S0-BMG-660

THE ROLE OF RNA IN REPRODUCTION
AND DEVELOPMENT

American Association for the Advancement of Science.

THE ROLE OF RNA IN REPRODUCTION AND DEVELOPMENT

Proceedings of the A.A.A.S. Symposium, December 28-30, 1972

Edited by

M.C. NIU

Department of Biology, Temple University, Philadelphia, U.S.A.

and

S.J. SEGAL

Population Council, The Rockefeller University, New York, U.S.A.

1973

NORTH-HOLLAND PUBLISHING COMPANY – AMSTERDAM • LONDON
AMERICAN ELSEVIER PUBLISHING CO., INC. – NEW YORK

Library of Congress Catalog Card Number: 73-81529

North-Holland ISBN: 0 7204 4135 8
American Elsevier ISBN: 0 444 10539 5

PUBLISHERS:
NORTH-HOLLAND PUBLISHING COMPANY – AMSTERDAM
NORTH-HOLLAND PUBLISHING COMPANY, LTD. – LONDON

SOLE DISTRIBUTORS FOR THE U.S.A. AND CANADA:
AMERICAN ELSEVIER PUBLISHING COMPANY, INC.
52 VANDERBILT AVENUE, NEW YORK, N.Y. 10017

PRINTED IN THE NETHERLANDS

Preface

Serious pursuit of the function of RNA in living cells first began in the late nineteen-thirties when ultraviolet microspectrophotometry was introduced for the study of intracellular localization of RNA. The early observations led Caspersson to conclude that embryonic cells and other fast growing cells, including microorganisms, are rich in RNA, which presumably plays some direct role in the growth process. When an abundance of RNA-containing particles was demonstrated cytochemically in the chordamesoderm of the amphibian gastrula, a role in development was suggested for this substance. As gastrulation progresses, the RNA particles gradually decrease in the invaginated chordamesoderm and concomitantly increase in the overlying ectoderm. Brachet proposed that this apparent transfer of RNA particles is causally related to the differentiation of the ectoderm into the central nervous system. Thus, RNA appeared to be involved in the two primary aspects of development – growth and differentiation, processes particularly characterized by the synthesis of new protein.

Subsequently, in the 1940's and early 1950's, cytochemists and geneticists amassed a body of qualitative evidence that correlated RNA and protein synthesis. With the introduction of more specific chemical methods for detection and measurement of RNA, quantitative data emerged. As a result of the accelerating pace of work during this period, it was established that RNA synthesis is linked to protein synthesis in a time sequence, with the former preceding the latter.

At the same time developmental biologists who were studying the nature of embryonic inductor substances discovered the importance of RNA in connection with the classical embryonic "organizer" phenomenon. For example, one of us (Niu) found that muscle differentiation occurs in a medium conditioned by developing myoblasts under conditions that suggest that released RNA carries the muscle information and can induce ectodermal differentiation into muscle cells.

Parallel to the study of RNA as an inducer, biochemical analysis of tobacco mosaic virus was climaxed in 1966 by the experiment of Gierer and Schramm showing that the RNA isolated from tobacco mosaic virus by a very mild procedure (treatment with phenol) is infectious. When such isolated RNA enters into tobacco leaves, it not only (reproduces itself but also synthesizes the viral protein components, thus yielding whole virus of the strain from which the RNA was initially isolated. Fraenkel-Conrat confirmed this role of RNA by using reconstituted virus particles in which RNA from one strain was encapsulated in the protein of another. He found that the lesions produced by the "hybrid" viral particles were characteristic of the strain from which the RNA was isolated; moreover, the progeny of the "hybrid" virus was composed of both the protein and the RNA characteristic of the RNA donor strain.

By the end of the 1950's the concept of messenger RNA emerged, and Jacob and Monod proposed a model for gene regulation based on operon gene control of structural genes. Activation of a structural gene by its operon results in transcription of the gene into newly formed messenger RNA. The new RNA migrates into cytoplasm and there acts as template for protein synthesis. By 1961, synthetic polynucleotides were being used to study RNA-programmed synthesis of specific proteins in vitro. Polyuridylic acid was used in a cell-free system by Nirenberg and found to give rise to the synthesis of polyphenylalanine. Extensive studies of this kind led to elucidation of the genetic code.

During the past decade, the focus of research on the function of RNA has again enlarged from in vitro systems to include higher biological levels of organization. The intense research efforts devoted to this subject coupled with the advent of several fundamental improvements in research methodology has led the search into diverse areas of biological investigation, many of which bear directly on reproductive physiology and development. They include RNA metabolism at organ and organismic levels; RNA-induced in vitro and in vivo syntheses of specific proteins; RNA-mediated transfer of biological phenomena, e.g. immunity and hormone action; and RNA-induced genetic changes and the "control" functions at the level of the genome.

Although research in these areas has progressed very rapidly and fruitfully, the accumulating body of information has not been assembled in one volume, nor have the active workers in the field been brought together for discussion and exchange of ideas. To meet this need we decided to organize an international symposium entitled "The role of RNA in reproduction and development". The meeting was organized under the auspices of The Division of Developmental Biology, American Society of Zoologists. Scientists from eleven countries contributed to the symposium, which was held in conjunction with the annual meetings of The American Association for The Advancement of Science, December 28–30, 1972, in Washington, D.C.

This symposium was supported by a grant (No. HD-07247-01) from The Research Grant Division, National Institute of Child Health and Human Development, National Institutes of Health, Bethesda, Maryland. We are grateful to all who have assisted in making both the symposium and this volume possible.

March 1973

M.C. Niu
S.J. Segal

List of participants

Vincent G. Allfrey
The Rockefeller University, New York, New York, 10021

Mirko Beljanski
Institut Pasteur, 25, Rue du Docteur Roux, Paris, XVe, France

James Bonner
California Institute of Technology, Division of Biology, Pasadena, Calif. 91109

Mario Burgos
University of Cuyo, Mendoza, Argentina

Richard Croissant
Department of Biochemistry, USC School of Dentistry, 925 W. 34th Street, Los Angeles, California 90007

A.K. Deshpande
Department of Biology, Temple University, Philadelphia, Pa. 19122

N. Dupont
Laboratory of Nuclear Medicine, The Free University of Brussels, 115, Boulevard de Waterloo, Brussels, 1-Belgium

Audrey Evans
Department of Radiology, Case Western Reserve University, Cleveland, Ohio 44106

J.-E. Edström
Department of Histology, Karolinska Institutet, S-104 01 Stockholm, Sweden

Marvin Fishman
Public Health Research Institute, New York, New York 10016

R.A. Flickinger
Department of Biology, State University of New York, Buffalo, New York 14214

John H. Frenster
Division of Oncology, Stanford University School of Medicine, Stanford, Calif. 94305

P. Galand
Biology Group Institut de Recherce Interdisciplinaire en Biologie Humaine et Nucleaire (LMN) Faculty of Medicine, Free University, Brussels, Belgium

A.W. Galston
Department of Biology, Yale University, New Haven, Conn.

Ajit Goswami
Boston Hospital for Women, Harvard Medical School, Boston, Mass. 02115

Paul R. Gross
Department of Biology, University of Rochester, Rochester, New York 14627

J.B. Gurdon
Medical Research Council, Laboratory of Molecular Biology, Hills Road, Cambridge, England

Paul R. Herstein
Division of Oncology, Stanford University School of Medicine, Stanford, Calif. 94305

David S. Holmes
California Institute of Technology, Division of Biology, Pasadena, Calif. 91109

Rufus Ige
University of Ibadan, Ibadan, Nigeria

C.-Y. Kang
McArdle Laboratory, University of Wisconsin, Madison, Wis. 53706

R. Kaur-Sawhney
Department of Biology, Yale University, New Haven, Conn.

J.S. Knowland
Medical Research Council, Laboratory of Molecular Biology, Hills Road, Cambridge, England

S.S. Koide
The Population Council, The Rockefeller University, New York, New York 10021

N.C. Kostraba
Biology Department, State University of New York, Buffalo, New York 14214

R.A. Laskey
Imperial Cancer Research Fund, Lincoln's Inn Fields, London WC2A 3PX, England

Philip Leder
Laboratory of Molecular Genetics, National Institute of Child Health and Human Development, National Institutes of Health, Bethesda, Maryland

Barry E. Ledford
Biology Division, Oak Ridge National Laboratory, Oak Ridge, Tennessee

H. Lee
Department of Biology, Rutgers University, Camden, New Jersey

N.C. Mishra
The Rockefeller University, New York, New York 10021

Jui-yun Mu
Department of Physiology, National Defense Medical Center, Taipei, Taiwan

Marshall Nirenberg
National Heart and Lung Institute, National Institutes of Health, Bethesda, Maryland 20014

L.C. Niu
Department of Biology, Temple University, Philadelphia, Pa. 19122

M.C. Niu
Department of Biology, Temple University, Philadelphia, Pa. 19122

Raphael Palacios
Department of Pharmacology, Stanford University, Stanford, California 94305

John Papaconstantinou
Biology Division, Oak Ridge National Laboratory, Oak Ridge, Tennessee

Michel Plaweck
Institut Pasteur, 25, Rue du Docteur Roux, Paris, XVe, France

Robert E. Rhoads
Department of Pharmacology, Stanford University, Stanford, California 94305

Naoi Sasaki
Embryological Laboratory, Biology Dept., Kuyushu University, Fukuoka, Japan

Sheldon J. Segal
The Population Council, The Rockefeller University, New York, New York 10021

Robert T. Schimke
Department of Pharmacology, Stanford University, Stanford, California 94305

Harold C. Slavkin
Department of Biochemistry, USC School of Dentistry, 925 W. 34th Street, Los Angeles, California 90007

Drew Sullivan
Department of Pharmacology, Stanford University, Stanford, California 94305

G. Szabo
The Biology Institute, Medical University, Debrecen, Hungary

E.L. Tatum
The Rockefeller University, New York, New York 10021

H.M. Temin
MacArdle Laboratory, University of Wisconsin, Madison, Wis.

Pentti Tuohimaa
University of Tampere, Tampere, Finland

Claude A. Villée
Department of Biochemistry, Harvard University Medical School, Boston, Mass. 02115

Arthur H. Whiteley
Department of Zoology, University of Washington, Seattle, Washington

S.F. Yang
Department of Biology, Temple University, Philadelphia, Pa. 19122

Table of contents

PREFACE V

LIST OF PARTICIPANTS VII

First session: RNA metabolism in developing embryos and organs . 1

Chairman: Arthur Whiteley

Chairman's introduction 3

Gene transcription and gene expression during sea urchin development, *by Paul R. Gross and Kenneth W. Gross* 4

Unbalanced growth and cell determination in frog embryos, *by Reed A. Flickinger* 11

Ovalbumin mRNA, complementary DNA and hormone regulation in chick oviduct, *by Robert T. Schimke, Robert E. Rhoads, Raphale Palacios and Drew Sullivan* 26

Regulation of albumin synthesis in cultured mouse hepatoma cells, *by John Papaconstantinou and Barry E. Ledford* 43

Second session: RNA programmed protein synthesis in cell-free systems

Chairman: Philip Leder

RNA mediated protein synthesis in eucaryotes, *by R.C. Huang*

Steroid hormone induction of specific translatable mTNAs, *by Gary Rosenfeld, A.R. Means and Bert W. O'Malley*

Translation and reverse transcription of purified mRNA, *by Philip Leder*

(Manuscripts of the three papers in Session Two were not received.)

Third session: RNA effects on in vivo synthesis of specific proteins

Chairman: Vincent G. Allfrey

Chairman's introduction 59

A hormone-controlled RNA fraction regulating enzyme development in plant cells, *by R. Kaur-Sawhney and A.W. Galston* 61

Effects of exogenous RNA on steriod metabolism in adrenals and gonads, *by Dorothy B. Villee and Ajit Goswami* 73

Thyrotropin-like activity of thyroid RNA in vitro, *by Jui-yun Mu* 86

In vivo uptake of RNA and its function in castrate uterus, *by M.C. Niu, L.C. Niu and S.F. Yang* 90

Injection of messenger RNA into living cells and its application to the study of gene action in *Xenopus Laevis, by John S. Knowland, John B. Gurdon and R.A. Laskey* 110

Fourth session: Transfer of Tissue Specificity

Chairman: Seldon J. Segal

Chairman's introduction 125

The role of macrophage RNA in the immune response, *by Marvin Fishman* 127

Studies on biological potentiality of testis-RNA I. Induction of axial structures in whole and axcised chick blastoderms, *by H. Lee and M.C. Niu* 137

Biological activity of RNA from estrogen-stimulated uterus, *by Paul Galand and N. Dupont* 155

Effects of exogenous polynucleotides on uterine enzymes, *by Claude A. Villee* 167

The role of RNA in the differentiation of presumptive ectoderm from urodele embryos, *by Naoi Sasaki and M.C. Niu* 183

Fifth session: Nucleic acid-induced chances in living systems

Chairman: Marshall Nirenberg

Transforming RNA as a template directing RNA and DNA synthesis in bacteria, *by M. Beljanski and M. Plawecki* 203

RNA mediated transformation in *Pneumococcus, by Audrey Evans* 225

Requirement of informational molecules in heart formation, *by Amrut K. Desphande, L.C. Niu and M.C. Niu* 229

Intercellular communication during odontogenic epithelial-mesenchymal interactions: isolation of extracellular matrix vesicles containing RNA, *by Harold C. Slavkin and Richard Croissant* 247

Nucleic acid-induced changes in *Neurospora, by Nawin C. Mishra, G. Szabo and Edward L. Tatum* 259

Specific and heterospecific transfer of hormone action by mRNA, *by Sheldon J. Segal, R. Ige, M. Burgos, P. Tuohimaa and S.S. Koide* 270

Sixth session: Mechanism of RNA action

Chairman: M.C. Niu

Chairman's introduction 287

Appearance and decay of ribonucleic acids in the cytoplasm of salivary gland cells of *chironomus tentans, by J.E. Edstrom* 289

Sequence composition and organization of the genome and of the nuclear RNA of higher organisms: an approach to understanding gene action, *by David S. Holmes and James Bonner* 304

Nonhistone proteins as gene derepressor molecules, *by T.Y. Wang and N.C. Kostraba* 324

RNA in gene de-repression, *by John H. Frenster and Paul R. Herstein* 330

RNA-directed DNA synthesis in viruses and normal cells: a possible mechanism in differentiation, *by C.-Y. Kang and Howard M. Temin* 339

INDEX 349

Session One

RNA Metabolism in Developing Embryos and Organs

Chairman

ARTHUR H. WHITELEY

Department of Zoology, University of Washington, Seattle, Washington

Chairman's introduction

When an egg is activated at fertilization, processes are started which lead in an orderly way to an elaboration of form, the differentiation of tissues and the synthesis of specific molecules. These processes are regulated by a series of interlocking control mechanisms. Some are extrinsic, others intrinsic. Movements bring cells and tissues into new relationships permitting inductive interactions. Hormones act on their targets. Blastomeres divide producing many cells; cells fuse to form new complexes. Proteins are made and assemble into higher order organelles, and form and structure emerge. In order that its development may proceed, sooner or later an embryo must have the information encoded in its genome. The genome invariably is expressed through the mediation of molecules of RNA, and the purpose of this symposium is to explore a number of ways in which these transcripts are expressly involved with development and differentiation.

Transcription is a simplistic term, encompassing the formation of many kinds of RNA from several segments of the genome. Transfer and ribosomal RNAs provide the cytoplasmic machinery for construction of polypeptides. The specificity of amino acid sequences is provided by messenger RNA. Most of these molecules are adenylated for transport to the cytoplasm, though some, notably the messages for histones, lack poly A stretches. Some are short lived; others are encapsulated with protein to form particles, informasomes, which are very stable. The transcriptional units may be huge molecules from which the functional molecule is processed prior to its transport out of the nucleus. Other molecules of RNA are proposed to remain in the nucleus to interact with the chromatin and regulate its activity.

Beyond the role of transcription in the expression of the genome in the form of proteins which characterize a differentiated tissue, there are a number of intriguing instances where specific transcripts from one differentiated tissue cause a phenotypic transfer when provided to tissues of alternative states of differentiation.

In the course of this symposium, there will be an exploration of these and other facets of genomic expression, and of the involvement of RNA in the regulated events of egg activation and differentiation. This discussion will open with an examination of gene transcription in two developing systems, proceed to a consideration of mechanisms of RNA involvement in protein synthesis, extend to a consideration of polynucleotide-mediated transfer of tissue specificity and induced changes in living systems, and deal lastly with mechanisms of regulation of the genome and RNA transport to the cytoplasm.

Arthur H. Whiteley

Niu and Segal (eds.). The role of RNA in reproduction and development
North-Holland Publ. Co., 1973

Gene transcription and gene expression during sea urchin development

Paul R. GROSS and Kenneth W. GROSS
Department of Biology, University of Rochester, Rochester, N.Y. 14627, USA, and
Department of Biology, Massachusetts Institute of Technology, Cambridge, Mass. 02139, USA

1. *Introduction*

RNA and protein synthesis begin or are very greatly accelerated within a few minutes after fertilization of the sea urchin egg. Since the ripe unfertilized egg of sea urchins has a pronucleus, macromolecule syntheses following fertilization are related to development proper, rather than to meiosis and other processes associated with maturation. There is no obligatory period of ripening after spawning of the eggs, so that the small amount of protein synthesis characteristic of the spawned unfertilized egg does not reflect production of proteins necessary for further development. Protein synthesis that follows fertilization, however, is essential. If it is stopped, development stops within a single intermitotic interval, which can be very short – less than an hour. Expression of genetic information is thus a continuous process in sea urchin embryos. Some gene expression is "immediate", in the sense that newly-transcribed RNA enters the cytoplasm and functions there at once, and some part is "delayed", in the sense that a part of the protein synthesis of early development is directed by messenger RNA (mRNA) synthesized weeks to months earlier – during oogenesis. General arguments surrounding this position have recently been reviewed [1] and will not be repeated here.

The idea that oogenetically-transcribed RNA messages are stored untranslated in the egg in "masked" or otherwise unavailable form has been supported by strong, but essentially indirect evidence. Such evidence is available not only for eggs of invertebrates, but also for those of vertebrate animals (e.g. [2]). It concerns "informational" RNA in a general way, and also putative "maternal" messages for specific proteins, such as the monomers of microtubules [3]. Lacking until very recently, however, has been an example of so-called maternal mRNA identified by the most direct test, which is its translation in a cell-free system and a demonstration of identity between the proteins synthesized therein and those made naturally in the embryo.

We report here on our recent identification by this method of the maternal component of histone mRNA. This group of messages, probably five in number, offers some conveniences for experimental work, as outlined below. The experiments on histone mRNA permit, therefore, a final proof of the maternal mRNA hypothesis. They show, in addition, that these messages reside in a compartment of the cytoplasm outside the protein synthetic machinery. Establishment by example of the correctness of the maternal mRNA hypothesis has important consequences for the analysis of animal development, some of which we mention below.

2. *"9 S" RNA synthesized on embryonic genomes*

During the period of cleavage, polyribosomes acquire a very prominent, rapidly-labeled 9 S RNA whose presence is closely correlated with the synthesis of histones. A body of such correlative evidence led to the conclusion that 9 S RNA contains messages for histone production and that these are among the quantitatively most important of the "immediate" class of gene expressions during cleavage [4]. There remained the necessity to demonstrate that 9 S RNA can direct the synthesis of sea urchin histones in a cell-free, heterologous system. This has recently been accomplished [5]. 9 S RNA prepared from mid-cleavage polyribosomes contains sequences that direct synthesis of all five histones, and furthermore, the best preparations are remarkably free of other translatable sequences. Whereas, for example, purified globin mRNA stimulates incorporation of tracer tyrosine and tryptophan into proteins when added to the Krebs mouse ascites system, 9 S RNA from embryo polyribosomes stimulates tyrosine but *not* tryptophan incorporation. The tyr/trp mole ratios are the expected ones for globin, and, of course, the absence of tryptophan is characteristic of histones. Presence of significant contamination by other messages in 9 S RNA is therefore ruled out. Peptide fingerprints of the histones made in vitro are indistinguishable from those of marker embryo histones.

Among the varieties of indirect evidence used earlier to identify labeled 9 S RNA as histone mRNA was the observation that inhibition of histone synthesis via the arrest of DNA replication with hydroxyurea specifically and selectively stops the entry of labeled 9 S RNA into polyribosomes [6]. This result shows that the *labeled* 9 S RNA, rather than solely unlabeled, is histone mRNA.

3. *"Maternal" histone mRNA*

Blocking transcription with actinomycin brings about a major reduction in histone synthesis during cleavage, but not a total loss of it [7]. Recent and more detailed studies by Ruderman [7] reinforce this observation and show that the residual histone synthesis includes all five major classes. Survival of histone synthesis in the absence of immediate transcription products places histones in the class of proteins for which there is probably a maternal provision of mRNA. This provision would account for about half of the normal histone production during the peak period of synthesis at mid-cleavage [8].

The foregoing thus argues for histone mRNA as representative of *both* classes of informational RNA functioning during development – the immediate and the delayed. What might be the purpose for such a dual distribution is an interesting question, but it would divert us from our immediate objective to discuss it here.

Suffice it to say that 9 S RNA contains immediately transcribed (labeled) and immediately translated mRNA for histones, plus, on the basis of indirect evidence from inhibitor studies, some maternal histone mRNA. Having in hand an assay system for these messages, however, we were in a position to make the desired direct test for maternal mRNA, i.e., to attempt to extract such polyribonucleotides from the cytoplasm of unfertilized eggs, in which the rate of protein synthesis is very low indeed, and whose histone synthesis reflects at most a small fraction of that already low rate.

4. Histone mRNA activity in egg fractions

Eggs of *Lytechinus pictus* were homogenized in a variety of media and fractionated according to several alternative schedules. Experimental details and a full account of results are to be published elsewhere [9]. Since the relevant conclusions are the same for several alternative kinds of preparation, only typical results will be reported below. Homogenization of carefully washed eggs in TKM buffer (0.2 M KCl, 5 mM $MgCl_2$, 20 mM Tris buffer, pH 7.6) yields cell-free preparations in which the protein synthetic machinery is minimally disturbed (e.g., undergraded polyribosomes are obtained with this medium for embryo homogenates). A schedule of differential centrifugations was employed with TKM homogenates to produce frac-

TABLE 1
Fractionation scheme for TKM homogenates of unfertilized sea urchin eggs

Relative centrifugal force and time	Fraction contents
200 *g*, 1.5 min, pellet	Unbroken cells, debris, nuclei
1,500 *g*, 10 min, pellet	Large yolk particles and pigment granules
5,000 *g*, 10 min, pellet	Yolk
12,000 *g*, 15 min, pellet	Mitochondria plus scant "microsomes"
12,000 *g*, 15 min, supernatent	Ribosomes and solubles
12,000 *g* supernatant, on 20–40% sucrose gradient, 19 hr @ 26,000 rev/min, pellet	Ribosomes (polyribosomes)
Regional fractions from gradient	**Sedimentation velocity ranges**
No. 1	60–80 S
No. 2	40–60 S
No. 3	±40 S
No. 4	20–40 S
No. 5	<20 S solubles

tions whose contents were identified by microscopy or by appropriate chemical tests. This schedule is shown in table 1.

RNA was purified from all fractions by repeated partitionings with phenol-chloroform-isoamyl alcohol. The purified RNA in bulk or fractions of it repurified from sucrose gradients was tested in the mouse ascites cell-free system as has been described [5]. Incubations were carried out at 30°C for 1 hr with appropriate controls for non-specific stimulation of incorporation. Following incubation, samples were DNAase-treated and then dialyzed against 0.1% sodium dodecyl sulfate–0.2% mercaptoethanol. The proteins were separated by electrophoresis on 12.5% polyacrylamide gels according to the method of Laemmli [10]. Incubation mixtures contained ^{3}H-phenylalanine, ^{3}H-lysine, or ^{3}H-aspartic acid, according to their purpose. ^{3}H-phe was used for identification of the tryptic peptides of histones, while the other two amino acids provided tracers for newly-synthesized proteins on the gels. Marker histones were present on all gels, labeled with ^{14}C-leucine in vivo during mid-cleavage.

Post-ribosomal supernatants from 12,000 *g* supernatant preparations were further fractionated on sucrose gradients (40%–20%) for 19 hr at 26,000 rev/min (Spinco SW 27 rotor). These gradients were pumped through the flow cell of a recording spectrophotometer and fractions were collected and pooled for RNA extraction. The RNAs were tested in the cell-free system as described above.

All the particulate fractions, taken together, account for about 10% of the total RNA of homogenates, and this RNA either inhibited incorporation in the cell-free system (e.g., RNA from pigment granules and yolk) or stimulated it slightly ("nuclei", which contain some unbroken cells and undispersed cytoplasm, and mitochondria plus "microsomes"). The bulk of the RNA and most of the stimulatory activity is recovered from the post-mitochondrial supernatant. From this fraction, stimulatory activity was further fractionated as shown in the table, the conditions of gradient centrifugation having been chosen so as to pellet a portion of the ribosomes. In a typical experiment, neither the pelleted ribosomes (plus any polyribosomes present) nor the bottom-most gradient fraction stimulated the cell-free system significantly. This latter fraction contained a large amount of RNA from unpelleted ribosomes. Likewise, the fraction from the top of the gradient, representing a range of sedimentation velocities up to about 20 S, failed to stimulate significantly.

Efficient stimulation of protein synthesis in vitro was obtained with the intermediate fractions of the post-ribosomal supernatant: the range of sedimentation velocities is approximately 20 S–60 S. Three such fractions were normally collected by pooling: the middle one contained the absorbancy peak, at 260 nm, of the 40 S ribosomal subunit. These fractions stimulated protein synthesis 3- to 8-fold, depending upon the amino acid used as tracer and the specific scheme of fraction pooling employed.

One of these fractions, with equivalent sedimentation velocities of 20 S–40 S, stimulated lysine incorporation much more efficiently than it did aspartic acid incorporation. The other fractions did the reverse. Fraction No. 4 in this scheme (20 S–40 S) gave a 7.6X increment over background in lysine incorporation, and a 3.6X increment in aspartic acid incorporation, whereas other fractions (40 S–60 S, approximately, and 60 S–80 S) gave greater aspartate stimulation than lysine stimulation. The ratio of lys to greater stimulation was greater than 2 for fraction 4 and less than one of all others.

Electrophoresis of the products showed that fraction 4 RNA stimulated synthesis of a group of proteins co-electrophoretic with marker histones. Their phepeptide fingerprints were those of natural histones. Fraction 4 directed synthesis of little other protein (ca. 20% or less of the total). The remaining postribosomal fractions directed synthesis of an heterogeneous array of proteins, some of them of high molecular weight and rich in aspartic acid, and the one adjacent to and with higher sedimentation velocities than fraction 4 provided RNA capable of stimulating some histone synthesis (ca. 20% of the total).

Quantitative data on stimulation, the kinetics of incorporation in stimulated systems, and the electrophoretic patterns for proteins made by RNA from all fractions are shown in detail in the forthcoming paper [9]. There is no question about the presence of RNA in eggs capable of stimulating protein synthesis in vitro, but much more important, the rigorous product identification steps described earlier [5] leave no doubt now that among the products of stimulation are proteins indistinguishable from those made during embryogenesis. These proteins are, furthermore, among those that can be made even when new transcription is prevented. This supports the conclusion based upon inhibitor experiments: the unfertilized egg contains histone mRNA, and furthermore, this mRNA sediments in homogenates with a class of particles distinct from ribosomes (or polyribosomes) and other, larger particulates. Maternal mRNA is, therefore, present in a compartment of the cytoplasm from which translational machinery is absent.

5. The relevance of maternal mRNA to development

Direction of the protein synthesis upon which development and differentiation depend is accomplished by new as well as old mRNA, "new" implying very recent transcription upon the genes of blastomeres and embryonic cells, "old" implying transcription during oogenesis. The observation that mRNA for histones exists in the cytoplasm of unfertilized eggs provides us with an example and the most direct possible tests for maternal, or oogenetically-transcribed mRNA. It is an interesting, and probably unusual example, because histone mRNA is also transcribed continuously and for immediate translation during cleavage [11]. It is probably true that the maternal and embryonic mRNA complements are on the whole quite different [12]. There is in any case no valid reason to doubt that the sort of direct test accomplished for histone mRNA can be made successfully for other maternal messages, nor is there any reason to doubt that the program encoded upon them is a large one, comprising many proteins.

In a general way, these facts provide us with some insight into the economics of developmental information flow, and since so many other things needed for development are stored in the egg, it is perhaps not surprising that RNA messages are among them. The existence of such messages, probably in the form of ribonucleoprotein particles (although the latter is not *proven* by the experiments discussed above), raises a number of interesting questions about the mechanism by which they are selected for use and their translation controlled. Our ability to purify, at least partially, several distinct classes of maternal mRNA-bearing particles should help in the eventual analysis of that mechanism.

There is, however, an equally important issue, and one that has, perhaps, as much relevance to development proper as it has to the molecular biology of translation: this is the issue of maternal mRNA distribution in the egg and among the blastomeres arising during cleavage. It is clearly possible, and indeed likely unless proof to the contrary is obtained, that the entire population of maternal mRNAs is distributed symmetrically throughout the cytoplasm. If such proves to be the case, then oogenetic provision of mRNA can be viewed as another case of generalized storage for cellular "housekeeping", as, for example, seems to be the case with organelles such as mitochondria, and for large numbers of enzymes of intermediary metabolism. Symmetry of distribution would not lessen the interest of the translational control problem, since some central functions of embryonic metabolism are bound up with it.

The alternative possibility is, however, very much alive at this time, i.e., that the population of maternal mRNA is distributed asymetrically in the egg – spaced within the egg cytoplasm according to a pattern as simple as the gradient of yolk in the amphibian egg, or perhaps as complex as a large series of interdigitating and oppositely-directed gradients. The massive reorganization by flow that takes place in the cytoplasm after fertilization could establish or reinforce such asymmetries. If they were found to exist, then it is very likely indeed that the several classes of blastomeres formed in early cleavage, e.g., the macro-, meso-, and micromeres of the 16-cell stage, would acquire different sets of maternal mRNA in the course of their separation by cytokinesis. Differential protein synthesis, which is generally assumed to be the basis of stable cellular differentiation, would then be a consequence of cleavage geometry, the relevant transcriptions having occurred at a much earlier time. If among the proteins synthesized differentially on the maternal mRNA in blastomeres there were some capable of regulating transcription on the embryonic chromosomes, there would then come into operation a cascading system of divergence for protein and nucleic acid syntheses, i.e., for providing the essential molecular events in determination and eventually in terminal differentiation.

The technologies of nucleic acid hybridization and of cell-free protein synthesis are rapidly approaching a state of versatility within which it should be possible to test for asymmetry of maternal mRNA distribution. Obviously, it would be very hard to prove that it does *not* occur, although the general tendency can be elucidated. If it can be shown that it *does* occur, however, even in a few examples, then significant reinforcement will be given to an emerging view that a very large part of the story of development, even in the most "regulative" of embryos, is told in the maternal body, as the oocyte is being shaped and provided with its supply of materials for making a self-sustaining animal.

References

[1] P.R. Gross, In: M. Sussman, ed. Molecular genetics and developmental biology (Prentice-Hall, Inc., Englewood Cliffs, N.J., 1972).
[2] B.R. Hough and E.H. Davidson, J. Mol. Biol. 70 (1972) 491.
[3] R.A. Raff, H.V. Colot, S.E. Selvig and P.R. Gross, Nature 235 (1972) 211.
[4] L.H. Kedes and P.R. Gross, Nature 223 (1969) 1335.
[5] K. Gross, J. Ruderman, M. Jacobs-Lorena, C. Baglioni and P.R. Gross, Nature New Biol. 241 (1972) 272.
[6] L.H. Kedes, B. Hogan, G. Cognetti, S. Selvig, P. Yanover and P.R. Gross, Cold Spring Harbor Symposia 34 (1969) 717.
[7] J. Ruderman, Devel. Biol. (1973) in press.
[8] L.H. Kedes, P.R. Gross, G. Cognetti and A.L. Hunter, J. Mol. Biol. 45 (1969) 337.
[9] K.W. Gross, M. Jacobs-Lorena, C. Baglioni and P.R. Gross, Proc. Natl. Acad. Sci. U.S. (1973) in press.
[10] U.K. Laemmli, Nature 227 (1970) 680.
[11] B. Hogan and P.R. Gross, J. Cell Biol. 49 (1971) 692.
[12] R.O. Hynes and P.R. Gross, Biochim. Biophys. Acta 259 (1972) 104.

Niu and Segal (eds.). The role of RNA in reproduction and development
North-Holland Publ. Co., 1973

Unbalanced growth and cell determination in frog embryos

R.A. FLICKINGER
Department of Biology, State University of New York, Buffalo, N.Y. 14214, USA

The amount and number of kinds of DNA-like RNA (D-RNA) transcribed in nuclei decreases during early development of the amphibian embryo. A specific example is the restriction of transcription during development of messenger RNA(s) for collagen synthesis. Even though larvae synthesize much more collagen than neuralae, the injection of neurula nuclear RNA into *Xenopus* oocytes promotes collagen synthesis while an equal amount of larval nuclear RNA does not. However, a greater amount and number of kinds of D-RNA accumulate in developing embryos due to the formation of more cells in a similar volume of cytoplasm. This is an extreme example of unbalanced growth characterized by cell division and D-RNA synthesis with low levels of cellular protein synthesis. It is characteristic of all early stage embryos and is evidenced by decreasing cell size and increasing nuclear/cytoplasm ratios as early development proceeds. It is likely that it plays a role in cellular determination. For example, a greater number of cell divisions under conditions of unbalanced growth causes the accumulation of more RNA in gastrula ectoderm being induced to form neural tissue than in uninduced ectoderm. Evidence for an entrainment of transcription with DNA replication exists for the early frog embryo. Stages and parts of embryos that transcribe more kinds of D-RNA spend a greater part of their cell cycle in the S phase than those transcribing fewer kinds of D-RNA.

Hybridization of nuclear and whole embryo RNA preparations to DNA of varying degrees of reiteration reveals that the percent reduction in the variety of D-RNA molecules transcribed is greatest for the least reiterated DNA during development. In turn this accounts for the greater percent accumulation in variety of D-RNA molecules transcribed from the more reiterated DNA. However, the absolute variety of D-RNAs transcribed from the more reiterated DNA is quite low. Many copies, but few kinds, of D-RNA are transcribed from the highly reiterated DNA and these D-RNAs accumulate early in development, although they continue to accumulate later as well. Fewer copies, but many more kinds, of D-RNA are transcribed from the less reiterated DNA and these accumulate later in development.

Reassociation of denatured labeled frog DNA of varying degrees of reiteration with denatured mouse DNA has revealed that the more reiterated DNA sequences are more conservative than the less reiterated sequences. It is hypothesized that the ontogenetic sequence of differentiations is due to a sequence of accumulation of D-RNA molecules present in fewer copies transcribed from DNA sequences which are progressively less reiterated and conservative, and this accumulation depends upon an increasing number of cell divisions under conditions of unbalanced growth.

1. Introduction

It is likely that qualitative differences in transcription in the various cells of developing embryos play a central role in the synthesis of different tissue-specific proteins which account for cell specialization. The role of nucleic acid synthesis in embryos in relation to cell differentiation has been reviewed recently [1]. The content of the present paper will deal with possible mechanisms that control the synthesis and accumulation of DNA-like RNA (D-RNA) during the early development of the amphibian embryo and the relations to cell differentiation.

2. *Quantitative pattern of D-RNA synthesis during development*

In developing *Xenopus laevis* embryos there is little accumulation of D-RNA from fertilization to late cleavage but from gastrulation to an early larval stage there is a doubling of the content of newly synthesized D-RNA each time the DNA content of the embryo doubles [2]. In the late blastula the accumulation of high molecular weight non-ribosomal RNA occurs earlier in the endoderm and inner mesoderm cells than in the ectoderm [3]. In terms of RNA synthesis per cell, incubation of isolated nuclei of *Xenopus laevis* embryos in the presence of labeled uridine has revealed a decrease in rate of RNA synthesis from early cleavage on through to later stages of development [4]. Injection of RNA precursors into developing embryos has show that after gastrulation there is a decreased synthesis of D-RNA per cell in developing *Xenopus* up to the larval stage and that the endoderm cells accumulate a greater amount of D-RNA per cell than do the ectoderm-mesoderm cells of the embryo [5].

The level of DNA accumulation in frog gastrula ectoderm which is induced to form neural tissue is greater than that of uninduced ectoderm but the levels of RNA accumulation per cell were similar in the induced and uninduced ectoderm [6]. However, if the calculation of the level of RNA accumulation in induced and uninduced ectoderm was based on equal amounts of total protein, then the accumulation of RNA was greater in the induced ectoderm. This latter result is ascribed to the presence of more cells in the induced ectoderm and is substantiated by the higher DNA/protein ratio found in the induced relative to the uninduced cells. The results of these experiments suggest that a stimulation of cell division under conditions of unbalanced growth in which there is an increase in the nuclear/cytoplasm ratio as the cells become smaller may be the primary response of a competent tissue to an induction stimulus. More D-RNA may be accumulating in the induced ectoderm since D-RNA accumulation is proportional to DNA accumulation [2] and this might account for determination.

3. *Qualitative pattern of D-RNA synthesis during development*

3.1. *Hybridization experiments with unfractionated DNA*

The RNA–DNA hybridization experiments with RNA isolated from developing amphibian embryos indicate an accumulation in whole embryos of more kinds of D-RNA transcribed from reiterated DNA as development proceeds [7, 8]. A different picture is presented when RNA prepared from nuclei isolated from early neurulae, tailbuds and larvae of *Rana pipiens* is labeled in vitro with ^{3}H-dimethylsulfate and hybridized to total DNA. Neurula, tailbud and larval nuclear RNA hybridize to 16, 6 and 2% of the DNA respectively, indicating that RNA representing 32, 12 and 4% of the reiterated genome is present in nuclei at these stages of development [9]. There is not only a quantitative reduction of D-RNA synthesis after neurulation in amphibian embryos [5] but also a qualitative one [9]. The increased quantitative and qualitaitve accumulation of D-RNA in developing amphibian embryos may be explained by the great increase in the number of nuclei in virtually the same volume of cytoplasm during this period of development. Even though DNA transcription is being restricted the increase in the ratio of nuclei to cytoplasm with successive cell divisions as the cells become smaller gives an increase in the amount of D-RNA in the total embryo [10]. The conservation of more of the D-RNA on cytoplasmic polysomes also may be important in accounting for the increased number of kinds of D-RNA accumulating in embryos during this period of development.

3.2. Injection of nuclear RNA into Xenopus oocytes

The qualitative restriction of transcription of nuclear D-RNA during development from the neurula to the larval stage of *Rana pipiens* [9] has now been demonstrated for nuclear RNA coding for the synthesis of a specific protein, collagen. It is known that collagen synthesis begins at gastrulation and increases markedly during development in embryos of *Xenopus laevis* [11] and *Rana pipiens* [12]. In the present work a messenger RNA fraction obtained from polysomes of frog larvae and equal amounts of RNA prepared from nuclei isolated from neurulae and larvae were injected into growing *Xenopus* oocytes according to the techniques of Gurdon et al. [13]. Messenger RNA was prepared by releasing ribonucleoprotein from polysomes with ethylenediaminetetraacetic acid and deproteinization with sodium dodecylsulfate [13]. The RNA was separated by sucrose density gradient centrifugation and the messenger RNA fractions located between 18 S and 4 S were pooled for the injections. Nuclei were isolated in 0.25 M sucrose–2 mm $MgCl_2$, followed by washing in 0.025 M citric acid to dissolve yolk platelets. Then RNA was prepared from the isolated nuclei [8]. Growing oocytes (1.8 mm diameter) were dissected free of connective tissue and incubated with labeled proline (20 μCi/ml) 2 hr before and 10 hr after the injections. Each oocyte was injected with 50–70 nl of messenger RNA or nuclear RNA at a concentration of 800 μg/ml. Column chromotography of protein hydrolysates using a 150 cm column containing MS fraction D, Aminex (Bio-Rad Co.) revealed the presence of labeled hydroxyproline indicative of collagen synthesis after injection of larval messenger RNA [14]. Neurula nuclear RNA also promoted collagen synthesis in the oocytes (fig. 1) while the injection of an equal amount of larval nuclear RNA did not promote collagen synthesis nor was

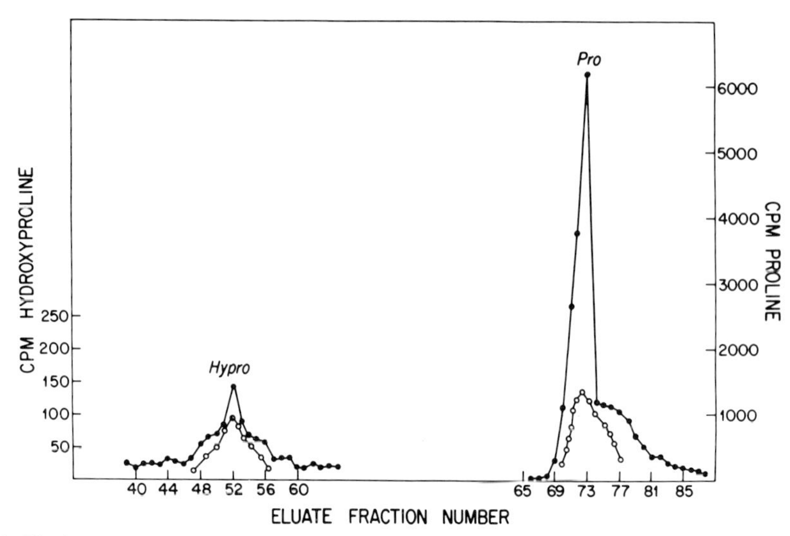

Fig. 1. Elution pattern of labeled hydroxyproline (Hypro) and proline (Pro) (●—●—●). The absorption of added authentic hydroxyproline and proline is indicated (○—○—○). Fifty-five growing oocytes were preincubated for 2 hr in 20 μCi/ml of ^{3}H-proline, followed by injection with 20 μg/ml (intracellular concentration) of neurula nuclear RNA and returned to the labeled incubation medium for 10 hr. Cpm Hypro (corr)/cpm Pro × 100 = 2.7%.

there any collagen synthesis in growing oocytes injected with saline. The corrected hydroxyproline/proline ratio × 100 was 0.041 after injection of the messenger RNA fraction of larvae and 0.027 after injection of neurula nuclear RNA.

Since larvae are synthesizing considerable collagen, the detection of collagen synthesis after injection of the larval messenger RNA fraction was expected. However, the stimulation of collagen synthesis after injection of neurula nuclear RNA was surprising. It implies that the neurula nuclear RNA has collagen messenger activity, or that the heterogeneous nuclear RNA is processed in the *Xenopus* oocyte before it is translated. The absence of activity for collagen synthesis of the larval nuclear RNA is thought to be due to a lower quantity of the messenger or pro-messenger, not to its qualitative absence as equal amounts of nuclear RNA were injected. The level of proline radioactivity in the protein hydrolysates, indicative of general protein synthesis, was higher after injection of neurula nuclear RNA than for larval nuclear RNA. It is believed that the decreased ability of larval nuclear RNA to promote both collagen and generalized protein synthesis is due to the quantitative restriction of nuclear D-RNA synthesis from neurulation to the larval stage in developing amphibian embryos [5]. However, it is possible that the nuclear D-RNA might turn over more rapidly at the later stage.

3.3. Hybridization experiments with nuclear RNA of regions of embryos

Labeled nuclear RNA from the dorsal axial regions and from the belly endoderm regions of neurula and tailbud stages of a large number of developing *Rana pipiens* embryos was hybridized to DNA. Equal amounts of labeled nuclear RNA of dorsal axial ectoderm-mesoderm cells of neurulae and tailbuds hybridized to a greater percent of the reiterated DNA than did the labeled nuclear RNA of belly endoderm cells from these stages (fig. 2) [15]. The addition of successive DNA filters to saturating amounts of labeled nuclear RNA of neurala dorsal axial and belly regions revealed a more rapid decrease in percent hybridization for the labeled RNA of the dorsal axial regions than for the belly endoderm regions. This implies that those species of D-RNA that are rapidly depleted by the first few DNA filters are present in fewer numbers than those species of D-RNA that resist depletion. It appears that some of the D-RNA molecules of dorsal axial cell nuclei accounting for a higher level of hybridization are present in fewer copies and that endoderm nuclei do not possess many of these kinds of few copy D-RNA. A gradual approach towards saturation (fig. 2) also indicates the presence of D-RNA present in few copies in nuclei of the dorsal axial regions. Similar kinds of experiments have demonstrated that the new kinds of D-RNA molecules present in fewer copies accumulate at later stages of development in intact embryos [8], while they are present in greater quantity in early neurula nuclei than in nuclei of later stages [9]. Unbalanced growth, the increase in nuclei/cytoplasm ratio as cell size decreases with successive cell divisions, may account for the gradual accumulation in intact embryos of these D-RNA molecules present in few copies despite the restriction of their synthesis in nuclei during the same period. In other words, even though nuclei from later stages transcribe less D-RNA both quantitatively and qualitatively, the presence of more nuclei in relation to the same volume of cytoplasm accounts for an increase of D-RNA quantitatively and qualitatively in the embryo.

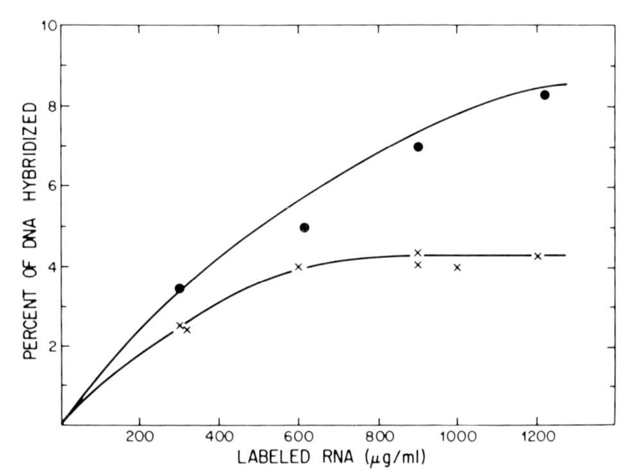

Fig. 2. Increasing amounts of labeled RNA from nuclei of frog neurula neural plates (•) and neurula belly endoderm regions (×) were added to nitrocellulose filters containing 1.24 μg of denatured frog blood cell DNA.

3.4. Hybridization experiments with DNA of different degrees of redundancy

In order to learn the relative degrees of transcription of frog highly repetitive, moderately repetitive and least repetitive DNA these fractions were prepared and hybridized to labeled nuclear and whole cell RNA from frog embryos [16]. Unlabeled frog red blood cell DNA was mixed with a small quantity of ^{14}C-labeled frog tailbud DNA and then sheared, denatured and separated into three fractions using hydroxyapatite columns. The repetition frequency of the highly repetitious fraction was 10^3–10^6, that of the moderately repetitious fraction 10–10^3 and that of the least repetitious fraction 1–10. These denatured DNA fractions were fixed on nitrocellulose filters and hybridized to ^{3}H-labeled RNA of nuclei of neurulae and larvae and that of intact gastrulae, neurulae, and larvae. The RNA was labeled in vitro with ^{3}H-dimethysulfate so that the RNA is uniformly labeled. Although much of the sheared DNA is lost from the filters during the course of the incubations, at the end of the incubations the amount of ^{14}C-label on the filters indicates the amount of DNA present, while the ^{3}H-activity indicated the amount of RNA hybridized to that DNA. More kinds of D-RNA transcribed from each of these DNA fractions accumulate in whole frog embryos during development, while there was a decrease in the kinds of D-RNA transcribed in the nuclei from the three classes of DNA from the early neurula to the larval stage (table 1). The percent restriction of transcription during development is greatest for the kinds of nuclear RNA transcribed from the least reiterated DNA, which probably accounts for the greater percent accumulation in whole embryos of kinds of RNA transcribed from the more reiterated DNA.

TABLE 1
Hybridization of labeled RNA to highly repetitious (HR), moderately repetitious (MR) and least repetitious (LR) frog DNA fractions. Results are the averages of 3 or 4 experiments with standard deviations

	Source of RNA	% Hybridization to each fraction			Total % hybridization [a]
		HR	MR	LR	
Embryo nuclei	Neurula nuclei (stage 14)	2.4 ± 0.3	48.0 ± 3.2	46.5 ± 2.1	38.4 ± 2.2
	Larval nuclei (stage 25)	1.3 ± 0.3	17.7 ± 2.3	10.0 ± 1.1	11.5 ± 1.4
Whole embryos	Early gastrulae (stage 10)	1.7 ± 0.2	3.6 ± 0.5	4.1 ± 0.4	3.4 ± 0.4
	Early neurulae (stage 14)	1.1 ± 0.2	5.9 ± 1.5	4.3 ± 0.3	4.3 ± 0.8
	Swimming larvae (stage 25)	5.5 ± 0.7	16.7 ± 1.1	6.5 ± 0.7	10.6 ± 0.9
Liver	Adult liver	2.1 ± 0.6	3.9 ± 0.7	4.8 ± 0.9	3.9 ± 0.8
	Adult liver nuclei	1.6 ± 0.2	3.1 ± 0.1	5.4 ± 0.7	3.7 ± 0.6

[a] Each percent hybridization was multiplied by the fraction of DNA in its repetition class. The total percent hybridization is the sum of these fractions. For example, the total percent hybridization of neurula nuclear RNA to frog DNA is $2.4 \times 0.20 + 48.0 \times 0.42 + 46.5 \times 0.38$.

The decrease in number of kinds of D-RNA transcribed from the three reiteration frequencies in nuclei and the increase in whole embryos with development is similar to the results obtained with unsheared DNA, which indicated hybridization only to total reiterated DNA [8, 9].

This lends further support to the validity of the data. There is only a very low level of transcription from highly reiterated DNA, but the levels of hybridization of neurula nuclear RNA to moderately and least reiterated DNA are near the expected maximum of 50% assuming only one strand of the DNA is transcribed! This suggests that saturation of the DNA is attained in the hybridizations with nuclear RNA. True saturation is not obtained with whole embryo RNA since the large amount of ribosomal RNA dilutes out the small quantity of D-RNA. However, this amount of ribosomal RNA is similar at all stages and hence the relative accumulation of D-RNA transcribed from each repetition class can be compared for the stages of development.

3.5. Depletion experiments

To determine if few-copy RNA hybridized chiefly to highly repetitious DNA or to moderately repetitious DNA, depletion experiments were performed with each DNA preparation under conditions of RNA saturation. Larval RNA was used because it has previously been shown to contain a population of D-RNA molecules present in few copies. Filters containing highly repetitious or moderately repetitious DNA were incubated 10 times longer than necessary for saturation and then the filters were withdrawn, washed and counted. New filters were added to the same incubation medium and incubated for the same length of time and this process was repeated 5 times.

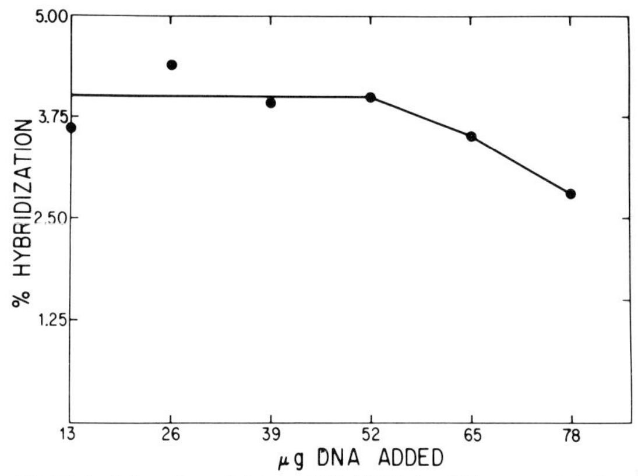

Fig. 3. Depletion of tritiated frog larval (stage 25) RNA by filters containing ^{14}C-labeled highly repetitious (HR) DNA. RNA was dissolved at 600 μg/ml in 1 M $NaHClO_4$–50% formamide in a volume of 0.05 ml. Filters (3 mm) containing 13 μg HR DNA were added to RNA solutions and incubated at 37°C for 10 min to an RNA Cot of 10. Each incubation contained one blank filter and one DNA filter. The average of duplicates is plotted.

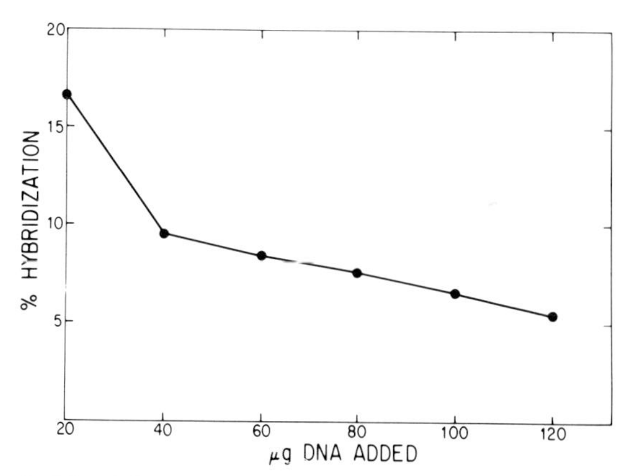

Fig. 4. Depletion of tritiated frog larval (stage 25) RNA by filters containing 20 μg ^{14}C-labeled moderately repetitious (MR) DNA. Conditions as in fig. 3 except that RNA was dissolved at 6.94 mg/ml and incubation time was 12 hr (to RNA Cot of 10,000).

Larval RNA showed only a slight decrease in percent hybridization after five filters with highly repetitious DNA had been added (fig. 3), while moderately repetitious DNA caused a striking decrease in percent hybridization after addition of only two DNA filters (fig. 4). These data indicate that the majority of D-RNA species present in few copies in frog larvae are homologous to moderately repetitious DNA. Previous work had shown that the kinds of D-RNA present in fewer copies accumulate later in development in whole embryos [8]. These depletion experiments with the highly repetitious and moderately repetitious DNA suggest that the kinds of D-RNA present in fewer copies which accumulate later in development in intact embryos are transcribed from less reiterated DNA than are the kinds of D-RNA that accumulate earlier in many copies and continue to accumulate later in development.

3.6. Degree of divergency in each DNA reiteration class

The degree of relatedness between the three frog DNA fractions and whole mouse DNA was determined by examining the thermal stability of these heterologous hybrids. Trace amounts of denatured tritium-labeled frog DNA of each reiteration frequency were added to a large excess of sheared, denatured mouse DNA so as to insure that the labeled frog DNA would be hybridized to mouse DNA. Double-stranded material was then adsorbed on hydroxyapatite and eluted by stepwise temperature increase. The melting profiles of the labeled frog-mouse DNA hybrids of each reiteration frequency were compared to that of the reassociated mouse DNA and the extent of mismatch estimated. A 1°C difference in melting temperature is equivalent to a 1.5% mismatch in base sequences. The results show that the degree of divergency between mouse and frog DNA is greater for the less

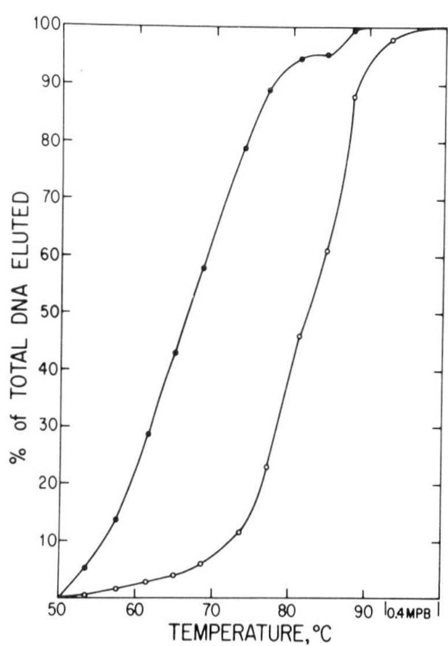

Fig. 5. Integral plot of melt of the least repetitious frog DNA-mouse DNA hybrids (•) and reassociated mouse DNA (○). Frog least repetitious ^{3}H-DNA (0.21 μg) was mixed with 40,000 P.S.I. sheared mouse DNA (432 μg) in a combined volume of 0.15 ml of 1 M phosphate buffer (PB) containing 2.5 mM EDTA. The mixture was denatured and incubated at 65°C for 6.8 days to a Cot (mouse) of 50,000. The reassociated material was isolated on hydroxyapatite at 50°C in 0.14 M PB, then melted in 3–4°C steps. DNA remaining on the column at 93°C was eluted with 0.4 M PB.

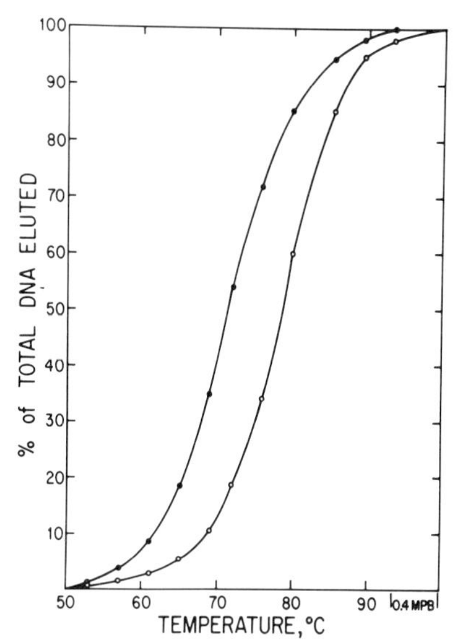

Fig. 6. Conditions as in fig. 3 except that moderately reiterated frog ^{3}H-DNA was added. The incubation was for 1.4 days to a mouse DNA Cot of 10,000. Frog-mouse DNA hybrids (•); mouse DNA duplexes (○).

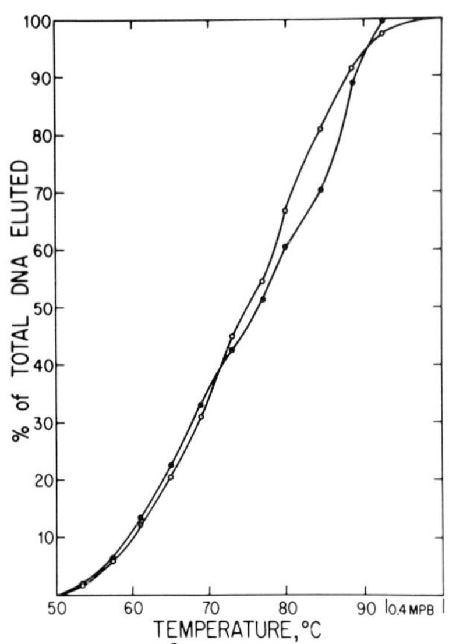

Fig. 7. Melt of frog highly repetitious ^{3}H-DNA mouse DNA hybrids (•) and of reassociated sheared mouse DNA (○). Conditions as in fig. 2 except that the ratio of mouse to frog DNA was 1000 : 1. DNA was dissolved in 0.14 M PB and incubated to a mouse DNA Cot of 10.

reiterated DNA sequences (figs. 5, 6 and 7). In other words, the more reiterated DNA sequences of mouse and frog DNA are more conservative than the less reiterated sequences.

The less repetitious DNA sequences are less conservative and transcribe D-RNAs present in fewer copies which accumulate later in development in intact embryos. These data suggest that the sequence of accumulation of D-RNA reflects the scale of redundancy of the DNA sequences. The kinds of D-RNA transcribed from more reiterated and conservative DNA sequences accumulate before those transcribed from less reiterated and conservative DNA sequences. This may be the means by which the sequence of differentiations in ontogeny reflects the sequence of gene acquisition during phylogeny [10, 16].

The idea that the more conservative features of development are expressed before the more specialized features was put forth by Van Baer during the last century. The similar appearance of the embryos of an early stage of development of such diverse organisms as fish and mammals illustrates this point; their basic pattern of early development is homologous and conservative. In this regard it is of interest that DNA reassociation experiments reveal that mouse and salmon DNA have common reiterated DNA, but little or no non-repeated DNA in common [17]. More specialized organs, containing less conservative proteins, appear later in development during the time when D-RNA present in fewer copies and transcribed from DNA sequences which are less reiterated and less conservative is accumulating. In several cases it has been shown that highly conservative molecules are coded for by reiterated DNA. Ribosomal RNA, transfer RNA and 5 S RNA [18], as well as the D-RNA coding for histone synthesis [19], are homologous to reiterated DNA.

It is possible that conservative DNA sequences account for the synthesis of D-RNA coding for the active sites of the numerous enzymes common to quite diverse organisms. In the instance of common proteins that play a role in the early differentiation of homologous organs in diverse organisms, once again the DNA sequences are likely to be conservative, although perhaps less conservative than for common metabolic "housekeeping" functions. It is possible that the DNA sequences which are even less conservative may control the synthesis of more specialized proteins appearing later in development, with these accounting for individual differences being the least conservative.

If the DNA sequences which are more conservative are also more reiterated (figs. 5, 6, 7), then it appears that D-RNA transcribed from more reiterated DNA accumulates before that transcribed from less reiterated DNA although it continues to accumulate later in development as well. According to this idea the course of development is regulated by the redundancy of the genes. However, it is clear that some highly repetitive DNA containing similar families of DNA sequences, such as the various satellite DNAs, are of recent evolutionary origin [20]. A greater number of cell divisions under conditions of unbalanced growth may be needed in a developing embryo in order for the transcripts from the less reiterated DNA sequences to accumulate in the cytoplasm. While this might only be true for early embryos in which cells are becoming smaller as the nuclear/cytoplasm ratio increases, it could be a factor in any cells undergoing some type of unbalanced growth.

The least repetitious DNA contains the greatest variety of base sequences and therefore transcribes the most kinds of D-RNA. Fewer kinds of D-RNA are transcribed from more reiterated DNA, although more copies are produced. Highly reiterated DNA, which may play a structural sole in the chromatin, is an exception to this!

4. Possible control mechanisms for transcription

4.1. Gene dosage

The number of DNA templates available for transcription is one way in which RNA synthesis may be regulated. The best example of this is the amplification of the genes for ribosomal RNA synthesis in developing oocytes [18]. The doubling of amounts of D-RNA with a doubling of DNA in developing amphibian embryos [2] also has a parallel in synchronized cultures of HeLa cells [21]. The synthesis of each kind of RNA, including D-RNA, increases two-fold during the early part of the S period when most of the DNA replicates.

It is interesting that treatment of gastrula ectoderm of amphibian embryos with sodium or lithium ions [22, 23, 24] can alter the prospective fate of these cells.

These treatments are accompanied by a change in the number of cell divisions, with sodium stimulating cell division while lithium inhibits division [25].

A change in the number of cell divisions under these conditions of decreasing cell size might operate as a gene dosage type of control. Successive cell divisions under conditions of unbalanced growth are essentially filling the same volume of cytoplasm with more copies of the genome, hence affording a greater opportunity for cytoplasmic accumulation of D-RNA molecules which could account for cell determination [10, 26].

Another possible means by which the number of genes might regulate the amounts of D-RNA transcribed is that a greater degree of reiteration of certain DNA sequences in the genome might allow more transcription from those sequences. While it seems clear that some of the most highly reiterated sequences undergo little transcription, e.g., mouse satellite DNA [27], and may play a structural role in the chromatin [28], this is not true of all reiterated DNA sequences. In *Xenopus* oocytes most of the D-RNA is transcribed from reiterated DNA [29]. The depletion experiments, in which successive filters with denatured highly reiterated or moderately reiterated DNA are added to labeled larval RNA, reveal a relation between the degree of reiteration and number of copies (not kinds of D-RNA) transcribed. The moderately reiterated DNA depletes those kinds of D-RNA molecules present in fewer copies in the labeled larval RNA preparation, while the highly reiterated DNA filters barely reduce the level of hybridization (figs. 3, 4). This implies that the kinds of D-RNA present in many copies are transcribed from the more highly reiterated DNA and the kinds of D-RNA present in fewer copies are transcribed from the moderately reiterated DNA. The presumption is that even fewer copies of D-RNA are transcribed from the least reiterated DNA. However, these assumptions relate to the quantitative aspect of transcription. The least repetitious DNA contains the greatest variety of those sequences and although its transcripts accumulate later in development in fewer copies, they represent a greater variety, or number of kinds of D-RNA.

4.2. Restriction of transcription during developments

4.2.1. Late-replication of DNA

After gastrulation in amphibian embryos when the nuclear synthesis of D-RNA is restricted quantitatively [5] and qualitatively [9] more of the DNA sequences of frog embryos become late-replicating [30, 31]. Explants of dorsal axial ectoderm-mesoderm and belly endoderm regions of neurulae and tailbuds were partially synchronized by FUdR, followed by unlabeled thymidine release [31]. At two hour intervals during the S and G_2 periods the explants were given two hour pulses of ^{3}H-deoxyguanosine. The DNA synthesis is expressed as cpm DNA/ cpm acid soluble pool/μg total DNA and represents the levels of DNA synthesis on a per cell basis. At both the neurula and tailbud stages there are two peaks of DNA synthesis during the S period (fig. 8). The shaded areas represent the relative amounts of

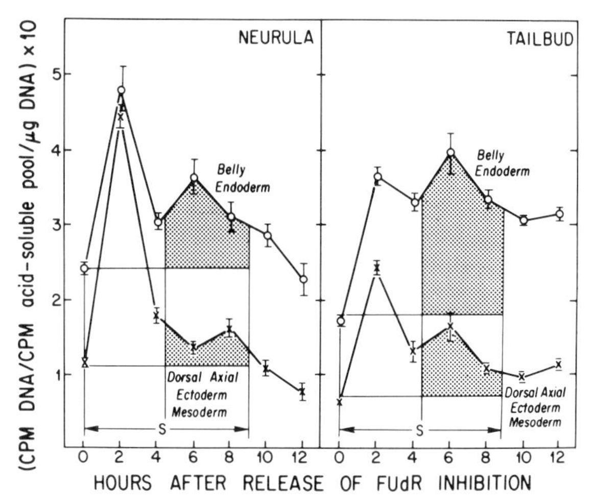

Fig. 8. Changes in the rate of DNA synthesis during the S period. At two hour intervals during the S and G_2 periods partially synchronized neurula and tailbud explants were incubated in 2 μCi/ml ^{3}H-deoxyguanosine for 2 hr. The levels of incorporation in unsynchronized control cultures remained relatively constant during the period of incubation. The shaded areas represent late-replicating DNA.

DNA that replicate during the last 4½ hr of the 9 hr S period. It is clear that more of the total DNA has become late-replicating by the tailbud stage, as compared to early neurulae. Measurements of DNA-like RNA synthesis in these partially synchronized explants has revealed a restriction of synthesis in the last half of the S period for tailbud embryos [31] and this may be due to the increased amount of late-replicating DNA. Further, the levels of labeled D-RNA/labeled pool÷total DNA are increased in the partially synchronized cells in which more cells are in the S phase and decreased during G_2. Apparently the act of replication of DNA somehow stimulates transcription.

The possibility that the more highly reiterated DNA sequences preferentially become late-replicating has been examined [32]. DNA of partially synchronized frog embryo explants was labeled during either the first or last half of the S period. The DNA was isolated, sheared, denatured and the renaturation rates compared to that of unlabeled DNA of adult red blood cells. No differences in the proportion of DNA of any reiteration frequency were detected when comparing labeled early or late-replicating DNA. In other words, the late-replicating DNA did not contain more of the highly reiterated DNA sequences.

4.2.2. Restriction of transcription and redundancy

In considering the possible mechanisms by which transcription is restricted there is one common feature, namely, the DNA sequences affected. If restriction affects DNA of varying degrees of reiteration equally, then the less reiterated DNA sequences would be affected most since a greater degree of repetition would allow similar sequences a better chance to avoid inactivation. Depletion experiments of labeled nuclear RNA with successive DNA filters show that the transcription of kinds of D-RNA molecules present in fewer copies undergoes the greatest degree of restriction at later stages [9]. The depletion experiments reported here with DNA of different degrees of redundancy suggest that the kinds of D-RNA present in fewer copies are transcribed from less reiterated DNA (figs. 3, 4). This suggests that restriction of transcription is greatest for the less reiterated DNA. The later accumulation of these kinds of D-RNAs present in fewer copies in intact developing frog embryos is thought to be due to the great increase in number of nuclei in a similar volume of embryo cytoplasm.

The incorporation of 5-bromodeoxyuridine into DNA can prevent the expression of the characteristics of differentiated cells without affecting their growth and division [33]. An equal level of incorporation of 5-BUdR into DNA of each reiteration frequency would result in a higher proportion of the analog in the less reiterated DNA sequences. It is possible that this could restrict transcription of the D-RNAs present in fewer copies that account for differentiation. The analog would be present in a lower proportion in the more conservative and reiterated DNA sequences and the transcription of D-RNAs accounting for growth and cell division might be unimpaired since there would be more of the repeated sequences that had not incorporated 5-BUdR.

4.3. Replication of DNA in populations of cells and transcription

As restriction of transcription occurs from the neurula to the larval stage in *Rana pipiens* embryos, fewer of the cells are in the DNA synthetic period (S), as evidenced by a decreasing percent of labeled nuclei after a short pulse with ^{3}H-thymidine [34]. In an asynchronous population of cells such as this a greater percent of nuclei in the S period indicates that the S period occupies a greater percent of the generation period [35]. Nuclei of more rapidly dividing ectoderm-mesoderm cells of frog neurula and tailbuds also spend a greater percent of their cell cycle in the S period than do the more slowly dividing endoderm cells [25]. Not only do neurula nuclei transcribe more kinds of D-RNA than tailbud nuclei [9], but ectoderm-mesoderm nuclei at both stages transcribe a greater variety of hybridizable D-RNAs than do endoderm cells [15]. In each case the cells spending a greater part of their cell cycle in the S period transcribe a greater variety of D-RNA molecules. It is possible that the qualitative control of D-RNA synthesis is linked to DNA replication. RNA polymerase has a greater binding affinity for denatured DNA, although less RNA is transcribed from denatured DNA in vitro [36]. If short sequences of DNA have more strand separation during DNA replication, this might facilitate the binding of RNA polymerase. Therefore, if more of the individual cells in a *population* of cells are in S at any given time, this might promote binding of RNA polymerase and account for transcription of more kinds of D-RNA. Removal of bound protein during DNA replication might also increase transcription.

References

[1] J.B. Gurdon, Essay Biochem. 4 (1967) 26.
[2] D.D. Brown and E. Littna, J. Mol. Biol. 20 (1966) 81.
[3] R. Bachvarova and E.H. Davidson, J. Exptl. Zool. 163 (1966) 285.
[4] W.C. Claycomb and C.A. Villee, Exptl. Cell Res. 69 (1971) 430.
[5] H.R. Woodland and J.B. Gurdon, J. Embryol. Exptl. Morphol. 19 (1968) 363.
[6] R.A. Flickinger, Roux Archiv. 171 (1972) 256.
[7] H. Denis, J. Mol. Biol. 22 (1966) 285.
[8] R.F. Greene and R.A. Flickinger, Biochem. Biophys. Acta 237 (1970) 447.
[9] J.C. Daniel and R.A. Flickinger, Exptl. Cell Res. 64 (1971) 285.
[10] R.A. Flickinger, Devel. Biol. Suppl. 4 (1970) 12.
[11] H. Greene, B. Goldberg, M. Schwartz and D. Brown, Devel. Biol. 18 (1968) 391.
[12] J. Klose and R.A. Flickinger, Collagen synthesis in frog embryo endoderm cells, Biochim. Biophys. Acta 232 (1971) 207.
[13] J.B. Gurdon, C.D. Lane, H.R. Woodland and G. Marbaix, Nature (London) 233 (1971) 177.
[14] J.W. Rollins and R.A. Flickinger, Science 178 (1972) 1204.
[15] R.A. Flickinger and J.C. Daniel, Roux Archiv. 169 (1972) 350.
[16] R.A. Flickinger, J.C. Daniel and R.A. Mitchell, Exptl. Cell Res. (1972) in press.
[17] R.J. Britten and D.E. Kohne, Carnegie Inst. Year Book Washington 65 (1965) 78.
[18] D.D. Brown and C.S. Weber, J. Mol. Biol. 34 (1968) 661.
[19] L.H. Kedes and M.L. Birnstiel, Nature New Biol. 230 (1971) 165.
[20] R.J. Britten and D.E. Kohne, Science 161 (1965) 529.
[21] S.E. Pfeiffer, J. Cell Physiol. 75 (1968) 91.
[22] L.G. Barth and L.J. Barth, J. Morphol. 110 (1962) 347.
[23] Y. Masui, J. Embryol. Exptl. Morphol. 15 (1966) 371.
[24] D.O.E. Gebhardt and P.D. Nieuwkoop, J. Embryol. Exptl. Morphol. 12 (1969) 317.
[25] R.A. Flickinger, M.R. Lauth and P.J. Stambrook, J. Embryol. Exptl. Morphol. 23 (1970) 571.
[26] R.A. Flickinger, In: Developmental aspects of the cell cycle, Eds. I.L. Cameron, G.M. Padilla and A.M. Zimmerman (Academic Press, New York, 1971) 161–190.
[27] W.G. Flamm, P.M.B. Walker and M. McCallum, J. Mol. Biol. 40 (1969) 423.
[28] M.L. Pardue and J.G. Gall, Science 168 (1970) 1356.
[29] B.R. Hough and E.H. Davidson, J. Mol. Biol. 70 (1972) 491.
[30] P.J. Stambrook and R.A. Flickinger, J. Exp. Zool. 179 (1970) 101.
[31] J.A. Remington and R.A. Flickinger, J. Cell Physiol. 77 (1971) 411.
[32] R.A. Mitchell and R.A. Flickinger, Life Sci., Part II, 11 (1972) 1011.
[33] J. Abbott and H. Holtzer, Proc. Natl. Acad. Sci. U.S. 59 (1968) 1144.
[34] R.A. Flickinger, M.L. Freedman and P.J. Stambrook, Devel. Biol. 16 (1967) 457.
[35] V. Defendi and L.A. Manson, Nature (London) 198 (1968) 359.
[36] M. Chamberlin and P. Berg, J. Mol. Biol. 8 (1964) 708.

Niu and Segal (eds.). The role of RNA in reproduction and development
North-Holland Publ. Co., 1973

Ovalbumin mRNA, complementary DNA, and hormone regulation in chick oviduct*

Robert T. SCHIMKE, Robert E. RHOADS,
Raphael PALACIOS and Drew SULLIVAN
Department of Pharmacology, Stanford University, Stanford, Calif. 94305, USA

Chick oviduct differentiation is controlled by various steroid hormones, including both estrogen and progesterone. Administration of estrogen to chicks results in differentiation of tubular gland cells, which synthesize differentiated cell products, including ovalbumin and conalbumin. Ovalbumin can constitute as much as 60% of the protein synthesized by oviduct in the fully stimulated chick. Methods have been developed that allow for the quantitative assay of ovalbumin mRNA, the specific isolation of ovalbumin synthesizing polysomes by use of immunologic techniques, as well as the synthesis of a DNA probe complementary to the ovalbumin mRNA. With these techniques we have found that the rate of ovalbumin synthesis is proportional to the content of ovalbumin mRNA. Hybridization studies using the DNA probe indicate that there are only two copies of the ovalbumin gene per genome, and that there is no amplification of the ovalbumin genes to account for the large amount of ovalbumin synthesized.

1. Introduction

Estrogens and progesterone regulate the differentation and function of chick oviduct [1, 2, 3, 4]. Our attention has been focused on hormonal regulation of ovalbumin synthesis, since this single polypeptide comprises 50–60% of protein synthesized in the fully differentiated oviduct. These features allow for the isolation of the molecular elements involved in specific protein synthesis, including specific polysomes mRNA and genes, and for an analysis of various regulatory steps in the transcription and translation of specific mRNAs as affected by developmental and hormonal variables. In this report we describe some of our more recent studies on the regulation of ovalbumin synthesis in this system, specifically analysis of ovalbumin mRNA content, methodology for isolation of ovalbumin polysomes by immunologic techniques, the synthesis of a DNA probe complementary to the ovalbumin mRNA, and an analysis of gene amplification in this differentiated system.

* Supported by Research Grants from the National Institutes of Health, GM 14931, the Medical Scientists Training Program, GM 1922, and the American Cancer Society, B66, and a Public Health Service International Research Fellowship (FO5RW 1601).

2. Hormonal regulation of ovalbumin synthesis

Fig. 1 depicts the effects of estrogens and progesterone on oviduct development and function [2]. Estrogen administration to immature chicks results in cytodifferentiation of tubular gland cells which synthesize the major egg white proteins including ovalbumin, conalbumin, and lysozyme. When estrogen administration is stopped, the cells persist but ovalbumin synthesis stops. We have previously shown that reinitiation of ovalbumin synthesis in the "inactive" cells (restimulation) can

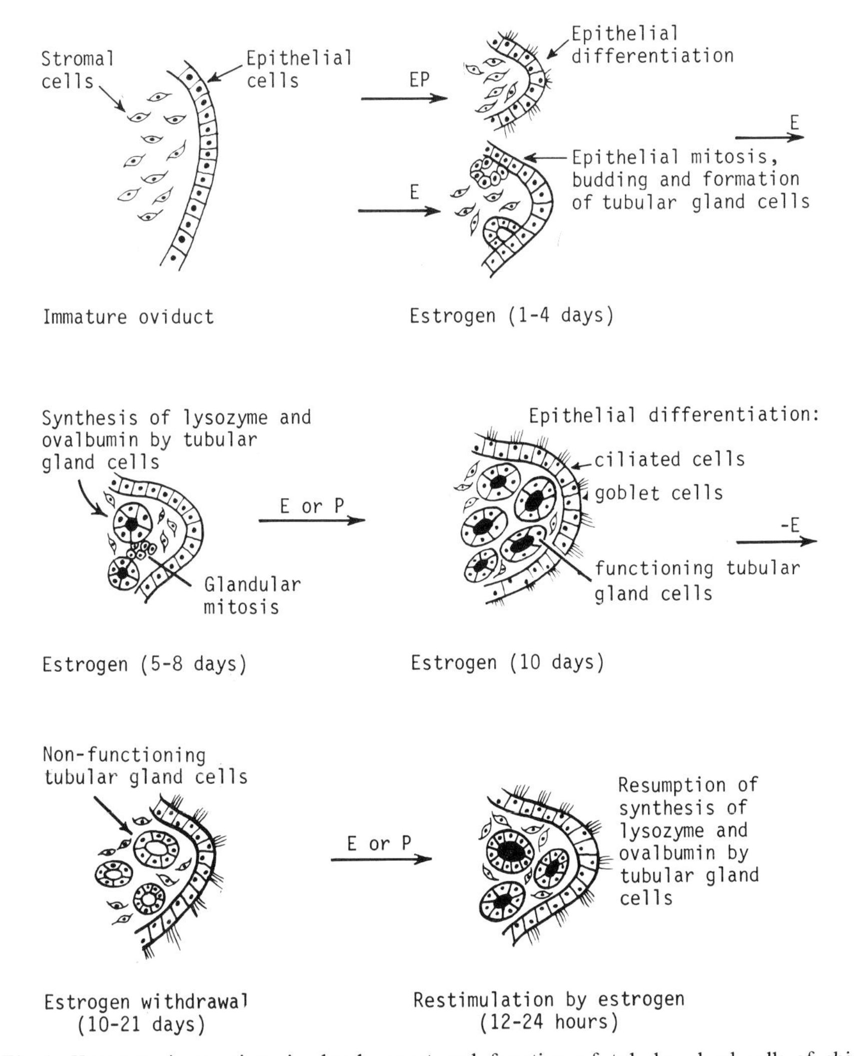

Fig. 1. Hormone interactions in development and function of tubular gland cells of chick oviduct. E, estrogen; P, progesterone.

occur on pre-existing ribosomes, and that both estrogen and progesterone can produce this effect [5]. In contrast concommitant administration of progesterone with estrogen to immature chicks prevents the typical cytodifferentiation [2, 4].

Rather an abortive synthesis of ovalbumin occurs in the surface cells, but such cells fail to undergo typical changes in including proliferation and formation of tubular gland cells [4].

Fig. 2 shows the percent of protein synthesis that ovalbumin comprises during different stages of estradiol administration and withdrawal. In the immature chick essentially no ovalbumin synthesis can be detected [4].

The capacity to synthesize ovalbumin increases such that by 10 days of daily estradiol administration, as much as 50% mf the protein syntheiszed is ovalbumin. Upon cessation of hormone treatment, ovalbumin synthesis decreases rapidly. The initial lag in capacity to synthesize ovalbumin results from the necessity for cytodifferentiation of the tissue (day 0–3), wheras the rapid increase of ovalbumin synthesis

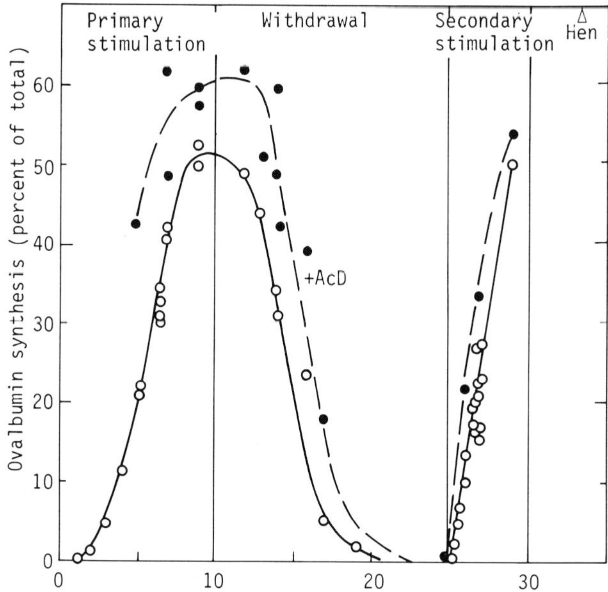

Fig. 2. Effect of estrogen and actinomycin D on the relative rate of synthesis of ovalbumin in chick oviduct during primary stimulation, withdrawal, and secondary stimulation. Immature chicks 4-day-old were injected intramuscularly with 1 mg estradiol benzoate daily (primary stimulation), and after 10 days without estrogen administration (withdrawal) administration was resumed (secondary stimulation). Chicks in groups of 2–4 were injected with actinomycin D (5 mg/kg) for 4–5 hr prior to isolating oviducts. Fragments of oviduct were then incubated in Hanks' salt solution for one hour with [^{3}H] amino acids (10 μCi/ml). Following homogenization and centrifugation at 100,000 *g* for 1 hr, ovalbumin was precipitated from the supernatant using a specific antibody. Results are presented as percent of total acid-precipitable radioactivity in supernatant that is precipitated immunologically with anti-ovalbumin antibody. Details are given in ref. [9]. ○——○, Estrogen ●——●, estrogen plus actinomycin D 4 hr before killing.

during restimulation results from stimulation of preexisting tubular gland cells. Administration of actinomycin D concomitant with initial restimulation abolishes the subsequent increase in ovalbumin synthesis [6].

In contrast, when actinomycin D is administered to intact chicks actively synthesizing ovalbumin and the capacity for ovalbumin synthesis is determined in fragments isolated 4 hr later, ovalbumin has become an increased proportion of the protein synthesized. This phenomenon is analogous to "superinduction" as studied in other systems [7]. The mechanisms we proposed for superinduction in the oviduct system differs from that proposed by Tomkins et al. [8], and are discussed in a paper by Palmiter and Schimke [9].

Central to understanding of hormonal control of cytodifferentiation and modulation of cell function is the ability to dissect each of the potential regulatory steps that may be involved in specific protein synthesis, i.e., ovalbumin. Thus does gene amplification occur to explain the large amount of ovalbumin synthesized? Is there a direct relationship between the content of mRNA and the rate of specific protein synthesis? Are there rate-limiting steps between synthesis of ovalbumin mRNA, its potential packaging, and transport from nucleus to cytoplasm? What roles do hormones play in regulating the rate of mRNA translation, or the stability of this mRNA? Critical to these questions is the ability to quantitate and to isolate ovalbumin mRNA.

3. Quantitation of ovalbumin mRNA

We are able to synthesize ovalbumin in a rabbit reticulcyte lysate system, employing various crude and partially purified oviduct RNA fractions [10]. The ovalbumin synthesized is isolated by immunoprecipitation. Fig. 3A shows that no radioactivity is incorporated into immunospecific protein as displayed on SDS acrylamide gels in the absence of added oviduct RNA. Marker ^{14}C ovalbumin is added to show its electrophoretic properties (open circles). Fig. 3B shows radioactivity precipitated by anti-ovalbumin antibody when RNA isolated from total yen oviduct polysomes is added to the lysate. Fig. 3C shows that authentic ovalbumin (synthesized in oviduct fragments; open circles) differs slightly from the immunospecific product synthesized in the rabbit reticulocyte lysate. We tentatively attribute the more rapid migration of the lysate product to the lack of the 7 carbohydrate molecules which exist on authentic ovalbumin, and which presumably are not added when ovalbumin is synthesized by the lysate. Fig. 3D displays the electrophoretic pattern of total labeled lysate protein in the absence of added oviduct RNA and shows that the vast majority of incorporated radioactivity occurs in globin chains, and that no protein with a mobility of ovalbumin is synthesized. More rigorous proof that the product is ovalbumin comes from studies (not presented) which show that the tryptic peptides of the lysate product co-chromatograph with those obtained from authentic ovalbumin [10].

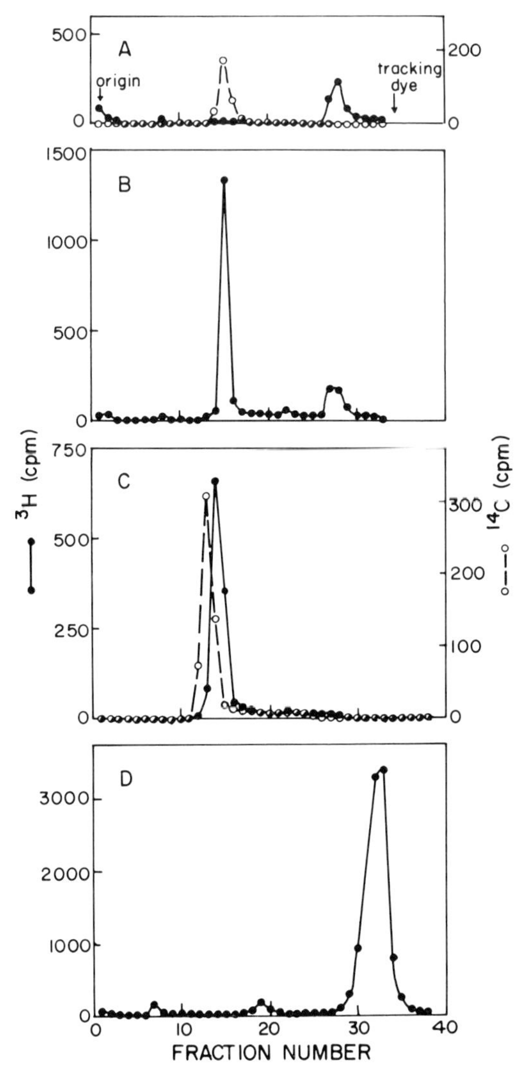

Fig. 3. SDS-acrylamide gel electrophoresis of anti-ovalbumin-precipitable radioactivity (A to C) and total reaction mixture (D). Antibody precipitates from 200 μl of reaction mixture were subjected to electrophoresis. Migration toward the anode was from left to right. (A) Reaction contained no oviduct RNA and [^{14}C] ovalbumin [6] was added as marker. (B) Reaction contained 40 μg per ml of oviduct polysomal RNA. The radioactive peak at Fraction 28 in Gels A and B probably results from the fact that the washing procedure for antibody precipitates used for these gels was less efficient than that used for Gel C. (C) Reaction contained 400 μg per ml of oviduct total nucleic acid. Before antibody precipitation the reaction mixture was combined with [^{14}C] leucine-labeled ovalbimin. (D) 200 μg of trichloroactic acid-precipitated protein from total reaction mixture containing no oviduct RNA . (See ref. [10] for details.)

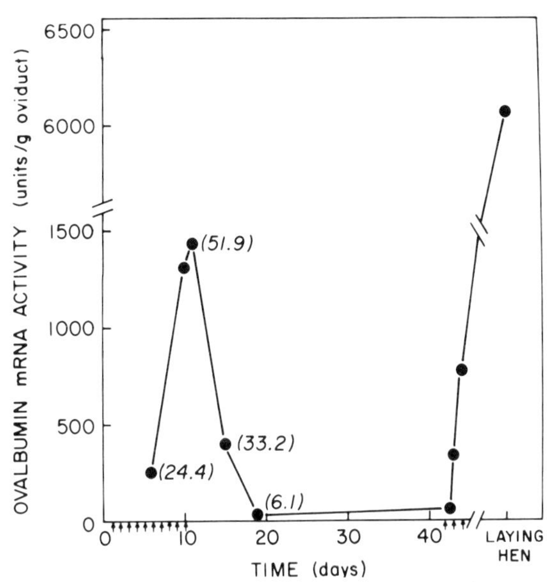

Fig. 4. Ovalbumin mRNA content of chick oviduct tissue during primary estrogen stimulation (1–10 d), withdrawal (11–41 d), and secondary stimulation (42–44 d). Estrogen (1 mg) was administered to chicks on days designated with arrows. Oviducts were removed and total mRNA activity determined. At several points portions of the same tissue used for the preparation of nucleic acid were incubated in culture for 1 hr with tritiated amino acids to determine the relative rate of ovalbumin synthesis. This is expresses as a percentage of total protein synthesis (numbers in parentheses). (See ref. [24] for details.)

The assay can be used as a quantitative measure of mRNA content and allows us to answer the fundamental question: is the content of mRNA directly related to the rate of ovalbumin synthesis? This experiment is depicted in fig. 4, which shows that the content of assayable ovalbumin mRNA changes in direct proportion to the rate of ovalbumin synthesis (also compare figs. 2 and 4). Thus apparently ovalbumin synthesis is directly related to ovalbumin mRNA content, not to an "unmasking" of previously existing (and translatable) mRNA.

4. Isolation of ovalbumin synthesizing polysomes

The ability to obtain ovalbumin synthesizing polysomes is necessary for the isolation of specific mRNA, its complementary DNA (gene), and for the assessment of possible involvement of specific regulatory proteins involved in transcriptional and translational control of ovalbumin synthesis. We have developed methods for the immunologic isolation of ovalbumin polysomes, dependent on the ability of an antibody against ovalbumin to react with the specific nascent chains of ovalbumin on polysomes. Fig. 5 indicates the specificity of binding of ^{125}I antiovalbumin to polysomes [11].

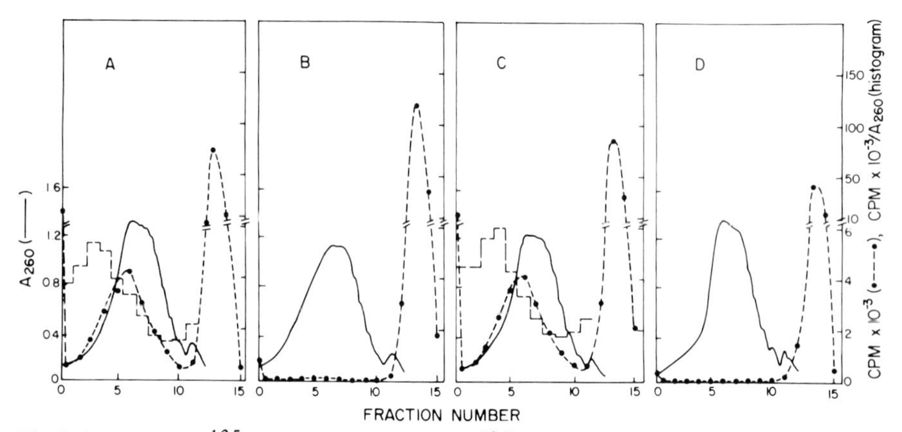

Fig. 5. Binding of [^{125}I] anti-ovalbumin and [^{125}I] anti-BSA to hen oviduct polysomes. Polysomes (10 A_{260} units in 1.0 ml) were incubated at 4°C with (a) 30 μg of [^{125}I] anti-ovalbumin for 30 min; (b) 500 μg of unlabeled anti-ovalbumin for 30 min followed by 30 μg of [^{125}I] anti-ovalbumin for 30 min more (c) 500 μg of unlabeled anti-BSA for 30 min followed by 30 μg of [^{125}I] anti-ovalbumin for 30 min; and (d) 30 μg of [^{125}I] anti-BSA for 30 min. After the incubation the polysomes were layered over a continuous sucrose gradient and centrifuged. Fractions (1.0 ml) were collected to measure specific activity and radioactivity (See ref. [20] for details.)

Fig. 5A shows the binding of 30 μg of ^{125}I antiovalbumin, showing binding to the polysomes with approximately 12 ribosomes. This is the major size class of polysomes and that an appropriate size for polysomes synthesizing a protein of the molecular weight of ovalbumin (42,000). That the binding sites are saturable is shown in fig. 5B, where the polysomes were first incubated with 500 μg of unlabeled anti-ovalbumin before addition of 30 μg of labeled antibody. In this case no binding of labeled antibody occurs in the polysome region. On the other hand, if polysomes are first incubated with 500 μg of bovine serum albumin antibody, the specific binding by anti-ovalbumin is not altered (fig. 5C). Fig. 5D shows that labeled bovine serum albumin antibody does not bind to polysomes. Also labeled anti-ovalbumin antibody does not bind to polysomes from hen brain or liver. In addition the binding is not the result of adsorbed ovalbumin from the supernatant, since mixing of liver polysomes with hen oviduct supernatant, which contains large amounts of ovalbumin, prior to polysome preparation did not result in binding of antibody to the liver polysomes.

The degree of binding can be used as a measure of the relative amount of polysomes engaged in the synthesis of ovalbumin. Fig. 6 shows the binding of labeled antibody to a constant amount of polysomes isolated from chicks at different times during restimulation (see fig. 2). There is a good correlation between the extent of binding of antibody to polysomes and the percent synthesis of ovalbumin.

Fig. 7 shows the method currently being used for the isolation of specific oval-

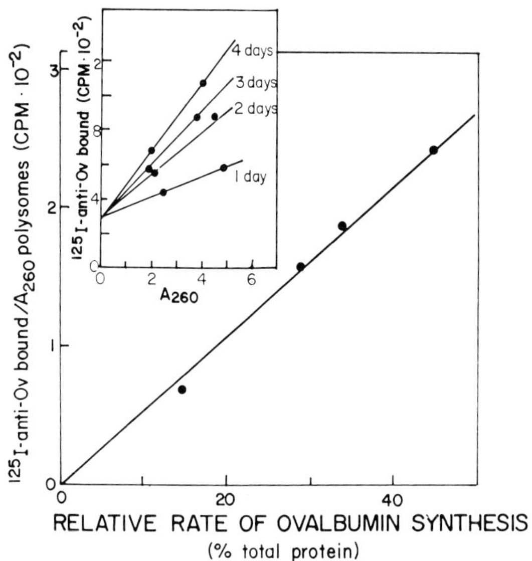

Fig. 6. Binding of ^{125}I-anti-Ov to oviduct ribosomes from chicks given 1 to 4 days of secondary stimulation with estrogens. Chicks were treated with estrogen (2 mg per day, secondary stimulation). Oviduct magnum was isolated and total ribosomes were prepared. Different amounts of ribosomes from each preparation were incubated with ^{125}I-anti-Ov for 1 hr at 4°C (inset), and amount bound determined. ^{125}I-anti-Ov bound per A_{260} of polysomes was determined from the slope of the line in inset and corrected for the percentage of total ribosomes which sediment as polysomes (determined by analyzing an aliquot of homogenate) on sucrose gradients; in all preparations 70 to 80% of the ribosomes sedimented as polysomes. The relative rate of ovalbumin was determined by incubating pieces of oviduct magnum in Hanks medium with ^{3}H-amino acids and the determining the percentage of the total protein synthesized which was precipitable with anti-Ov. (See ref. [20] for details.)

bumin polysomes. Following the binding of antibody to polysomes, a matrix of glutaraldehyde-fixed ovalbumin is added to the polysome antibody mixture. This matrix contains ovalbumin sites that react with the antibody molecules that has been previously bound to the nascent chains. The specific polysomes are thereby bound to the matrix. Fig. 7 shows washing of the polysome-ovalbumin matrix to remove the non-reacted polysomes (open bars), followed by elution of the specific polysomes with a buffer that contains EDTA (dashed bars). When each of the fractions i.e., original polysomes, non-reacted, and reacted polysomes, are assayed for ovalbumin mRNA content, there is an approximate 7-fold enrichment of ovalbumin mRNA in the immunoadsorbed fraction. This degree of enrichment is consistent with the fact that the polysomes were taken from oviducts of chicks in which 17% of the protein being synthesized was ovalbumin. The recovery of ovalbumin mRNA is essentially quantitative in this experiment. The amount of nonspecific precipitation of polysomes has been assessed by adding uridine-labeled hen liver polysomes to the immunoadsorption system, and determining what percent of

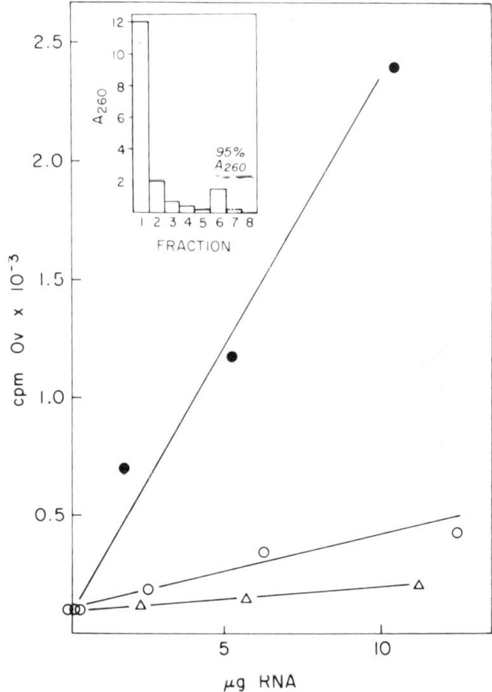

Fig. 7. In vitro synthesis of ovalbumin by RNA extracted from immunoadsorbed chick oviduct polysomes. Chicks received a secondary stimulation with estrogen for 18 hr. Oviduct magnum was isolated and the relative rate of ovalbumin synthesis was measured by incubating explants of the oviduct. Ovalbumin was 17% of the total protein synthesized. Polysomes prepared from the same oviduct (20 A_{260} units) were treated for immunoadsorption. The A_{260} of the different fractions during the treatment was measured as in fig. 1 and is presented in the inset. Shaded fractions represent the eluted polysomes. RNA was extracted from total (○——○), nonadsorbed (△——△), and adsorbed polysomes (● ——— ●), and was assayed at different concentrations in the reticulocyte lysate system for the synthesis of ovalbumin. (See ref. [12] for details.)

the radioactivity is eluted in the EDTA washes. Only 2% of labeled hen liver polysomes were adsorbed, whereas 20% of the hen oviduct polysomes were saved and eluted from the matrix [12].

The fraction of RNA isolated by the immunoadsorption technique contains specific ovalbumin mRNA, as well as considerable amounts of ribosomal RNA. The ovalbumin mRNA can be further purified by adsorption to nitrocellulose filters (millipore filters) by the technique of Brawerman et al. [13], or by the use of poly[dT]-cellulose columns as described by Nakazato and Edmonds [14]. Therefore we presume that this specific mRNA, like globin mRNA [15, 16], contain relatively long sequences of poly [rA]. It has been proposed that many mRNAs of animal tissues contain poly[A] sequences, presumably at the 3′ end of the message [13, 14, 17].

5. Synthesis of a complementary DNA, and an analysis of gene amplification for ovalbumin genes

We have used the RNA dependent DNA polymerase of Rous Sarcoma Virus [18], to synthesize a nucleic acid sequence complementary to ovalbumin mRNA. This enzyme uses as a template single stranded RNA with short primer region of double stranded nucleic acid at the 3′ end. By addition of oligo [dT] to the purified ovalbumin mRNA preparation, we convert the mRNA molecules, containing the putative poly[A] sequence, are converted to a double stranded primer for the RNA dependend DNA polymerase reaction.

As shown in table 1, the ovalbumin mRNA fraction that has been specifically immunoprecipitated and selectively adsorbed on millipore filters is active as a template for RNA-dependent DNA polymerase in a system that contains Rous Sarcoma Virus (RSV) reverse transcriptase and oligo[T] [19]. That reaction is dependent on RNA is shown by the fact that RNAse treatment of the RNA completely abolishes incorporation of deoxynucleoside triphosphate into an insoluble form. Likewise the reaction is essentially totally dependent on oligo[T] addition.

The mRNA fraction used as template contains approximately 75% ribosomal RNA [20]. To show that "reverse transcriptase" cannot use ribosomal RNA as a template ribosomal RNA isolated from a monosome fraction from hen oviduct catalyzes only 1/40 th of the incorporation (table 1). Moreover part of this incorporation, 950 cpm, presumably results from viral template RNA present in the

TABLE 1
Template activity of RNAs

RNA added to polymerase system	Acid precipitable ^{3}H cpm/50 μl reaction
+ ovalbumin messenger fraction (1.3 μg/ml)	79,000
+ ovalbumin messenger fraction + RNase	250
polymerase system alone (no added RNA)	950
polymerase system alone + RNase	200
+ ovalbumin messenger fraction, no oligo dT	900
+ RSV 70 S RNA (2 μg/ml)	51,000
+ RSV 70 S RNA + RNase	380
+ monosomal RNA (1.2 μg/ml)	2,000
+ monosomal RNA + RNase	300
+ monosomal RNA (1.2 μg/ml) + ovalbumin messenger fraction (1.3 μg/ml)	84,000

In those samples described as "+RNase" the added RNA, if any, was preincubated thirty minutes at 37°C with 100 μg/ml boiled pancreatic RNase before this mixture was added to the polymerase reaction. (See ref. [19] for details.)

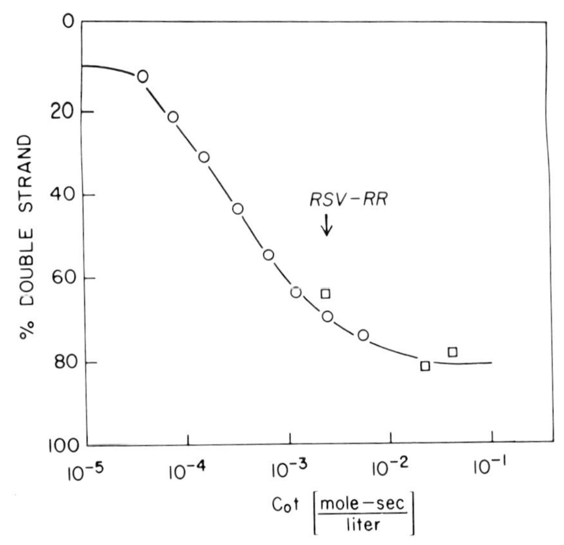

Fig. 8. Double-stranded probe DNA (3 ng/ml) was melted, allowed to reassociate and assayed for secondary structure on hydroxylpatite. The two symbols (○, □) represent two separate determinations. The arrow, labeled RSV–RR, indicated the $C_0t_{1/2}$ of Rous Sarcoma Virus rapidly reannealing DNA, with a complexity of = 3×10^3 base pairs (Faras et al. [18]) reassociated under identical conditions. (See ref. [19] for details.)

enzyme preparation. The remaining incorporation could be accounted for by the small amount of mRNA is present in the monosome preparation. Hence we conclude that ribosomal RNA is not being read in this system.

The RSV "reverse transcriptase" first synthesizes a DNA complementary to the added mRNA. It will then use the DNA–RNA hybrid to synthesize a second strand of DNA complementary to the DNA strand synthesized initially. If, on the other hand, actinomycin D is present during the reaction, only the single stranded DNA, complementary to the mRNA is synthesized [21].

Fig. 8 shows a C_0t curve of the reanealing of the double stranded product to itself. The kinetics of the DNA–DNA hybridization indicate a single second-order reaction, in which at least 85% of the DNA participates. The arrow indicates the $C_0t_{1/2}$ for the hybridization of complementary DNA synthesized from RSV template RNA. From the complexity of this DNA, it can be estimated that the complexity of the DNA product synthesized from the ovalbumin mRNA fraction is approximately 400 nucleotides. This number is consistent with the size of the probe estimated from sedimentation on sucrose gradients [19]. We estimate that the messenger for ovalbumin should contain approximately 1400 nucleotides, and hence the RSV "reverse transcriptase" does not appear to be reading the entire messenger.

An experiment to indicate that the DNA probe is complementary to ovalbumin mRNA is shown in fig. 9. In this experiment the single stranded DNA probe (i.e.,

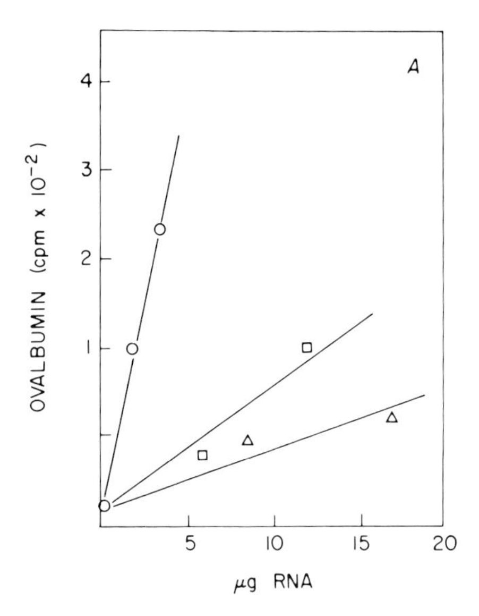

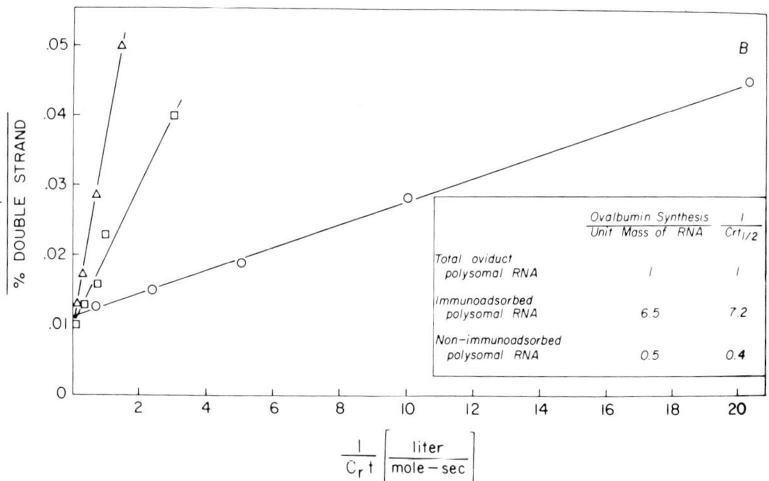

Fig. 9. Chicks were given secondary stimulation with estrogen, sacrificed 18 hr later, and polysomes were isolated from their oviducts. Approximately 15% of the oviduct polysomes from chicks in this state of secondary stimulation are specific for ovalbumin. An aliquot of these polysomes was separated by the ovalbumin-anti-ovalbumin system into an immunoadsorbed and a non-immunoadsorbed fraction and the RNA was then isolated from these polysomes. RNA from total oviduct polysomes (□), immuno-adsorbed polysomes (○) and non-immunoadsorbed polysomes (△) was assayed at different concentrations in the reticulcyte lysate system for ability to synthesize ovalbumin (panel A). Aliquots of the same RNAs were tested for rate of hybridization with single stranded DNA probe (panel B). The hybridization reaction has 5 ng/ml single stranded probe DNA, and either 27 μg/ml (for total oviduct polysomal RNA, ○) 4 μg/ml (for immunoadsorbed polysomal RNA), or 64 μg/ml (for non-immunoadsorbed polysomal RNA △) of polysomal RNA. Double strandedness was scared by resistence to S_1 nuclease. In the table (panel B, ovalbumin synthesizing ability/mass of RNA and the inverse $C_r t_{1/2}$ of each of these three RNAs are compared. Values are expressed relative to the value determined for total oviduct polysomes. (See ref. [19] for details.)

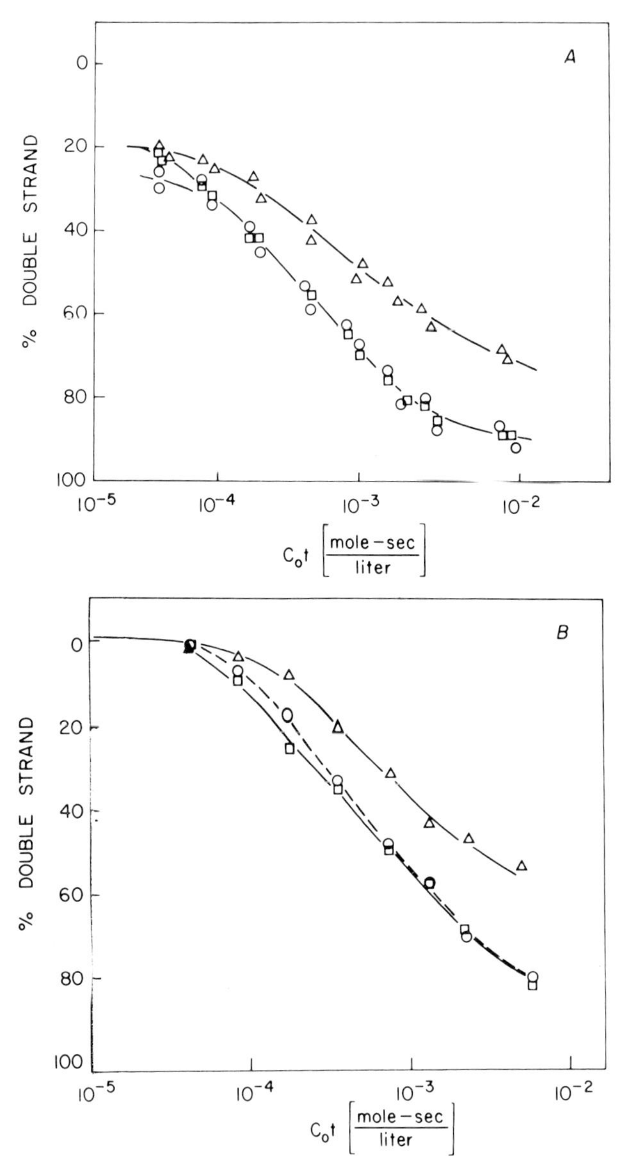

Fig. 10. Double stranded probe (5 ng/ml) and liver (○——○) or oviduct (□——□), DNA, prepared by the Marmur procedure, or salmon sperm (△——△) DNA (all at 5 mg/ml) were melted and reannealed. At different times aliquots were taken and assayed for secondary structure by either hydroxyapatite (panel A) or S_1 nuclease (panel B). (See ref. [19] for details.)

synthesized in the presence of actinomycin D) has been employed in DNA–RNA hybridization, again using resistence to S_1 single strand specific nuclease as a measure of the secondary structure. The RNA used in these studies was fractionated by immunoadsorption, as described previously, such that three fractions, original polysomes, immunoadsorbed polysomal RNA, and non-adsorbed polysomal RNA, were obtained. The inset in fig. 9B of data indicates relative amounts of ovalbumin mRNA as quantitated by the lysate system (fig. 9A). Fig. 9B shows kinetics of DNA–RNA hybridization, here presented as double reciprocal plots [22].

Note that the data fit well to a single straight line, indicating the probe is homogeneous to a single RNA species in the population of molecules, and that the different RNA preparations hybridize with the probe at different RNA preparations hybridize with the probe at different rates. The inset table in fig. 9B indicates that for all three species tested the amount of ovalbumin synthesized and the rate of hybridization with the specific DNA probe are inversely proportional. Thus we conclude that the DNA probe is homogeneous in its reaction with a single species of mRNA, and that ovalbumin mRNA and the RNA sequences complementary to the DNA probe have been copurified by specific immunoadsorption. Hence we conclude that the DNA probe, is, indeed, hybridizing with ovalbumin mRNA.

The DNA probe can now be used to determine the number of ovalbumin genes, and thereby answer the question of whether there is differential gene amplification

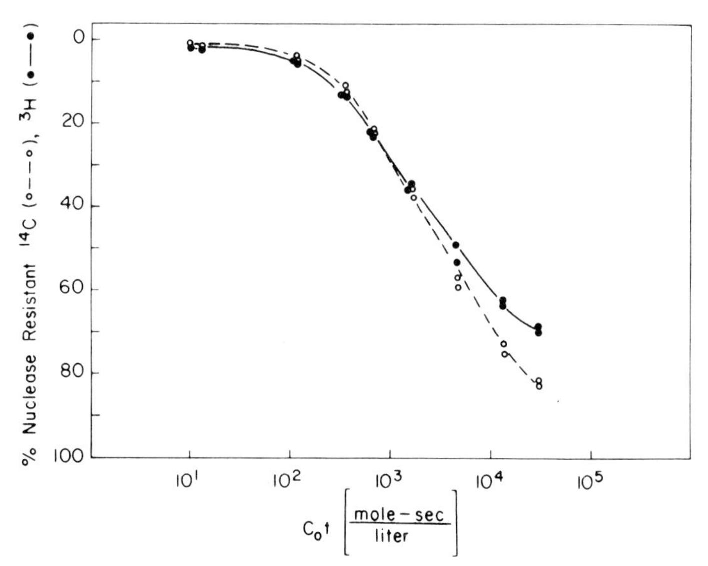

Fig. 11. Determination of the absolute copy number of ovalbumin sequences per chicken liver genome. Chicken liver DNA at 10 mg/ml (prepared by the technique of Varmus et al. [27], $[^{14}C]$ unique sequence chicken DNA at 30 µg/ml (○——○) and $[^{3}H]$ single stranded probe DNA at about 1 ng/ml (○——○) were mixed together, then melted and reannealed. At different times aliquots were taken and assayed for S_1 nuclease resistance. The data is plotted relative to the C_0t of the unlabeled liver DNA. (See ref. [19] for details.)

for ovalbumin. Fig. 10 shows an experiment using the technique of Gelb et al. [23], in which a constant amount of double stranded probe DNA (3 ng/ml) is reannealled with either chicken oviduct or chicken liver DNA (both at 5 mg/ml) or self annealled. To control for the effect of the high viscosity of the solution in the test reactions, the self annealling reaction was performed in the presence of 5 mg/ml salmon sperm DNA, with which the probe does not react. If the added cellular DNAs contain sequences complementary to the probe DNA, the apparent rate of reassociation of those probe sequences should be accelerated by an amount proportional to the number of complementary cellular sequences added. As seen in fig. 10, both liver and oviduct DNA accelerate the rate of hybridization to essentially the same degree when present at identical concentrations. Hence we conclude that the DNA from each tissue contains the same number of ovalbumin sequences per unit mass of DNA, i.e., that there is no differential amplification of the ovalbumin gene in oviduct, the only tissue in which ovalbumin is synthesized in detectively amounts.

To determine the absolute number of ovalbumin genes in the chick genome, the experiment shown in fig. 11 was performed. A large amount of unlabelled chicken liver DNA was melted and allowed to reanneal in the presence of trace amounts of ^{14}[C] labeled unique sequence chicken DNA (open circles) and ^{3}[H] labeled ovalbumin specific DNA (closed circles). The rate of reassociation of the labeled DNAs should depend on the concentration of complementary sequences in the unlabeled mass DNA. Fig. 11 clearly indicates that both ovalbumin specific and unique sequence DNA reassociate at the same rate, therefore ovalbumin sequences are only present at a concentration of 1 copy per haploid genome in chicken DNA.

6. Discussion

The results presented here are not meant to provide complete answers to the question of which the multiple regulatory steps is rate limiting for the synthesis of the major egg-white protein, ovalbumin. Nor does this discussion attempt to review the literature in the field. We can presently state that the content of translatable ovalbumin mRNA appears to be one of the important rate-limiting reactants in the system. Likewise we can state that differential gene amplification is not a factor.

Most importantly, perhaps, is the description of various techniques utilized in our laboratory that provide the tools for answering fundamental questions, including mRNA assay and the isolation of mRNA (polysomes) by immunologic techniques. With these, and in particular with the DNA probe, we can now begin an assessment of the roles of gene transcription and mRNA transport in the regulation of protein synthesis.

An answer to the question of why the oviduct synthesizes such great proportion of its protein ovalbumin remains somewhat unclear. We believe that a major determinant is the stability of this mRNA [9, 24, 6].

The exact mechanisms that determine differential mRNA stability will require the isolation of a number of pure mRNAs and a study of their physical and biological properties. We believe that the immunologic approach offers such an opportunity.

The finding of a single ovalbumin genes per haploid genome raises a difficult problem concerning the number of binding sites in the nucleus for a specific hormone. Although data is not available for the number of binding sites in oviduct nucleus for estrogens, in other steroid target tissues this number approximates 4000 to 10,000 per nucleus [25, 26, 27]. In one assumes a simple regulatory system in which a nuclear binding protein–steroid complex binds at a specific regulatory site on the DNA to allow transcription of single structural genes (such as ovalbumin), then the number of binding sites and proteins appears enormous relative to the probable number of proteins whose synthesis is regulated by estrogen. It may well be, then, that steroid hormone binding proteins are heterogeneous with respect to molecular structure and to function. Thus the potential role of such protein–steroid complexes as modifiers of either RNA polymerases or specific transport molecules must also be considered.

References

[1] T. Oka and R.T. Schimke, J. Cell Biol. 41 (1969) 816.
[2] T. Oka and R.T. Schimke, J. Cell Biol. 43 (1969) 123.
[3] B.W. O'Malley, W.L. McGuire, P.O. Kohler and S.G. Korenman, Rec. Progr. Horm. Res. 25 (1969) 105.
[4] R.D. Palmiter and J. Wrenn, J. Cell Biol. 50 (1971) 598.
[5] R. Palmiter, A.K. Christensen and R.T. Schimke, J. Biol. Chem. 245 (1970) 833.
[6] R.D. Palmiter, T. Oka and R.T. Schimke, J. Biol. Chem. 246 (1971) 724.
[7] G.M. Tomkins, B.B. Levinson, J.D. Baxter and L. Dethlefsen, Nature New Biol. 239 (1972) 88.
[8] G.M. Tomkins, T.D. Gelehrter, D. Granner, D. Martin, H.H. Samuels and E.B. Thompson, Science 166, 3912 (1969) 1474.
[9] R.D. Palmiter and R.T. Schimke, J. Biol. Chem. (1973) in press.
[10] R.E. Rhoads, G.S. McKnight and R.T. Schimke, J. Biol. Chem. 246, 23 (1971) 7407.
[11] R. Palacios, R.D. Palmiter, R.T. Schimke, J. Biol. Chem. 247 (1972) 2316.
[12] R. Palacios, D. Sullivan, N.M. Summers, M.L. Kiely and R.T. Schimke, J. Biol. Chem. (1973) in press.
[13] G. Brawerman, J. Mendecki and S.Y. Lee, Biochemistry 11 (1972) 637.
[14] H. nakazato, and M. Edmonds, J. Biol. Chem. 247, 10 (1972) 3365.
[15] J. Ross, H. Aviv, E. Scolnick and P. Leder, Proc. Natl. Acad. Sci. U.S. 69, 1 (1972) 264.
[16] I.M. Verma, G.F. Temple, H. Fan and D. Baltimore, Nature New Biol. 235, 58 (1972) 163.
[17] M. Adesnik, M. Salditt, W. Thomas and J.E. Darnell, J. Mol. Biol. 71 (1972) 21.
[18] A.J. Faras, J.M. Taylor, J.P. McDonnell, W.E. Levinson and J.M. Bishop, Biochem. 11, 12 (1972) 2334.
[19] D. Sullivan, R. Palacios, J. Stavnezer, J.M. Taylor, A.J. Faras, M.L. Kiely, N.M. Summers, J.M. Bishop and R.T. Schimke, J. Biol. Chem. (1973) submitted.

[20] R. Palacios, R.D. Palmiter and R.T. Schimke, J. Biol. Chem. 247 (1972) 3296.
[21] J.P. McDonnell, A.-C. Garapin, W.E. Levinson, N. Quintrell, L. Fanshier and J.M. Bishop, Nature 228 (1970) 433.
[22] M.L. Birnstiel, B.H. Sells and I.F. Purdom, J. Mol. Biol. 63 (1972) 21.
[23] L.D. Gelb, D.E. Kohne and M.A. Martin, J. Mol. Biol. 71 (1971) 21.
[24] R.T. Schimke, R.E. Rhoads and G.S. McKnight, Methods of enzymology (1973) in press.
[25] P. Rennie and N. Bruchovsky, J. Biol. Chem. 245 (1972) 1546.
[26] S. Fang, K.M. Anderson and S. Liao, J. Biol. Chem. 244 (1969) 6584.
[27] D. Toft, G. Shyamala and J. Gorski, Proc. Natl. Acad. Sci. U.S. 57 (1967) 1740.
[28] H.E. Varmus, W.E. Levinson and J.M. Bishop, Nature New Biol. 233 (1971) 19.

Niu and Segal (eds.). The role of RNA in reproduction and development
North-Holland Publ. Co., 1973

Regulation of albumin synthesis in cultured mouse hepatoma cells*

John PAPACONSTANTINOU and Barry E. LEDFORD**
*Biology Division, Oak Ridge National Laboratory,
Oak Ridge, Tennessee, USA*

1. Introduction

The differentiation of cells into specific tissues is usually accompanied by the acquisition of morphological and biochemical specificity. In most cases these highly differentiated cells can synthesize large quantities of tissue specific structural proteins. This is clearly seen in the synthesis of crystallins in the lens [1, 2], globin in the reticulocyte [3], albumin in the liver [4], and other highly differentiated tissues.

In recent years cultured cells have been used extensively to study the regulation of synthesis of tissue specific proteins. The development of new techniques and culture media for the maintenance of differentiated functions has facilitated studies on the regulation of tissue specific protein synthesis under more chemically defined conditions. Although some of these cells can be maintained in continuous culture, it is usually the long-term primary cultures such as the chick myoblasts [5], pigment cells [6], thyroid cells [7], or organ cultures such as the mammary gland [8, 9] which maintain their biochemical specificity. These culture systems have been extremely useful in studies on the regulation of protein and nucleic acid synthesis in vitro. Similarly it has been observed in many laboratories that tumors of highly differentiated tissues will grow indefinitely in culture and will maintain their biochemical specificity. Thus, melanomas will synthesize large amounts of pigment [10], hepatomas synthesize serum proteins [11–15], and the induction of tissue specific enzyme and protein synthesis is observed in hepatomas [16, 17], neuroblastoma [18–20] and leukemia cells [21].

* Research sponsored by the U.S. Atomic Energy Commission under contract with the Union Carbide Corporation.
** Supported by U.S.P.H.S. Postdoctoral Fellowship GM51857.

It is apparent, therefore, that cells can be grown under well-defined chemical and physical conditions and that stimuli such as hormones or nutrients, which play a major role in the regulation of nucleic acid and protein synthesis, can be added to the medium to facilitate studies on the regulation of macromolecular synthesis.

An interesting cell line that we have chosen to use in our studies on the regulation of tissue specific protein synthesis is the mouse hepatoma derived by Bernhard, Darlington and Ruddle [22]. It has been shown by Bernhard et al. that these cells synthesize albumin which is secreted into the culture medium. In the studies to be presented below we have confirmed that these cells synthesize and secrete proteins into the culture medium, the major one being mouse serum albumin. We are able to show that the rate of synthesis is determined by the generation time, and are proposing that this relationship may be explained by discontinuous synthesis of the albumin mRNA during the cell cycle [23].

2. Results

2.1. Identification of albumin synthesis by Hepa cells

Electrophoretic analyses were carried out to establish an assay for mouse serum albumin. The pattern shown in fig. 1a represents the resolution of mouse serum

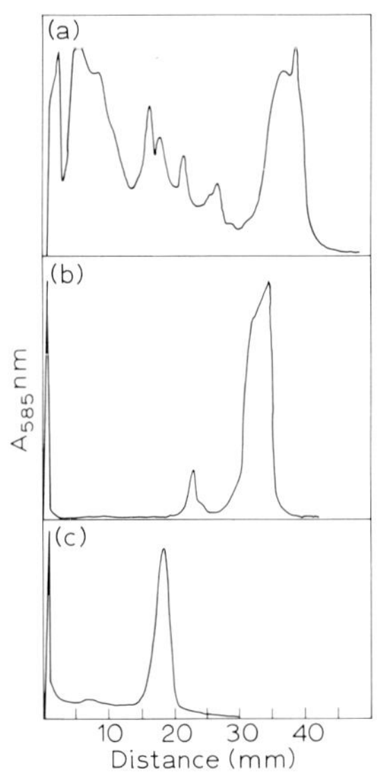

Fig. 1. An electrophoretic analysis of (a) mouse serum proteins. (b) Mouse serum albumin extracted from whole serum by the ethanol-TCA procedure [24]. The fast moving component represents albumin. (c) SDS-acrylamide gel electrophoresis of the ethanol-TCA extracted albumin after reduction with 10^{-2} M dithiothreotol. Standard 7% acrylamide gels, pH 8.9, were used in (a) and (b). A 10% acrylamide, 0.1% SDS gel was used in (c).

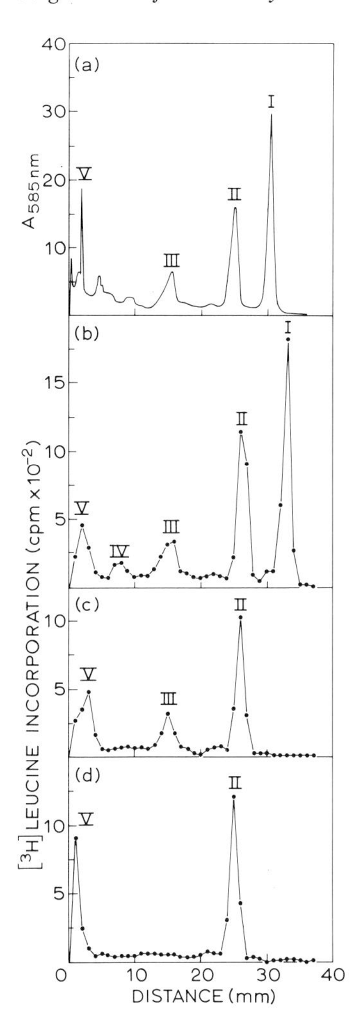

Fig. 2. An electrophoretic analysis of the proteins secreted by the hepatoma BW7756 in tissue culture: (a) Acrylamide gel electrophoresis of the secreted proteins. The gels were stained with Coomasie Blue and scanned at 585 nm. (b) Acrylamide gel electrophoresis of radioactive serum proteins secreted by Hepa cells in the presence of $[^3H]$leucine. (c) Electrophoresis of the secreted radioactive proteins after the mixture was treated with anti-mouse albumin anti-serum. The elimination of peak I by the specific antibody identifies this protein to be albumin. (d) Electrophoretic analysis of the secreted radioactive proteins after the mixture was treated with antibody to whole adult mouse serum. The pattern indicates that peaks II and V are not found in adult serum. Conditions of electrophoresis are the same as in fig. 1.

proteins. Albumin was specifically extracted from the serum by the ethanol-TCA extraction procedure [24]. An electrophoretic pattern of the ethanol-TCA extract is shown in fig. 1b. This pattern shows that albumin is the leading band in the electrophoresis. The minor, trailing band is probably the dimer form of albumin.

After reduction with dithiothreitol, SDS acrylamide gel analysis of the ethanol-TCA extract results in formation of the albumin monomer form which migrates as a single band (fig. 1c).

The mouse hepatoma cells (Hepa) used in these studies were established from the hepatoma BW7756 by Bernhard et al. [22]. These cells have been shown to secrete significant quantities of protein into serum-free culture medium. An electrophoretic analysis of the proteins secreted by the Hepa cells is shown in fig. 2a. Approximately five protein bands could be resolved, and on the basis of electrophoretic mobility the albumin fraction is indicated by I. A similar analysis was done after incubating the Hepa cells in serum-free medium with [^{3}H]leucine. It can be seen from the pattern in fig. 2b that the radioactivity follows the optical density pattern. Treatment of this protein mixture with antiserum to mouse albumin resulted in the specific elimination of albumin (peak I). In addition, when the secreted protein mixture is treated with antiserum to mouse serum proteins, peaks I, III and IV were precipitated from the mixture. Thus, peak II, which is a major component of the proteins secreted into the medium, is not a component of adult mouse serum.

2.2. Analysis of the rate of albumin synthesis by Hepa cells

Having shown that one of the major proteins secreted by the Hepa cells is albumin, an experiment was done to determine the rate of secretion of albumin into the

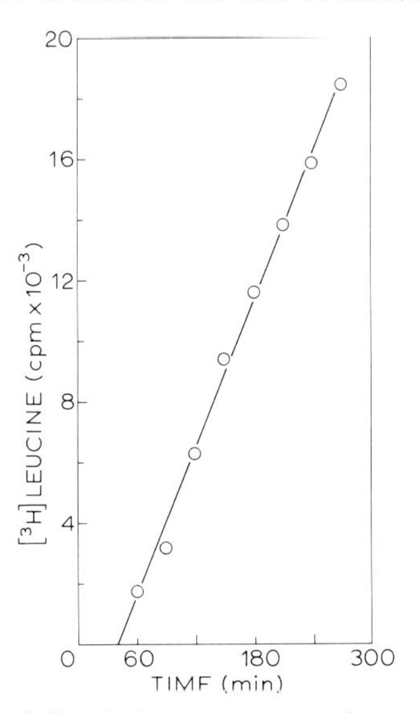

Fig. 3. The rate of secretion of albumin by mouse hepatoma cells in tissue culture. The cells were grown in Dulbecco's modified MEM medium supplemented with 10% calf serum. 10 μCi/ml [^{3}H]leucine was added at zero time. Aliquots (0.01 ml) of the medium were taken, albumin extracted and counted. Counts were corrected to indicate incorporation by the entire culture.

medium. The mixture of secreted proteins was treated with ethanol-TCA to extract albumin, and the rate of secretion of albumin and nonalbumin protein was determined. The data for the rate of albumin secretion are shown in fig. 3. A lag period of approximately 45 min occurs before albumin can be detected in the medium. This lag period presumably represents the time required for synthesis of the protein and movement through the secretory apparatus. Our recent data indicate that the rate of secretion is the same as the rate of synthesis of albumin. Thus, the data presented in fig. 3 represent true rates of albumin synthesis.

Hepa cells have been successfully cultured in Waymouth's MAB 87/3 medium and in Dulbecco's modified MEM in our laboratory. Significant differences in the rate of replication of these cells were observed in medium with and without serum. For example, the generation time of Hepa cells in Waymouth's medium supplemented with insulin and fetal calf serum was 15 hr (fig. 4a), whereas in medium without serum the generation time increased to 26 hr (fig. 4b). Analysis of the rate of albumin synthesis by these cells, under different growth conditions, was undertaken. The results of these experiments are also shown in fig. 4. In the presence of calf serum, the rate of albumin synthesis remains at a constant level during the logarithmic growth phase (fig. 4a). Thus, for the first four days of culture, the rate of albumin synthesis is at a level of 7.5×10^{10} molecules/min/culture. Between days 4 and 5, when the cells are in a stationary phase, there is an abrupt increase in

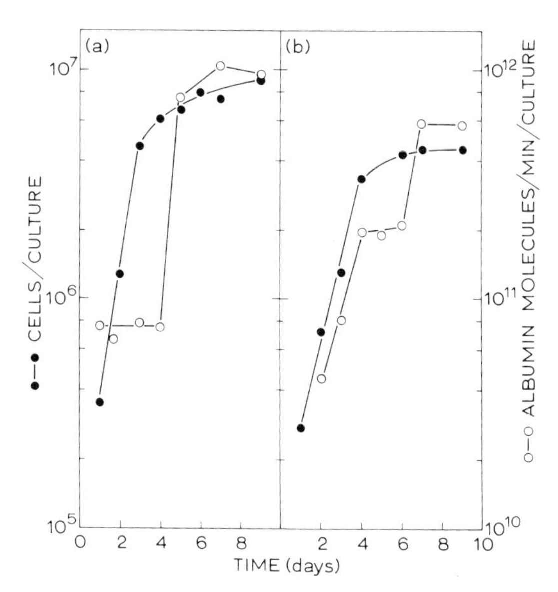

Fig. 4. The effect of fetal calf serum on the rate of growth and rate of albumin synthesis by Hepa cells. (a) Hepa cells grown in Waymouth's medium supplemented with insulin and fetal calf serum; (b) cells grown in Waymouth's medium plus insulin. Rates of albumin synthesis were determined from incorporation rates as in fig. 3.

the rate of synthesis to a new level of 9.8×10^{11} molecules/min/culture. These data show that there are two phases of albumin synthesis in cells grown under optimal conditions. During the logarithmic growth phase, the rate of albumin synthesis is at a relatively low level. The highest level of albumin synthesis occurs in the stationary phase, suggesting that the control of albumin synthesis is linked to a specific phase of the cell cycle, namely, that in which the Hepa cells are arrested.

A similar pattern of synthesis is seen in the cells grown in the absence of serum (fig. 4b). Under these conditions the growth of the cultures is slowed down to a generation time of 26 hr. It can be seen that during the log phase (from days 2–4) there is an increase in the rate of albumin synthesis from 4.5×10^{10} molecules albumin/min/culture to 2×10^{11} molecules albumin/min/culture. When the cells enter the logarithmic growth phase, the rate of albumin synthesis parallels the growth curve; as the cells approach the stationary phase a lag in the rate of albumin synthesis is seen, and finally in the stationary phase a second increase in the rate of synthesis is seen. Although there is a biphasic curve in the slower growing serum free cultures, the increase in albumin synthesis is seen when the cells are not replicating (i.e. in the stationary phase).

The experiments described above have shown that Hepa cells synthesize albumin in Waymouth's medium. Similar studies were done to determine whether these cells will synthesize albumin in Dulbecco's medium. These data are shown in fig. 5. During the logarithmic growth phase, the rate of albumin synthesis per culture

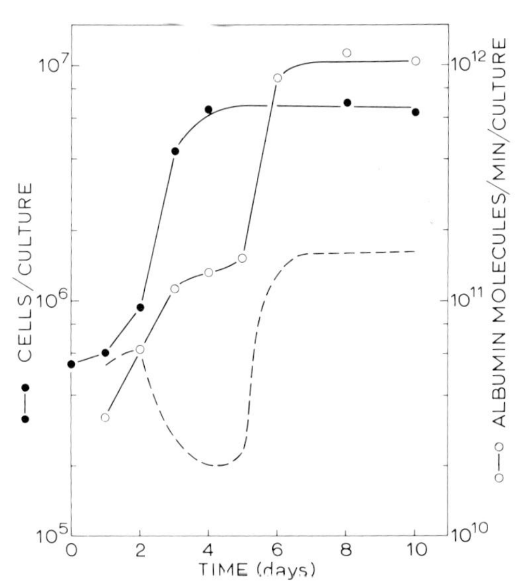

Fig. 5. The rate of albumin synthesis and cellular replication of Hepa cells in Dulbecco's medium plus fetal calf serum. ●——●, cells/culture; o——o, albumin molecules/min/culture; – – –, albumin molecules/min/cell. Synthetic rates were determined from incorporation rates as shown in fig. 3.

increases; this rate of synthesis plateaus as the cells approach stationary phase, and after approximately 24 hr in stationary phase there is an abrupt increase in albumin synthesis per culture. When the data are expressed as albumin synthesis/cell (dotted line in fig. 5), it can be seen that during the log growth phase each cell makes less albumin and that the increase in albumin/culture is due to the increase in the cell population. As the cells enter stationary phase, albumin synthesis/cell levels off, and 24 hr later the cellular synthetic rate increases abruptly reaching its highest level in stationary cells. Thus, albumin synthesis is inversely proportional to the rate of cell replication. A similar pattern of albumin synthesis is seen in cells grown in Waymouth's medium (fig. 4).

2.3. The effect of actinomycin D on albumin synthesis in stationary and log phase cells

The data in fig. 5 show that albumin synthesis is at its highest level in stationary phase cells. On this basis it would appear that the mRNA for this protein is being actively synthesized in these cells. To test for this, cells were incubated in the presence of [^{3}H]leucine and actinomycin D (1 μg/ml) for a period of 24 hr, and the rate of albumin synthesis was determined. The results of this experiment are shown in fig. 6. These data show that albumin synthesis remains constant for a period of 24 hr under conditions in which 96% of the RNA synthesis is inhibited. These observations indicate that the albumin mRNA is very stable and that it is no longer being synthesized in stationary phase cells. This is consistent with earlier

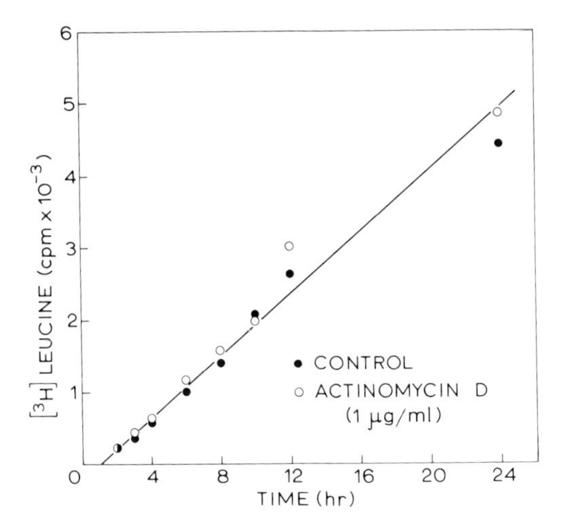

Fig. 6. The effect of actinomycin D on the synthesis of albumin by mouse hepatoma cells in stationary phase. The cells were grown in Dulbecco's medium supplemented with 10% fetal calf serum. Actinomycin D (1 μg/ml) o——o. Control •——•. [^{3}H]leucine and actinomycin D were added to the culture medium at time zero.

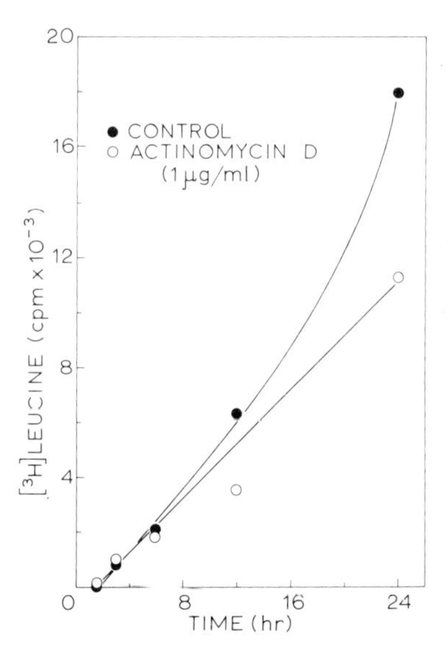

Fig. 7. The effect of actinomycin D on the synthesis of albumin by mouse hepatoma cells in logarithmic growth phase. The conditions of culturing are the same as in fig. 6. Control ○——○. Actinomycin D (1 μg/ml) ●——●.

observations of the intensitivity of albumin synthesis to actinomycin D in vivo [25, 26].

A similar series of experiments was done using Hepa cells in the logarithmic growth phase. These data are shown in fig. 7. In the presence of actinomycin D the rate of albumin synthesis remains constant over a period of 24 hr, indicating that the albumin template is stable in a dividing population of cells as well as in stationary cells (fig. 6). The rate of synthesis in the control cells also remains constant during the first 12 hr of the incubation period. However, from 12–24 hr there is a significant increase in the rate of synthesis of albumin. This increase in albumin synthesis by control cells reflects the slow increase observed in logarithmically growing cells shown in fig. 5. This increase is blocked by treatment with actinomycin D suggesting that synthesis of albumin mRNA is responsible for the observed increase in the rate of albumin synthesis.

3. Discussion

In these studies we have shown that albumin is one of the major proteins secreted by the mouse hepatoma line (Hepa). These results are in agreement with the observations reported by Bernhard et al. [22]. The other major proteins secreted by these cells have been tentatively identified as α-feto-protein and haptoglobin. These studies show that Hepa, which was derived from the mouse hepatoma BW7756 [22] and which has been maintained in culture for many transfers, has retained several very specific biochemical characteristics of the liver parenchymal cell. Furthermore, our data show that the amount of albumin synthesized (and secreted) per cell weight is comparable to that observed in vivo [4].

One of the major goals of this project was to study the regulation of albumin synthesis in a dividing and resting population of cells. Our first series of experiments provided evidence that the regulation of albumin synthesis is linked to the replicative activity of the cell. The data showed that albumin synthesis is inversely proportional to the rate of cellular growth. Thus, during the logarithmic growth phase, the rate of albumin synthesis is less than 20,000 molecules/cell/min, and when the cells enter stationary phase the rate increases to over 100,000 molecules/cell/min. These observations indicate that the regulation of albumin synthesis may be linked to a specific phase of the cell cycle. We propose, therefore, that when the generation time of the cell is minimal, the period of albumin synthesis would be minimal, resulting in the synthesis of a low number of mRNA molecules, this level of template concentration is expressed in the low level of albumin synthesis seen in the log phase cells. When the cells go into stationary phase, the generation time increases. (This is usually achieved by an extension of the G_1 phase of the cell cycle, but remains to be shown experimentally for the Hepa cells.) If we tentatively assign the period of albumin mRNA synthesis to the G_1 phase, then an extension of this phase in stationary cells could result in an extension of the period of albumin mRNA synthesis, thus resulting in an increase in the number of mRNA molecules. This in turn would account for the increase in rate of albumin synthesis observed in stationary phase cells. We are presently attempting to synchronize these cells to determine whether the synthesis of mRNA is actually localized in a specific phase of the cell cycle.

Further evidence which suggests that albumin mRNA synthesis may be discontinuous arises from the observation that actinomycin D will inhibit the increase in albumin synthesis in log phase cells. During a 24-hr incubation in the presence of actinomycin D (1 μg/ml), there is no significant inhibition for the first 12 hr. However, from 12–24 hr, the increase in albumin synthesis normally seen in control cells is blocked. During this 24-hr period there is an increase in the cell population, and the daughter cells would pass through a period of albumin mRNA synthesis. This period of transcription does not occur in actinomycin treated log phase cells.

Finally, our data support the conclusion that albumin mRNA in Hepa cells is very stable. This is indicated by the observation that actinomycin D has no effect on albumin synthesis by stationary cells. This high degree of stability is also seen in log phase cells where albumin synthesis remains linear for a period of 24 hr in the presence of actinomycin. The high degree of stability of this template has been shown to occur in vivo [25, 26]. It has been suggested that the formation of membrane bound albumin polysomes may play an important role in the stabilization of this template [27].

In conclusion, we propose the following sequence of events is involved in the regulation of albumin synthesis in Hepa cells: (a) albumin mRNA is synthesized during a specific phase of the cell cycle, we tentatively assign this period to the G_1 phase; (b) increasing the generation time increases the duration of the period of mRNA synthesis, thereby increasing the rate of albumin synthesis; (c) albumin mRNA shows a high degree of stability in cultured cells, and this stability may be achieved by the formation of membrane bound polysomes.

Acknowledgements

We wish to thank Drs. H.P. Bernhard, G.J.Darlington, and F.H. Ruddle for having made the mouse hepatoma cells available to us for these studies. The excellent technical assistance of Mrs. Anna M. Henley is gratefully acknowledged.

References

[1] J. Delcour and J. Papaconstantinou, J. Biol. Chem. 247 (1972) 3289.
[2] K. Yoshida and A. Katoh, Exp. Eye Res. 11 (1971) 184.
[3] K. Scherrer and L. Marcaud, J. Cell Physiol. 72 (suppl. 1) (1968) 181.
[4] T. Peters, Jr. and J.C. Peters, J. Biol. Chem. 247 (1972) 3858.
[5] I.R. Konigsberg, Science 140 (1963) 1273.
[6] R.D. Cahn and M.B. Cahn, Proc. Natl. Acad. Sci. U.S. 55 (1966) 104.
[7] B.S. Spooner, J. Cell Physiol. 75 (1970) 33.
[8] R.W. Turkington, In: Biochemical actions of hormones, Vol. 2. Ed. G. Litwack. (Academic Press, New York, 1971) 315-355.
[9] R.W. Turkington, D.H. Lockwood and Y.J. Topper, Science 156 (1967) 945.
[10] G.E. Moore, Exptl. Cell Res. 36 (1964) 422.
[11] A.H. Tashjian, Jr., F.C. Bancroft, I.U. Richardson, M.B. Goldlust, F.A. Rommel and P. Ofner, In Vitro 6 (1970) 32.
[12] S.H. Ohanian, E.B. Taubman and G.J. Thorbecke, J. Nat. Cancer Inst. 43 (1969) 397.
[13] H.-M. Rotermund, G. Schreiber, H. Maeno, U. Weinssen and K. Weigand, Cancer Res. 30 (1970) 2139.
[14] G. Schreiber, H. Rotermund, H. Maeno, K. Weigand and R. Lisch, European J. Biochem. 10 (1969) 355.
[15] U.I. Richardson, A.H. Tashjian, Jr. and L. Levine, J. Cell Biol. 40 (1969) 236.
[16] F.T. Kenney, J.R. Reel, C.B. Hager and J.L. Wittliff, In: Regulatory mechanisms for protein synthesis in mammalian cells. Eds. A. San Pietro, M.R. Lamborg and F.T. Kennety. (Academic Press, New York, 1968) 119.
[17] G.M. Tomkins, T.D. Geleherter, D. Granner, D. Martin, Jr., H.H. Samuels and E.B. Thompson, Science 166 (1969) 1474.
[18] A. Blume, F. Gilbert, S. Wilson, J. Farber, R. Rosenberg and M. Nirenberg, Proc. Natl. Acad. Sci. U.S. 67 (1970) 786.
[19] G. Augusti-Tocco and G. Sato, Proc. Natl. Acad. Sci. U.S. 64 (1969) 311.
[20] D. Schubert, S. Humphreys, F. De Vitry and F. Jacob, Dev. Biol. 25 (1971) 514.
[21] C. Friend, W. Scher, J.G. Holland and T. Sato, Proc. Natl. Acad. Sci. U.S. 68 (1971) 378.
[22] H.P. Bernhard, G.J. Darlington and F.H. Ruddle, Dev. Biol. (1973) in press.
[23] B.E. Ledford and J. Papaconstantinou, Fed. Proc. 32 (1973) 615 abs.
[24] J.R. Debro, H. Traver and A. Korner, J. Lab. Clin. Med. 50 (1957) 728.
[25] S.H. Wilson and M.B. Hoagland, Biochem. J. 103 (1967) 556.
[26] S.H. Wilson, H.Z. Hill and M.B. Hoagland, Biochem. J. 103 (1967) 567.
[27] T. Tanaka and K. Ogata, Biochem. Biophys. Res. Commun. 49 (1972) 1069.

Session Two

RNA Programmed Protein Synthesis in Cell-Free Systems

Chairman

PHILIP LEDER

Laboratory of Molecular Genetics,
National Institute of Child Health and Human Development,
National Institutes of Health, Bethesda, Maryland

(Manuscripts of the papers in this session were not received.)

Session Three

RNA Effects on in vivo Synthesis of Specific Proteins

Chairman

VINCENT G. ALLFREY

Rockefeller University, New York, New York

Chairman's introduction

In considering RNA mediation of specific protein synthesis, a clear distinction can be made between the role of RNA in translational mechanisms, emphasizing its messenger function, and the effects of exogenous RNA's on intact cells, leading to alterations in cell function and development. Earlier presentations in this symposium have dealt with the translation of specific mRNAs in vitro. The papers now to be presented emphasize the problem of how intact cells utilize and respond to RNA molecules. This approach, testing the capacity of exogenous RNAs to "transform" cell functions, to modify cell structure, or to influence programs of development, has its origins in the pioneering work of Dr. M.C. Niu, who in 1956 reported the first experiments in this important area.

Inevitably, the impact of these findings and their interpretations are strongly influenced by the history of bacterial cell transformations with exogenous DNA, and by the clear evidence for transfer of genetic information by transducing phages. There are clear alternatives in RNA action which warrant presentation; the first possibility is that ingested RNA is used *directly* in protein synthesis. For example, experiments by Dr. Claude Villee (cited below) indicate that synthetic polynucleotides such as polyuridylic acid and polycytidylic acid do increase the incorporation of phynylalanine and proline, respectively, in the target cells. Thus, exogenous RNAs can be translated. Since other workers have shown that exogenous *tRNAs* can be utilized for amino acid transfer in the recipient cells, it is clear that the potential for *direct* mediation of exogenous RNAs in different stages of protein synthesis is a real one.

Other studies suggest that exogenous RNAs may modify the transcriptional capacity of the cell and thus affect its future development. Indeed, many of the results described in this symposium are best explained on that basis, especially if one considers that the effects observed usually require considerable time for their expression. While the mechanisms through which transcription may be modified by RNAs remain largely conjectural, the possibility that low-molecular weight chromosomal RNAs may affect the specificity of transcription in chromatin has received much attention, notably by Dr. James Bonner and coworkers.

In comparison, studies of transcriptional control and of cell transformation by proteins are far along. For example, the isolation of the *lac* repressor in *E. Coli* and the C_1 gene repressor in *lambda* bacteriophage has established that proteins alone have the capacity to bind to specific polynucleotide sequences in DNA and to suppress their role in transcription. Conversely, the cAMP-receptor protein in *E. coli* exerts a *positive* control on the expression of the *lac* or *gal* genes. Similar specificities have emerged in studies of the nuclear phosphoproteins of eukaryotic cells which bind DNA selectively and stimulate the synthesis of specific RNA species.

The alterations in cell function induced by steroid hormones – usually evident in the induction of enzyme systems (such as cortisol induction of liver tyrosine amino transferase) or the production of characteristic cell products (e.g. ovalbumin induction by estrogens) are now interpreted in terms of hormonal interaction with specific *protein* receptors which enter the nucleus and modify patterns of transcription.

The discovery of the importance of cyclic AMP has also led to an emphasis on the role of proteins in genetic control, as cAMP has been found to modify the activity of the protein kinases affecting the structure of histones and other nuclear proteins. There is little doubt that such changes influence the structure and function of the genetic material.

Similarly, the existence of proteins which modify the function and development of intact cells is well documented. Examples include the "transformation" of lymphocytes by phytohemagglutinin or Concanavalin A, the response of lymphocytes to protein antigens, the induction of nerve proliferation by nerve growth factors, and the stimulation of capillary growth by the tumor angiogenesis factor.

The main emphasis of this symposium, however, is on *alternative* control systems which may or may not interlock with the protein-controlled systems of the genetic apparatus. The range of effects of exogenous RNAs is far-reaching, and it includes responses affecting hormonal regulation of genetic activity, as well as steroid metabolism in the target tissue.

Clearly this is an area of great potential, complementary to other avenues of investigation of genetic control by DNA-associated proteins. The program now to be presented represents the latest in support of the view originally and persistently advocated by Dr. M.C. Niu, that exogenous RNAs have the capacity to enter cells and to influence their form and function.

Vincent G. Allfrey

Niu and Segal (eds.). The role of RNA in reproduction and development
North-Holland Publ. Co., 1973

A hormone-controlled RNA fraction regulating enzyme development in plant cells

R. KAUR-SAWHNEY and A.W. GALSTON*
Department of Biology, Yale University, New Haven, Conn., USA

Indoleacetic acid (IAA) which represses de novo isoperoxidase formation in excised tobacco pith, also increases the titre within the tissue of a macromolecular peroxidase repressor. This substance can be extracted from donor tissue and vacuum-infiltrated into receptor cells, where it causes repression of isoperoxidase formation.

The repressor factor (RF) is extractable by phenol, precipitable with cold ethanol, and is completely degraded by pronase and ribonuclease used successively. It thus presumably contains both protein and RNA components.

The bulk of the RF activity coincides with a 25 S RNA fraction isolated by preparative electrophoresis on acrylamide gel slabs. IAA fed to tissue stabilizes part of RF to in vitro degradation by pancreatic ribonuclease, which specifically hydrolyzes single-stranded RNA. The protective effect of fed IAA can be reversed by heating and cooling RF in vitro, but not by exposure to pronase. This indicates that IAA may control the titre of RF in tissue by increasing either double-strandedness or the quantity of a nonprotein protective coat around RF.

1. Introduction

Fresh tobacco pith tissue has low peroxidase activity, resolvable into two anionic isozymes by starch-gel electrophoresis at pH 8.4. After excision and aseptic incubation for ca 12 to 24 hr, its peroxidase activity rises dramatically, due to the formation of several new isoperoxidases, both cationic and anionic [3, 5]. In the intact plant the low in situ peroxidase activity is correlated with high titre of a macromolecular repressor fraction (RF) present in partially purified phenolic RNA extracts. RF can be extracted, purified and vacuum-infiltrated into receptor tobacco pith, in which it represses peroxidase formation [9]. The titre of RF in pith cells is increased by incubation with $\geq 10\,\mu$M 3-indoleacetic acid, IAA [8], which also acts to repress peroxidase formation [5, 15, 19]. Conversely, kinetin, which promotes peroxidase formation, decreases the repressor activity of RF [8].

The activity of RF is not affected by DNase, is partially destroyed by RNase or

* With the expert technical assistance of Whitney Adams, Jr. and Liu-Mei Shih.

pronase, and is totally destroyed by pronase and RNase used sequentially [9, 10]. Thus we had tentatively concluded that the RF contains active RNA and protein. We now present further evidence on the specificity and chemical nature of RF, together with information on the mechanism through which IAA controls its titre and activity.

2. Materials and methods

2.1. Plant material

For isoperoxidase studies the pith from the upper internodes of preflowering tobacco plants *cv Wisconsin* 38 grown in a greenhouse were used. For RF studies the pith from the entire plant of preflowering and flowering tobacco plants were used. To study the effect of IAA on RF, fresh pith sections were vacuum-infiltrated with various concentrations of IAA and cultured aseptically for 24 hr in moist chambers.

2.2. Extraction and assay of RF

RF was extracted with phenol by a slight modification of the Loening and Ingle procedure [11]. The phenol reagent contained 0.1% by weight of hydroxyquinoline, 10% m-cresol and an equal volume of 10 mM Tris-HCl buffer, pH 7.4 containing 1% sodium lauryl sulfate, 12 mg/ml bentonite and 10 mM merceptoethanol. The aqueous layer was removed, reextracted twice with the phenol reagent, dialysed overnight against buffer and the macromolecular fraction (RF) precipitated several times with cold ethanol. The fraction thus obtained was examined spectrophotometrically for its RNA content. The isoperoxidase patterns were examined by vertical starch gel electrophoresis and analyzed quantitatively as reported earlier [2]. To study the action of RF, a solution in 10 mM Tris-HCl buffer, pH 7.4 was vacuum-infiltrated into pith sections; controls were similarly infiltrated with buffer. The pith sections thus treated were incubated aseptically for 24 hr, after which isoperoxidases were separated on starch gel and assayed. Complete repression of all new isoperoxidase formation was caused by 0.33 mg/ml RNA in the extract (as determined by optical density measurements at 257 nm, the peak of absorption of the RF). This concentration was used for all experiments except where otherwise stated. Repressor activity was calculated relative to the buffer control.

2.3. Fractionation of RF

Exclusion chromatography on Sepharose B columns and electrophoresis on 2.4% acrylamide gels [12] were used to separate the components of RF.

2.4. Enzymatic degradation

RF was exposed either to RNase A (Worthington Biochemical Co.), pronase (Cal. Biochem.) or both as described earlier [9]. RNase solution was heated for 5 min in a boiling water bath to denature DNase and protease activity prior to use. Pronase solution was self-digested for 2 hr at 37°C to inactivate nucleases prior to use. The digested RF in some cases was dialyzed, vacuum-infiltrated into fresh pith, aseptically incubated for 24 hr and the repression of isoperoxidases determined by densitometry of developed starch gels. RNase and pronase when infiltrated alone showed no effect on formation of new isoperoxidases.

3. Results and discussion

3.1. Specificity of RF

Similar phenol extracts containing identical quantities of RNA from tobacco leaves and floral buds, green pea stems and yeast RNA (Worthington Biochemical Co.) were investigated for their ability to repress new isoperoxidase formation in tobacco pith explants. The results (table 1) show that at the concentration employed, the phenol extract from tobacco pith repressed the new isoperoxidases completely, while the extract from buds and leaves caused only a slight repression. The extract from green pea stems was still less active, whereas the yeast RNA promoted isoperoxidase formation markedly, probably because of its high content of cytokinins [22].

As the ready entry of macromolecules into plant cells is somewhat unexpected, we also investigated the possibility that the tobacco pith RF might be viral in

TABLE 1
Effect of RF from different sources on repression of new isoperoxidase formation

Source of RNA	Percent repression of new isoperoxidases
Tobacco pith	100
Tobacco floral buds	16
Tobacco leaves	10
Green pea stems [b]	7
Yeast	(–114 [a])
TMV [c]	2
TMV [d]	(–5 [a])

[a] Promotion.
[b] RF was extracted from the apical internodes of 14-day-old plants.
[c] From A. Marcus, Institute for Cancer Research, Philadelphia.
[d] From A. Siegel, University of Arizona, Tucson.

origin. Samples of tobacco mosaic virus (TMV) from two sources were extracted with phenol in the usual fashion. The extracted RNA was ineffective in repressing the new isoperoxidase formation when applied at the usual concentration of 0.33 mg/ml (table 1, c and d).

Tobacco pith RF, effective in repressing de novo peroxidase formation in that tissue, is without effect on polyphenoloxidase formation as measured by chlorogenic acid (CA) or 3,4-dihydroxyphenylalanine (DOPA) oxidation [23]. Also, tobacco pith RF is ineffective in repressing isoperoxidase formation in excised bean hypocotyls or green pea stem tissue, and bean hypocotyl RF is effective in bean but ineffective in tobacco (unpublished results of J. Novak in this laboratory). These results demonstrate that the RF from tobacco pith possesses some specificity in repressing the formation of the new isoperoxidases of tobacco pith explants.

3.2. Hormonal control of RF activity

Earlier observations that IAA causes increased RNA synthesis [8, 21] and represses new isoperoxidase development in pith explants [5], together with the above observations, suggested that IAA might control the titre of RF which in turn controls peroxidase formation. To test this, fresh pith explants were vacuum-infiltrated with various concentrations of IAA and cultured aseptically for 12 to 36 hr. RF was then extracted from these donors and equal amounts of the extracts (based on O.D. at 257 nm) were vacuum-infiltrated into receptor pith explants which were in turn cultured for 24 hr to determine the effect of the infiltrated material on isoperoxidase formation. The results (table 2) show that as the concentration of IAA is increased, the content of RNA and activity of RF both increase. If a unit of repressor activity is defined as that amount which causes a 1% repression of new isoperoxidase formation relative to the buffer control (calculated from areas under the densitometer scan curves), then 1 mM IAA increased the number of RF units in the tissue by 144% (from 1065 to 2595) while increasing the RNA content only 35% (from 3.7 to 5.0). Thus, the RNA synthesized in the presence of IAA must have been enriched in RF activity. We also observed that when excised pith is

TABLE 2
Effect of IAA on RNA synthesis and RF activity

IAA (mM)	RNA content of pith (mg/50 g fr wt)	Percent repression of new isoperoxidases (relative to buffer control)	Repressor units per mg RNA	Total repressor units
0	3.7	46	288	1065
0.001	3.8	65	405	1539
0.01	4.2	69	432	1814
0.1	4.8	72	447	2145
1.0	5.0	83	519	2595

Concentration of RF used was 0.16 mg/ml in RNA equivalents.

incubated for prolonged periods in the absence of IAA, RF activity decays progressively with increasing incubation time; however, when IAA is added to the incubation medium, the rate of this decay is markedly reduced.

The IAA-induced increase in activity of the RF is not removed by prolonged dialysis, but it is sensitive to 0.3 N KOH. This indicates that the additional repressor effect is not due to IAA adhering to the extracted macromolecule. Further evidence that IAA increases the activity of macromolecular components of RF is obtained from detailed studies of residual RF activity after digestion with RNase and pronase. When the RF extracted from fresh pith or pith cultured in the absence of IAA is digested sequentially with RNase and pronase or vice versa, the repressor activity is completely destroyed (fig. 1). However, when the RF is extracted from IAA-treated pith and similarly digested, about 30% of the repressor activity remains after such enzymatic treatment. These observations suggest that the RF from IAA-treated pith is more resistant to combined pronase-RNase degradation than is the RF from untreated pith or from freshly excised pith.

To determine whether IAA affects the RNA or protein component of RF, the active fractions from control and IAA-treated pith were digested separately with RNase or pronase and their repressor activity subsequently determined. The results in table 3 show that following digestion with RNase, 47% of the RF activity in control pith and 67% in IAA-treated pith is resistant to RNase. Thus an additional 20% of the RF is stabilized to RNase degradation by prior incubation of the tissue

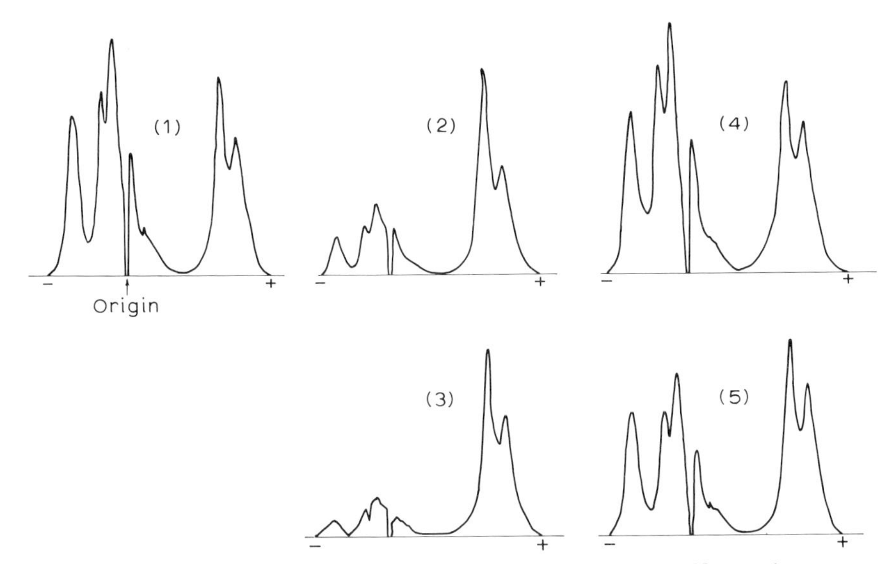

Fig. 1. Effect of IAA on the susceptibility of RF to degradation by RNase and pronase, as measured by the repression of new isoperoxidases. Peaks represent densitometric scans of isoperoxidases. Fresh pith sections were infiltrated with: (1) Buffer. (2) RF from control pith. (3) RF from 100 μM IAA-treated pith. (4) Like (2), but digested with RNase and pronase. (5) Like (3), but digested with RNase and pronase.

TABLE 3
Effect of IAA on the susceptibility of RF to degradation by RNase or pronase

New iso-peroxidases	Tissue infiltrated with			Tissue infiltrated with		
	Buffer	RF (−IAA)	RF (+IAA)	Buffer	RF (−IAA)	RF (+IAA)
		Hydrolyzed with RNase			Hydrolyzed with pronase	
C_3	73	40	32	48	41	33
C_2	65	24	19	37	14	14
C_1	92	48	21	52	32	35
A_1	21	14	9	19	12	15
A_2	30	24	12	26	29	33
Total	281	150	93	182	128	130
% Repression	0	47	67	0	30	29
Difference due to IAA			20			1

Numbers give peroxidase activity calculated from area under the densitometer scan curves. Concentration of IAA used in the incubation medium was 100 μM. RF from control and IAA-treated cultures gave 100% repression.

with IAA. A similar IAA-induced increase in stability of RNA to degradation by RNase in green pea stems has been reported previously [1]. Digestion with pronase, on the other hand, showed no difference in residual activity due to prior incubation with IAA. This indicates that the IAA-increased stability of RF is due not to the protein but to the RNA component of RF.

A kinetic study of the degradation of the repressor activity of RF by these enzymes (fig. 2) further shows that RF from IAA-treated pith is more stable to RNase degradation than RF from control pith, but that both are equally sensitive to pronase.

The data in table 3 show that RF from IAA-treated cultures represses formation of all 5 new isoperoxidases; the same has previously been shown to be true of IAA itself [3, 5]. This is further support for the view that the control of peroxidase formation by IAA is mediated by RF.

3.3. Partial characterization of RF

3.3.1. Fractionation on sepharose

3.3.1.1. Fresh pith. The RF from fresh pith was fractionated on Sepharose 4B columns. Optical density measurements at 257 nm, the absorption maximum of this fraction, revealed two peaks. The fractions corresponding to each peak were pooled, reprecipitated with 2½ volume of cold ethanol and their repressor activity

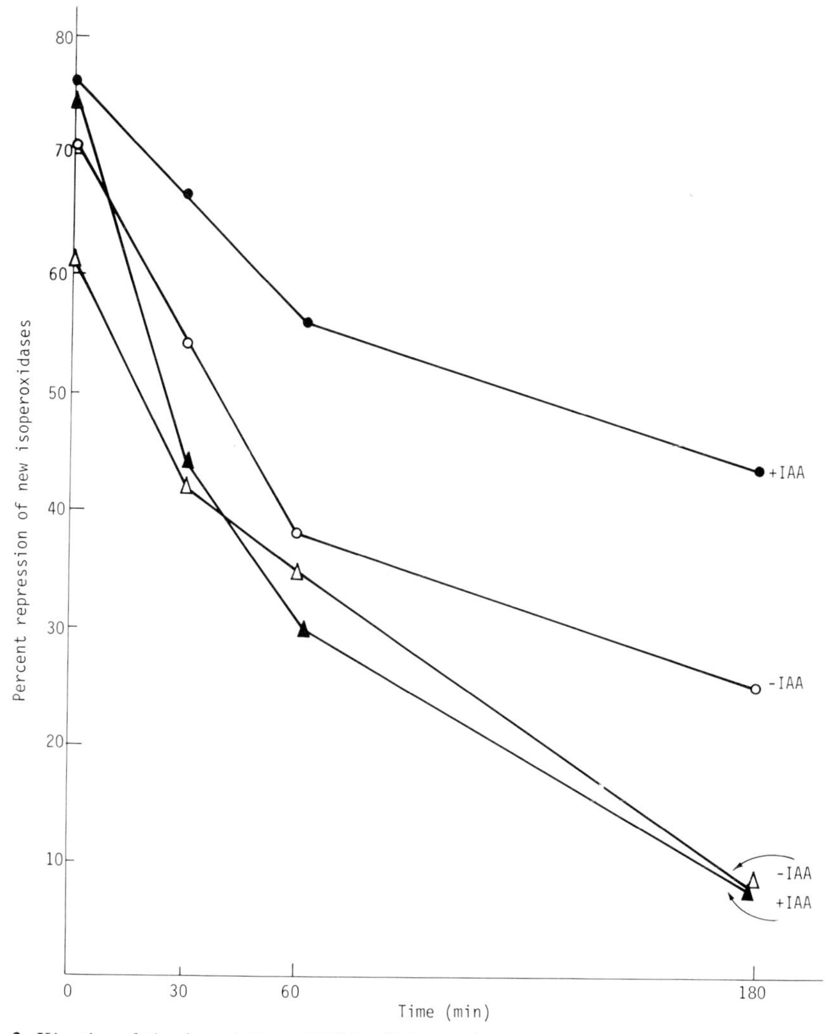

Fig. 2. Kinetics of the degradation of RF by RNase and pronase as affected by prior incubation of tissue with IAA. Rf extracted from control and 100 μM IAA treated-pith tissue was digested for 0.5 hr, 1 hr and 3 hr with RNase (10 μg/ml, ○) or pronase (60 μg/ml, △). The residual RF activities were determined by bioassay.

before and after digestion with RNase and pronase determined by bioassay on fresh pith. The results in table 4 demonstrate that there is more repressor activity in peak 1, more activity per unit RNA, greater susceptibility to RNase degradation and lower sensitivity to pronase. Clearly, in extracts of fresh pith, RF and RNA seem more concentrated in peak 1, while protein seems more abundant in peak 2.

TABLE 4
Fractionation of RF from fresh pith on a Sepharose column

	Peak 1	Peak 2
mg RNA	3.0	1.3
Total units RF	829	189
RF units/mg RNA	276	145
% activity destroyed by		
RNase	72	40
Pronase	39	60

Fresh pith sections were cut and frozen immediately. RF was extracted by the phenol method and dissolved in 10 mM Tris-HCl buffer, pH 7.4 containing 5 mM $MgCl_2$ and 50 mM NaCl and placed on a Sepharose 4B column (1 × 45 cm) and eluted with the same buffer: 1 ml fractions were collected and analyzed for absorbance at 260 nm. Fractions corresponding to each peak were collected and reprecipitated with $2\frac{1}{2}$ volumes of cold ethanol. The resulting precipitate was dissolved in 10 mM Tris-HCl buffer, pH 7.4. Suitable aliquots (0.33 mg/ml) were either not further treated or digested with RNase or pronase; they were then used for bioassays of repressor activity.

3.3.1.2. Pith incubated ± IAA. RF activity from pith cultured for 24 hr in the presence and absence of IAA resolved into 3 peaks on a sepharose column, the first two representing a splitting of peak 1 from fresh pith. The repressor activity and amount of RNA in the first two peaks were higher and again more resistant to RNase degradation in the RF from IAA-treated pith, but peak 3 showed no such differences (table 5). Thus, IAA treatment of tobacco pith increases the titre of RF-active, RNase-resistant RNA in fractions of peaks 1 and 2.

3.3.2. Fractionation on polyacrylamide gel

On electrophoretic analysis in a 2.4% polyacrylamide gel, RF from control and IAA-treated pith appeared heterogeneous, migrating as four peaks, two major and two minor (fig. 3). When RF was analyzed by a modified method of polyacrylamide gel slab electrophoresis [4, 16, 20], it migrated as 2 peaks, each representing a major and minor pair of fig. 3. The bulk of the RF from IAA-treated pith was found in the heavier peak 1, the size of which is in the 25 S range. These observations confirm the results obtained on sepharose fractionation, showing that the heavier peak 1 found in the presence of IAA is enriched in RF activity.

3.3.3. Effect of heating and cooling on RF

It is known that following heating and rapid cooling, double-stranded RNA which is resistant to degradation by pancreatic RNase separates into single-stranded RNA susceptible to the enzyme [13, 14]. RF from IAA-treated and control pith was heated in a boiling water bath, rapidly cooled and digested with RNase. The greater resistance of IAA-treated RF to RNase degradation disappeared completely after such heating and chilling treatment (table 6).

TABLE 5
Fractionation on a sepharose column of RF from pith incubated with and without IAA

	Peak 1 + 2		Peak 3	
	–IAA	+IAA	–IAA	+IAA
mg RNA	2.1	2.4	1.7	1.6
Repressor activity				
Total units	348	765	165	261
Units/mg RNA	165	319	97	163
Residual activity after RNase and pronase treatment	7	36	5	11

RF from control and 100 μM IAA treated pith was fractionated on a sepharose 4B column and treated as in table 4. Fractions corresponding to peak 1 and 2 were pooled, as were fractions corresponding to peak 3. Repressor activities before and after RNase and pronase treatment were determined by bioassay.

The effect of heat in removing the protective effect of IAA may be due either to the removal of a protective protein coating, thus permitting greater accessibility of the RF to RNase, or to separation of the RNA into single strands, which are then more susceptible to pancreatic RNase degradation. To help resolve this question, RF was digested with pronase prior to the heating and cooling treatment. The results in table 6 show that all RF activity from pith culture without IAA is destroyed by sequential treatment with pronase and RNase, while 25% of the RF activity from pith cultured with IAA remains after such treatment. However, interpolation of a heating and cooling treatment between the two enzymatic degradations results in a disappearance of this stabilized activity. It thus appears likely that the increased stability to RNase of RF derived from IAA-treated tissue is due not to

TABLE 6
Effect of enzyme and heat treatments on susceptibility to ribonuclease of RF from pith cultured with and without IAA

Treatment	% repression of new isoperoxidases	
	–IAA	+IAA (0.1 mM)
None	77	86
RNAse	43	62
Heat and cool	50	53
Heat, cool then RNAse	10	11
Pronase, then RNAse	(–1[a])	25
Pronase, heat and cool, then RNAse	(–2[a])	0

[a] Promotion.

RF was heated for 12 min in a boiling water bath and cooled immediately.

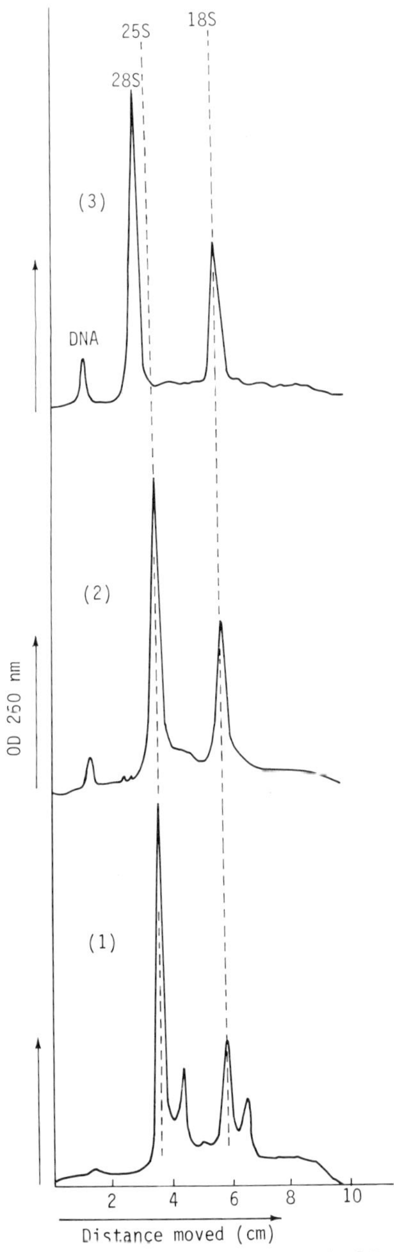

Fig. 3. Fractionation of RF by polyacrylamide gel electrophoresis. 25 μg of RF from control pith and pith incubated for 24 hr with 100 μM IAA were electrophoresed in a 2.4% polyacrylamide gel according to the method of Loening [12]. The migration profiles of RF from control and IAA-treated pith (1) were similar as detected by scanning the gels at 260 nm with a Gilford spectrophotometer equipped with a linear transport scanner. A similar gel separating markers of 25 S and 18 S from germinating axes of *Phaseolus vulgaris* (2) and 28 S and 18 S from *Xenopus laevis* (3) was electrophoresed at the same time.

a protective protein coat, but rather, to an increased double-strandedness in a portion of the RNA molecule.

3.4. Mode of action of RF

We have shown that IAA can be supplied to a specific tissue, causing an increase in the titre of a macromolecule (RF), presumably RNA in nature, with regulatory activity. The RF can be extracted from donor cells and when purified and reintroduced into receptor cells, it shows the same effect as the hormone itself, i.e., repression of peroxidase synthesis.

This finding raises several important questions. In the first place, all presently known repressors are proteins [6, 7, 17, 18]. Is RNA in RF active per se or must it first be translated in protein? Secondly, how does IAA affect RF titre? Is it by promoting the synthesis of a double-stranded RNA or through stabilizing the RNA to degradation by nucleases? Thirdly, is it the pronase-sensitive part of RF protein that is directly effective as repressor, and coded for by the RNA moiety which in turn is controlled by the hormone? Or is it simply a protective coat, which, once removed from around the RNA core of RF, permits degradation of the latter by endogenous nucleases? We hope to pursue these questions in the future.

Acknowledgment

We are indebted to Drs. S. Altman and D. Soll of Yale University for their useful suggestions and to the National Institutes of Health for generous support through a grant to the last author.

References

[1] F.E. Bendaña and A.W. Galston, Science 150 (1965) 69.
[2] H. Birecka and A.W. Galston, J. Exptl. Bot. 21 (1970) 735.
[3] H. Birecka, L.M. Shih and A.W. Galston, J. Exptl. Bot. 23 (1972) 655.
[4] R. de Wachter and W. Fiers, In: Methods in enzymology, Vol. XXI, Eds. L. Grossman and K. Moldave (1971) 167–178.
[5] A.W. Galston, S. Lavee and B.Z. Siegel, In: Biochemistry and physiology of plant growth substances, Eds. F. Wightman and G. Setterfield (Runge Press, Ottawa, 1968) 455–472.
[6] W. Gilbert and B. Muller-Hill, Proc. Natl. Acad. Sci. U.S. 56 (1966) 1891.
[7] F. Jacob and J. Monod, J. Mol. Biol. 3 (1961) 318.
[8] R. Kaur-Sawhney and A.W. Galston, In: Hormonal regulation in plant growth and development. Proc. Adv. Study Ist. Izmir, Eds. H. Kaldewey and Y. Vardar (Verlag Chemie, Weinheim, 1972) 27–35.
[9] Y. Leshem and A.W. Galston, Phytochem. 10 (1971) 2869.
[10] Y. Leshem, A.W. Galston, R. Kaur-Sawhney and L.M. Shih, In: Plant growth hormones, Ed. D.J. Carr (Springer Verlag, Heidelberg, 1970) 228.

[11] U.E. Loening and J. Ingle, Nature 215 (1967) 363.
[12] U.E. Loening, Fractionation of RNA on polyacrylamide gels; scanning of gels in the ultra-violet. The Joyce-Loebl Review (Joyce, Loebl and Co. Ltd., Princesway, Team Valley, Gateshead, England).
[13] K.-I. Miura, I. Kimura and N. Suzuki, Virology 28 (1966) 571.
[14] L. Montagnier and F.K. Sanders, Nature 199 (1963) 664.
[15] R. Ockerse, B.Z. Siegel and A.W. Galston, Science 151 (1966) 452.
[16] A.C. Peacock and C.W. Dingman, Biochem. 7 (1968) 668.
[17] M. Ptashne, Proc. Natl. Acad. Sci. U.S. 57 (1967) 306.
[18] A.D. Riggs and S. Bourgeois, J. Mol. Biol. 34 (1968) 361.
[19] R.W. Ritzert and B.A. Turin, Phytochem. 9 (1970) 1701.
[20] A. Sibatani, Anal. Biochem. 33 (1970) 279.
[21] J. Silberger., Jr. and F. Skoog, Science 118 (1953) 443.
[22] F. Skoog, D.J. Armstrong, J.O. Cherayil, A.E. Hampel and R.M. Bock, Science 154 (1966) 1354.
[23] H.A. Stafford and A.W. Galston, Plant Physiol. 46 (1970) 763.

Niu and Segal (eds.), The role of RNA in reproduction and development
North-Holland Publ. Co., 1973

Effects of exogenous RNA on steroid metabolism in adrenals and gonads

Dorothy B. VILLEE and Ajit GOSWAMI
Boston Hospital for Women, Harvard Medical School, Boston, Mass. 02115, USA

Preparations of RNA from one steroid-producing gland, administered either in vivo or in vitro, can markedly alter the pattern of steroid synthesis in another steroid-producing gland. The new pattern reflects the pattern of metabolism in the tissue from which the RNA was extracted. RNA from gonadotropin-stimulated testes has an effect similar to that of LH on androgen metabolism in immature testes. The effects of gonadotropin-induced RNA in androgen metabolism of testicular microsomes requires protein synthesis. Our results are in agreement with the conclusion that unique RNA molecules exist for different endocrine tissues and that these molecules function by increasing the synthesis of unique enzymes involved in the synthesis of specific hormones. The amount of this RNA is probably governed by the trophic hormone for that tissue.

1. Introduction

In the process of growth and differentiation of the nephrogenic ridge certain cells are distinguishable as steroid-producing cells. They occupy portions of the adrenal glands and the gonads; however, the steroid hormone formed differs for each of these endocrine glands. The adrenal forms primarily glucocorticoids such as cortisol and corticosterone. The testes form testosterone and the ovaries form estrogens and progesterone. Since testosterone is formed from 21C steroids, and is also a precursor of the estrogens, these tissues have many enzymes in common. Certain enzymes, however, would be unique for each tissue, such as the 11β-hydroxylase for the adrenal gland.

Because of the uniqueness of the products of these tissues, and because of the ease of isolation of the steroid hormones, effects of exogenous RNA on the function of specific endocrine tissues can be investigated. If the exogenous RNA can enter the cell and alter its pattern of steroid synthesis, it might be possible to determine if the synthesis of specific steroid enzymes is involved in the RNA effect.

2. Organ culture experiments

Adrenals from 1- to 3-week-old mice or fetal rats survive extremely well for short periods (24 to 48 hr) in organ culture in a chemically defined medium [1]. They remain histologically similar to freshly removed adrenals. The rates of oxygen consumption and lactate production over the first 29 hr in culture do not differ significantly from those of fresh tissue [1]. Also the incorporation of progesterone-4-^{14}C into either 11-deoxycorticosterone-^{14}C or corticosterone-^{14}C is linear over the first 24 hr. With this information, it seemed probable that the explants, under our conditions of organ culture, were viable and functionally differentiated.

Using these organ culture techniques we exposed adrenal and testicular explants to testicular and adrenal RNA respectively. The RNA was extracted with phenol and sodium dodecylsulfate and subsequently treated with deoxyribonuclease [2]. The absorption spectrum of the extracted RNA was compared to that of pure RNA.

The RNA preparations from testes were added to organ culture chambers containing 1 mg explants of mouse or rat adrenals on stainless steel rafts. 5 ml of Parker's medium 1066 was added to the chambers. After 24 hr culture at 26°C, the explants were removed, homogenized in buffer and added to incubation vessels containing an excess of pregnenolone-7α-^{3}H, NAD, NADP, and glucose-6-phosphate, and placed in a constant temperature shaking water bath (37°C). The radioactive steroids were isolated from the incubation mixture by carrier dilution and purification of the steroids involved the use of column chromatography, thin-layer chromatography, crystallization and derivative formation [2]. All steroids were purified to constant specific activity.

Adrenals which had been exposed for 24 hr to RNA from rat or mouse testis formed far more labeled testosterone than did control adrenals cultured in medium alone or in medium containing adrenal RNA. There was a concomitant decrease in labeled corticosterone. Thus the pattern of steroid synthesis in the adrenal was shifted toward one more nearly resembling that of rat testis (table 1).

Partial purification of the testicular RNA was attempted using methylated albumin kieselguhr columns [3]. Fractions of RNA were eluted with various molarities of saline. RNA which could not be eluted with 1 M saline was removed with brief exposure to 0.005 N NaOH followed by rapid precipitation of the RNA in the eluate with ethanol. This alkali-eluted RNA has a base ratio like that of DNA but does not show the colorimetric reactions of DNA. We therefore consider it to be a

TABLE 1
Effect of testicular RNA on adrenal steroid synthesis

	Control	+ Testicular RNA
Testosterone	8,480	12,500
Corticosterone	53,000	6,580

Radioactive steroid products, expressed as counts per minute per milligram of tissue, isolated from cell-free incubations of adrenals previously exposed to culture medium alone or culture medium + testicular RNA.

TABLE 2
Effects of saline-eluted RNA and alkali-eluted RNA from rat testes on mouse adrenal metabolism

	Mouse adrenals cultured for 24 hr in		
	Medium 1066	Medium 1066 + saline RNA	Medium 1066 + alkali RNA
Testosterone	984	4,170	16,000
Corticosterone	906	475	1,460

The numbers represent counts per minute of steroid isolated per milligram of adrenal in culture.

TABLE 3
Summary of experiments testing the biological activity of rat testicular RNA in organ culture

Experiment	Tissue in organ culture	Rat testicular RNA	% Increase in testosterone over control
1	Adult rat adrenals	Phenol extract DNAase treated	220
2	Human fetal adrenals	Phenol extract DNAase treated	100
3	25-day-old rat adrenals	alkali-eluted RNA	800
4	11-day-old mouse adrenals	saline-eluted RNA	320
4	11-day-old mouse adrenals	alkali-eluted RNA	1520

The substrate used for testosterone synthesis was pregnenolone-7α-^{3}H.
Abbreviation: DNAase = deoxyribonuclease.

TABLE 4
Effects of kidney RNA and testicular RNA on steroid synthesis in rat adrenals in organ culture

	Rat adrenals cultured for 24 hr in medium +		
	Adrenal RNA (209 μg)	Kidney RNA (1370 μg)	Alkali-eluted testicular RNA (145 μg)
Corticosterone	32,300 (27.9)	22,700 (23.6)	2,850 (3.0)
Deoxycorticosterone	3,160 (2.7)	3,730 (3.8)	3,730 (4.0)
Androstenedione	277 (0.2)	694 (0.7)	708 (0.7)
Progesterone	14,900 (12.8)	17,800 (18.4)	24,200 (26.2)
Testosterone	162 (0.1)	323 (0.3)	1,450 (1.5)

The numbers represent counts per minute of steroid isolated per milligram of tissue in culture. The numbers in parentheses represent the percentage of the total radioactivity associated with each steroid.

DNA-like RNA. The biological activities of the saline-eluted and alkali-eluted RNA are compared in table 2. The amount of labeled testosterone isolated from the incubations of adrenals previously exposed to alkali-eluted testicular RNA is considerably greater than that found in control tissues or even in tissues exposed to saline-eluted RNA. Compared to crude extracts of total RNA, alkali-eluted testicular RNA is more active in our organ culture system (table 3).

RNA from rat kidney (table 4) alters the pattern of steroid metabolism in rat adrenals to a small extent. The effect of kidney RNA is insignificant compared with the large increase in labeled testosterone and the decrease in labeled corticosterone found in incubations of adrenals exposed to alkali-eluted testicular RNA.

Comparable organ culture studies were performed using human fetal adrenals, ovaries and testes. In contrast to rodent adrenals human fetal adrenals have relatively little activity of the enzyme, 3β-hydroxysteroid dehydrogenase. This enzyme catalyzes the conversion of delta-5-3β-hydroxysteroids such as pregnenolone, 17-hydroxypregnenolone, and dehydroepiandrosterone, to delta-4-3-ketosteroids such as progesterone, 17-hydroxyprogesterone, and androstenedione. Therefore, human fetal adrenals incubated with pregnenolone would convert very little of this steroid to progesterone. Instead the pregnenolone would be 17-hydroxylated to 17-hydroxypregnenolone and the latter would be converted to dehydroepiandrosterone. The latter is the major product of the human fetal adrenal. The human fetal testis, from nine to fifteen weeks gestation, is rich in 3β-hydroxysteroid dehydrogenase and can convert pregnenolone to several delta 4-3-ketosteroids including the androgens, androstenedione and testosterone.

Human fetal adrenals exposed for 24 hr to human fetal adrenal RNA and subsequently incubated with pregnenolone-7α-^{3}H formed primarily labeled dehydroepiandrosterone (table 5). Thus the pattern of adrenal steroid metabolism is not altered by previous exposure to adrenal RNA. Adrenals exposed to testicular RNA formed far less dehydroepiandrosterone and greater amounts of progesterone and androstenedione (table 5). Using either labeled pregnenolone or labeled progesterone as substrate, human fetal adrenals previously exposed to testicular RNA formed

TABLE 5
Effect of human testicular RNA on the metabolism of pregnenolone in human fetal adrenals in organ culture

Radioactive products	Explants of adrenal exposed to	
	Adrenal RNA	Testicular RNA
Progesterone	197	754
Androstenedione	1760	3960
Dehydroepiandrosterone	9130	3920
Corticosterone	464	244

The numbers represent counts per minute of steroid isolated per milligram of adrenal explant.

TABLE 6
Effect of human fetal testicular RNA on testosterone synthesis in explants of human fetal adrenals

Fetus (crown-rump length)	Substrate			
	Pregnenolone-7α-^{3}H		Progesterone-4-^{14}C	
	0	+	0	+
4.8 cm	551	395	429	1300
5.8 cm	3930	5960	2130	6720
15.0 cm	245	480	94	342

The numbers represent counts per minute of testosterone-^{14}C or testosterone-^{3}H isolated from incubations of fetal adrenals previously cultured for 24 hr with or without fetal testicular RNA. The amount of RNA per organ culture chamber was 440 μg for the explants from the 4.8 cm and 5.8 cm fetuses and 680 μg for explants from the 15.0 cm fetus.

TABLE 7
Effect of adrenal RNA on steroid metabolism of fetal (8.5 cm) testicular explants

Radioactive products	Control	Adrenal RNA
Dehydroepiandrosterone	171,000	258,000
Androstenedione	83,000	33,400
Testosterone	4,800	3,100

The numbers represent counts per minute of steroid isolated per milligram of testis.

TABLE 8
Effects of adrenal RNA on the conversion of substrate to dehydroepiandrosterone

Explant	Substrate	Source of RNA: crown-rump length of fetus (cm)	Adrenal RNA (μg/ml)	% Increase over control
Human testis	Pregnenolone-7α-^{3}H	9.5	48	140
Human testis	Pregnenolone-7α-^{3}H	11.0	180	157
Human testis	Pregnenolone-7α-^{3}H	12.0	154	223
Human testis	Pregnenolone-7α-^{3}H	8.5	213	51
Mouse ovary	Pregnenolone-7α-^{3}H	8.5	236 (alkali-eluted)	105
Mouse ovary	17OH-Preg-7α-^{3}H	10.0	184	70
Human testis	17OH-Preg-7α-^{3}H	19.5	127	28
Mouse ovary	17OH-Preg-7α-^{3}H	19.5	137 (alkali-eluted)	286

Abbreviation: 17OH-Preg-7α-^{3}H = 17-hydroxypregnenolone-7α-^{3}H.

far more labeled testosterone than did control adrenals (table 6), with the exception of the 4.8 cm adrenals incubated with pregnenolone.

The reverse is also true. Human fetal testes exposed to human fetal adrenal RNA, and subsequently incubated with pregnenolone-7α-^{3}H formed 50% more labeled dehydroepiandrosterone and far less labeled androgens (table 7).

Comparing different RNA preparations and different substrates (table 8) we see that using either labeled pregnenolone or labeled 17-hydroxypregnenolone as substrate, fetal testes or mouse ovaries show enhanced formation of dehydroepiandrosterone after exposure to human fetal adrenal RNA. In a given experiment alkali-eluted RNA is far more active than crude RNA preparations from the same adrenals. We conclude from these experiments that the pattern of steroid metabolism in gonads or adrenals can be altered by exposure to RNA from adrenals or testes respectively. The new pattern resembles more closely that of the tissue donating the RNA.

3. The role of gonadotropins in steroid metabolism

We next turned to the question of the role of gonadotropins in producing testicular RNA which is biologically active in our organ culture systems. The pattern of androgen synthesis in Leydig cells from immature 21-day-old hypophysectomized rats and in mature 90-day-old male rats was studied. In order to have sufficient development of the Leydig cells so that they could be mechanically separated from the tubular elements of the testes the immature hypophysectomized rats were treated for five days with daily injections of LH (100 I.U.) followed by five days

TABLE 9
Steroid products isolated from Leydig cells incubated with cholesterol-7α-^{3}H

		Androstanediol	Androsterone	Testosterone
Immature	Control	48.1 ± 2.27	33.3 ± 2.06	10.3 ± 1.03
	+ RNA (100 μg)	51.3 ± 2.44 [b]	38.1 ± 1.99 [b]	13.6 ± 1.05 [a]
	+ LH (50 I.U.)	57.3 ± 2.55	47.1 ± 2.39	19.7 ± 2.96
Mature	Control	19.6 ± 1.06	10.5 ± 0.88	88.3 ± 3.60
	+ RNA (100 μg)	12.3 ± 0.98	7.8 ± 0.95 [b]	95.7 ± 4.29 [a]
	+ LH (50 I.U.)	13.1 ± 1.05	9.7 ± 1.36 [b]	98.8 ± 4.69

[a] Difference from control mean is of borderline significance ($P \simeq 0.05$).
[b] Difference from control mean not significant ($P > 0.1$).
Other means are significantly different from control mean ($P < 0.05$).

Each incubation dish contained 2.0 μCi (0.2 nmoles) of cholesterol-7α-^{3}H and Leydig cells suspended in medium 199 containing 10% fetal calf serum. Incubation was at 37°C for 24 hr. The reaction was terminated by freezing. The figures represent total counts per minute × 10^{-3} isolated and purified from the incubation medium. The data are expressed as mean ± S.E. of four experiments.

off LH. The animals were sacrificed and the Leydig cells were separated from the rest of the testes. The Leydig cells were cultured in suspension for 24 hr in the presence of excess cholesterol-7α-^{3}H (New England Nuclear Corp., S.A. 10 Ci/mM) or pregnenolone-7α-^{3}H (New England Nuclear Corp., S.A. 17 Ci/mM). At the end of the culture period the radioactive products were isolated by carrier dilution and purified by gas liquid chromatography after making suitable derivatives. The pattern of androgen synthesis from cholesterol is very different in mature Leydig cells as compared to immature cells (table 9). The immature Leydig cells show a preponderance of 5α-reduced androgens such as androstanediol (5α-androstan-3α,17β-diol) and androsterone (5α-androstan-3αol-17-one) over testosterone. The reverse is true of mature Leydig cells. If either LH or nuclear RNA from the testes of gonadotropin-treated rats is introduced into the culture medium for the 24 hr period, the amount of labeled testosterone isolated in the culture medium was increased over the control values (table 9). The effect of LH was statistically significant for both immature and mature cells, however, the effect of RNA was of borderline significance. The similarity of the effects of LH to those of RNA from gonadotropin-stimulated testes suggests that the effect of gonadotropins on rat testes may be mediated by RNA.

Similar effects were found using pregnenolone-7α-^{3}H as substrate (table 10). The deviations from control values are particularly significant in the mature Leydig cells. In all cases, however, the amount of labeled androgen was different in the LH-treated cultures compared to the controls. The RNA effects were always in the same direction as the LH effects, but the differences from control values were not always statistically significant.

TABLE 10
Steroid products isolated from incubations of Leydig cells with pregnenolone-7α-^{3}H

		Androstanediol	Androsterone	Testosterone
Immature	Control	91.4 ± 2.95	45.2 ± 1.65	29.6 ± 1.87
	+ RNA (100 μg)	97.5 ± 3.05 [b]	52.7 ± 1.57 [a]	32.0 ± 1.17 [a]
	+ LG (50 I.U.)	105.4 ± 3.16	59.0 ± 2.15	41.4 ± 2.33
Mature	Control	41.2 ± 1.95	25.6 ± 1.55	181.7 ± 5.24
	+ RNA (100 μg)	33.5 ± 1.86	18.1 ± 1.36	187.2 ± 5.82 [a]
	+ LH (50 I.U.)	30.1 ± 1.59	17.8 ± 1.09	192.0 ± 5.91

[a] Difference from control mean is of borderline significance ($P \simeq 0.05$).
[b] Difference from control mean not significant ($P > 0.1$).
Other means are significantly different from the control mean ($P < 0.05$).

Each incubation flask contained 1.7 μCi (0.1 nmole) of pregnenolone-7α-^{3}H and Leydig cells suspended in medium 199 containing 10% fetal calf serum. Incubation was at 37°C for 24 hr. The reaction was terminated by freezing. The figures represent total counts per minute × 10^{-3} of steroid isolated from the incubation medium. The data are expressed as the mean ± S.E. for four experiments.

If the testicular RNA from gonadotropin-stimulated rats is actually induced by gonadotropin, what kind of RNA is it? The RNA used in the Leydig cell experiments was prepared from the nuclei of testes from 21-day-old hypophysectomized rats treated for three days with 1 mg FSH (NIH-FSH-P1) and 100 I.U. of LH(NIH-LH-B8) daily. This nuclear RNA was characterized by polyacrylamide gel electrophoresis. Two distinct peaks were found, equivalent to 18 S and 28 S material. Could this RNA stimulate protein synthesis? This was tested using testicular microsomes and a protein-synthesizing system (table 11). The microsomes were isolated from testes of hypophysectomized 21-day-old rats, and incubated for 1 hr at 37°C in the presence of amino acids, glutathione, phosphoenol pyruvate, pyruvate kinase, GTP, NADH, NADPH, ATP, and leucine-^{14}C. In addition 0.1 ml of the 105,000 *g* supernatant fraction from testes of rats treated with FSH and LH was added to the incubation mixture. The incorporation of leucine-^{14}C into protein was compared in control incubations and incubations containing nuclear RNA from gonadotropin-treated rats or nuclear RNA plus cycloheximide. Microsomal incubations containing nuclear RNA had twice the protein specific activity as control microsomes. Cycloheximide not only abolished the RNA effect, but decreased the specific activity of the protein below control values.

Does this RNA-mediated protein synthesis involve a shift in steroid metabolism? Immature testicular microsomes have a pattern of androgen synthesis similar to that of immature Leydig cells (table 12), with a preponderance of 5α-reduced androgens. Immature microsomes incubated in the presence of nuclear RNA from gonadotropin-stimulated testes form more than double the amount of labeled testoste-

TABLE 11
Effect of RNA on polypeptide synthesis in microsomes from testes of immature rats

Additions	Radioactivity incorporated from leucine-^{14}C (cpm/mg microsomal protein)
0.1 M Tris	5,080
Nuclear RNA from gonadotropin-stimulated testes (100 μg)	10,860
Nuclear RNA + cycloheximide (150 μg)	1,540

Each incubation flask contained in a final volume of 2.3 ml: 0.1 ml of microsomal preparation from testes of hypophysectomized immature rats; 4.2 μmoles of $MgCl_2$; 3 μmoles of KCl; 14 μmoles of Tris-HCl buffer (pH 7.5); 0.1 μmoles of each of the 20 amino acids except leucine; 6.0 μmoles of GSH; 2.5 μmoles of phosphoenol pyruvate; 0.0625 μmoles of GTP; 25 μg of pyruvate kinase; 0.25 μmoles of NADH (diNA); 0.25 μmoles of NADPH (tetra Na); 0.5 μmoles of ATP; 0.05 ml of leucine-^{14}C (0.25 μCi; 0.1 μmole); 0.1 ml of the 105,000 g supernatant fraction from testes of rats treated with FSH and LH; and Tris or nuclear RNA as noted above. The mixture was incubated at 37°C for 1 hr. The reaction was terminated by the addition of 5% trichloric acid.

TABLE 12
Steroid products isolated from incubation of testicular microsomes with pregnenolone-7α-^{3}H

	Total cpm × 10^{-3}		
	Androstanediol	Androsterone	Testosterone
Complete system	39.3	29.9	5.13
+ nuclear RNA (100 μg)	32.1	17.3	12.00
+ nuclear RNA and cycloheximide (150 μg)	35.7	22.9	5.98
+ nuclear RNA minus lysine	40.2	27.3	6.89
– lysine	37.2	26.4	5.59

The incubation conditions were the same as those described in table 11 except that the amino acid cocktail was supplemented with DL-leucine and the radioactive substrate was pregnenolone-7α-^{3}H (1.7 μCi; 0.1 nmole).

rone as did the control microsomes. There is a concomitant decrease in 5α-reduced androgens. Addition of cycloheximide abolishes the RNA effect as does the elimination of one of the amino acids essential for protein synthesis. Thus the effect of RNA on androgen synthesis in testicular microsomes requires the synthesis of protein. The administration of gonadotropins to immature rats produces a testicular RNA which stimulates protein synthesis. This stimulation of protein synthesis results in a shift in androgen metabolism so that less 5α-reduced androgens and more testosterone accumulates in the incubation flask. The effect of RNA on protein synthesis may involve increased production of an inhibitor of 5α-reductase, thus decreasing the reduction of testosterone.

4. RNA effects on cells grown in tissue culture

Over the past three years we have been growing human fetal adrenal and testicular cells in tissue culture and studying the influence of exogenous RNA on monolayers of these cells. The endocrine tissue was first minced and the small fragments of tissue were spread in a coagulum on Petri dishes [4]. A medium containing 25% calf serum is added and the cultures are maintained at 37°C in room air containing 5% carbon dioxide. The cells grow out from the explants and over a period of weeks eventually cover the petri dish as a monolayer. The appearance of the cultures under the phase contrast microscope differs with the tissue. Adrenal cells form a confluent mass whereas testicular cells form circumscribed areas (fig. 1). We have maintained cultures for several months, during which period steroid hormones continue to be formed. Testicular cells continue to form labeled testosterone from labeled pregnenolone, progesterone or dehydroepiandrosterone without any appreciable decline from the level at confluency until cell death occurs. The amount of labeled testosterone isolated from the cultures depends on the gestational age of the

Fig. 1. Tissue culture of testicular cells from a 5 cm (crown-rump length) human fetus. These cells have been growing in culture for 94 days during which time they have continued to convert labeled dehydroepiandrosterone to labeled testosterone. The pattern of growth in well-delineated circumscribed areas is distinctly different from that seen in comparable cultures of fetal adrenal and pancreas. × 320.

fetus, since Leydig cells are present in large numbers only from 9 to 15 weeks gestation.

Fetal adrenal RNA added to testicular monolayers seems to alter the pattern of steroid metabolism in a manner similar to that found in our organ culture experiments (table 13). The amount of labeled dehydroepiandrosterone is increased and

TABLE 13
Effects of adrenal RNA on testicular cells grown in tissue culture

	12.4 cm fetus (8-day-old cultures)		12.7 cm fetus (22-day-old cultures)	
	0	+	0	+
Dehydroepiandrosterone	3,730	13,440	747	3,270
Testosterone	3,440	2,900	818	440

The numbers represent count per minute per day of steroid isolated from the culture medium. Substrate = pregnenolone-7α-^{3}H.
0 = culture medium alone; + = culture medium plus adrenal RNA.

TABLE 14
Effects of adrenal RNA on testicular cells from a 12.7 cm fetus grown in tissue culture

	After 24 hr exposure to adrenal RNA	Control period	After second 24 hr exposure to adrenal RNA
Dehydroepiandrosterone	1,260	728	1,507
Testosterone	724	628	396
Cortisol	631	220	1,320

The numbers represent counts per minute per day of steroid isolated from the culture medium. Substrate = pregnenolone-7α-^{3}H.

the amount of labeled testosterone is decreased after exposure of the cells to fetal adrenal RNA for 24 hr.

Since the testis possesses 3β-hydroxysteroid dehydrogenase, testicular cells could form cortisol if the adrenal enzymes 21-hydroxylase and 11β-hydroxylase could be provided. Testicular cells exposed to adrenal RNA for 24 hr were able to form significant amounts of labeled cortisol from labeled pregnenolone (table 14). If the cells are then grown for several days in fresh medium lacking RNA, and then tested again with labeled pregnenolone, the amount of labeled cortisol falls to very low values. The amount of labeled dehydroepiandrosterone declines also. Repeat exposure to adrenal RNA for 24 hr increases the conversion of pregnenolone to both dehydroepiandrosterone and cortisol and greatly reduces the amount of labeled testosterone.

Experiments using testicular cultures from five different fetuses are summarized in table 15. In each case the cells were maintained for several days in culture medium alone, then tested for their ability to metabolize labeled pregnenolone by adding the substrate to the culture medium and isolating the steroid products 24 hr later. The control period was followed by a 24 hr exposure to RNA extracted from the adrenals of the same fetus. In each case previous exposure to adrenal RNA resulted in enhanced synthesis of labeled cortisol and dehydroepiandrosterone over control values. There was no overlap in the values from control and RNA periods in culture. Thus the testicular cells are now capable of synthesizing an adrenal glucocorticoid, cortisol.

TABLE 15
Effect of human fetal adrenal RNA on steroid metabolism of fetal testicular cells in tissue culture

	Control	Adrenal RNA
Testosterone	630	540
Dehydroepiandrosterone	794	1,990
Cortisol	260	690

The numbers represent counts per minute per day of steroid isolated from the tissue culture medium (average of five experiments). Substrate = pregnenolone-7α-^{3}H.

TABLE 16
Effect of intraperitoneal injections of rat testicular RNA on steroid metabolism of rat ovaries and adrenals

	Ovary			Adrenal		
	Saline control	Crude RNA	Alkali-eluted RNA	Saline control	Crude RNA	Alkali-eluted RNA
Testosterone	106	812	823	251	251	173
Androstenedione	410	4,310	2,640	34,600	52,000	38,200
Estrone	895	545	505	0	0	0

100 μg of RNA (or an equal volume of saline) was injected every 10 hr × 3. The animals were then sacrificed and the adrenals and ovaries were homogenized and incubated for 1 hr at 37°C with dehydroepiandrosterone-7α-^{3}H, NAD, NADP, and glucose-6-phosphate. The numbers represent counts per minute of steroid isolated per milligram of tissue.

5. *Effect of RNA adiminstered in vivo*

More recently we have attempted to produce changes in steroid metabolism by administering RNA intraperitoneally to intact rats. Mature female rats were sacrificed after 3 injections of rat testicular RNA at 10 hourly intervals and the ovaries and adrenals were incubated with labeled dehydroepiandrosterone (table 16). Control animals were injected with an equal volume of physiologic saline. An eightfold increase in labeled testosterone and a comparable androstenedione increase was found in the ovaries from rats receiving either crude testicular RNA or alkali-eluted testicular RNA. These same ovaries showed a marked fall in labeled estrone compared to control rats. As was expected, no estrone could be isolated from the adrenal incubations. Crude testicular RNA appeared to increase 3β-hydroxysteroid dehydrogenase by increasing the conversion of labeled dehydroepiandrosterone to labeled androstenedione; however, no change in labeled testosterone was noted.

Though very few experiments have been performed, these preliminary results suggest that RNA administered in vivo may be capable of alterations similar to those we have demonstrated using RNA administered in vitro.

6. *Discussion*

Using a variety of experimental techniques we have shown that RNA administered either in vitro or in vivo is capable of altering the pattern of steroid metabolism, and that the new pattern reflects the origin of the RNA. In our original work [5] we showed that treating the RNA with ribonuclease abolished the effect; however, treating the RNA with pronase did not alter it. Thus we conclude that it is the RNA molecule itself that is responsible for the changes in steroid metabolism. We have found that partial purification of the RNA results in a DNA-like RNA which has greater biological activity in vitro than do the crude RNA preparations from the same tissue. This fact plus the shift in pattern of metabolism to that of the tissue donating the RNA suggests that the RNA may have messenger-like properties.

In assessing the role of gonadotropins in altering steroid metabolism, we have found that a biologically active RNA can be isolated from testes of rats treated with FSH and LH. This RNA is capable of increasing the uptake of labeled leucine into microsomal protein and of shifting the pattern of androgen metabolism toward the more sexually mature pattern. This RNA effect is abolished by cycloheximide and therefore presumably depends upon the synthesis of protein.

Our results are compatible with the conclusion that in the process of differentiation different species of RNA are formed in different tissues and that the specific products of the steroid-producing endocrine tissues are controlled by synthesis of unique enzymes under the direction of this RNA. The characteristic species of RNA for a given endocrine tissue is probably controlled by the specific trophic hormone for that tissue.

References

[1] D.B. Villee, In: Advan. Enzyme Reg. 4 (1966) 269.
[2] D.B. Villee, Science 158 (1967) 652.
[3] N. Sueoka and T.Y. Cheng, J. Mol. Biol. 4 (1962) 161.
[4] A.J. Milner and D.B. Villee, Endocrinology 87 (1970) 596.
[5] D.B. Villee, Proc. of the Fourth Intern. Congress on Pharmacology, Basel, 1969, Vol. 2, p. 149.

Niu and Segal (eds.), The role of RNA in reproduction and development
North-Holland Publ. Co., 1973

Thyrotropin-like activity of thyroid RNA in vitro*

Jui-yun MU
Department of Physiology
National Defense Medical Center,
Taipei, Taiwan,

1. Introduction

It has been shown that hormonally deprived target tissues respond to preparations of RNA extracted from hormonally-stimulated tissues, and that the RNA effect mimics the effect of the hormone in question [1–5]. The purpose of the present study was to learn whether or not RNA from TSH-stimulated thyroid glands has the ability to stimulate iodine metabolism of TSH-deprived thyroid tissue to the same extent as TSH itself.

RNA was prepared from the thyroid glands of male albino guinea pigs which had received 0.25 I.U. of bovine TSH at 12 and at 24 hr before sacrifice. Thyroid and liver RNA were prepared by a modification of the method of Kirby [6]. After precipitation from absolute ethanol, the RNA was treated with DNase [7] and was washed twice with ethanol and a mixture of ethanol–ether [8]. Denatured thyroid RNA was prepared by hydrolysis of the RNA with RNase. The thyroid glands which were used for organ culture were obtained from guinea pigs of the same strain but the secretion of TSH from the hypophysis and the function of the thyroid glands of these animals were suppressed by daily injections of 20 μg of triiodothyronine (T_3) for 1 week. The thyroid glands of the T_3-treated animals were removed under sterile conditions. Each pair of thyroid glands was placed on a sterilized glass pad in separate plastic petri dishes and incubated at 37°C in an atmosphere of 5% carbon dioxide in air. The medium used for culture was 10% v/v new born calf serum in Eagle's essential medium. After pre-incubation for 12 hr, the medium was changed with fresh medium which contained 0.2 μCi of carrier free ^{131}I per dish.

* Discussant's paper to the contribution by D.B. Villee and A. Goswami.

TABLE 1

Group No.	I	II	III	IV	V
Additions	–	TSH	Thyroid RNA	Denatured thyroid RNA	Liver RNA
Amount per dish	–	40 mUI	200 μg	200 μg	200 μg

TSH, thyroid RNA, denatured thyroid RNA, or liver RNA were added to groups II to V respectively (table 1). After an additional 24 hr period of incubation, the thyroid glands were removed from the dishes and were washed with Tris buffer (pH 8.0). After blotting, weighing, the glands were digested with 0.5% pronase in Tris buffer (pH 8.0). Iodine metabolism was studied using three different parameters:

(A) Twenty-four hour ^{131}I uptake per mg of thyroid tissue was calculated using the ratio of the radioactivity of a standard solution (0.2 μCi ^{131}I /liter of medium) to that of digested thyroid glands.

(B) Protein bound ^{131}I in the medium was determined by chromatography on IRA-400 resin [9].

(C) Incorporation of ^{131}I into amino acids in the thyroid tissue was determined by paper chromatography of the enzyme-digested thyroid tissue and radiochromatographic scanning.

Histological examination was performed on one pair of thyroid glands from each group. If necrotic change was observed, the data from that particular experimental group were discarded.

2. *Results*

The results showed that the 24 hr uptake of ^{131}I was significantly stimulated over control values by both TSH and thyroid RNA but not by liver RNA; RNase-treated thyroid RNA decreased ^{131}I uptake (table 2). Protein bound ^{131}I in the medium, which was used as an indicator of hormone release from the thyroid gland, was also significantly higher in both the TSH and the thyroid RNA groups compared to the control group. There was no significant difference between the control group and liver RNA of RNase treated groups (table 2).

Incorporation of ^{131}I into the amino acids of the thyroid tissue (table 3) showed that the percentage of radioactivity associated with T_3 and T_4 was increased over control values by exposure of the thyroid tissue to either TSH or to thyroid RNA. This increase was not found in the thyroid tissue exposed to liver RNA or to denatured thyroid RNA.

TABLE 2
24 hr uptake of ^{131}I and PB ^{131}I

Group	No. of guinea pigs	24 hr ^{131}I uptake per mg of thyroid tissue (%)[a]		24 hr PB ^{131}I per total added radioactivity (%)[b]	
		Mean ± S.D.	P[c]	Mean ± S.D.	P[c]
Control	36	0.32 ± 0.05		0.83 ± 0.11	
TSH	36	0.54 ± 0.10	< 0.001	1.43 ± 0.20	< 0.001
Thyroid RNA	24	0.40 ± 0.05	< 0.05	1.63 ± 0.24	< 0.001
RNase[d]	12	0.23 ± 0.01	< 0.05	0.90 ± 0.11	> 0.1
Liver RNA	12	0.29 ± 0.01	> 0.5	0.69 ± 0.11	> 0.1

[a] Expressed as $\frac{\text{cpm uptake/mg gland}}{\text{cpm added}} \times 100$.
[b] Expressed as $\frac{\text{cpm of PB } ^{131}\text{I}}{\text{cpm added}} \times 100$.
[c] Significance between control groups and others.
[d] RNase-treated.

The weight of the thyroid glands of T_3-treated guinea pigs was 23.7 ± 4.3 S.D. mg (N = 114). This is significantly lower than in the controls (35.7 ± 4.6 mg; N = 12). The total amount of thyroid RNA extracted from 12 pairs of TSH-treated thyroid glands (500–600 mg) was about 5–6 mg.

TABLE 3
The ^{131}I distribution in the iodoamino acids in the thyroid glands

Group	No. of guinea	Percent[a]				
		^{131}I	MIT	DIT	T_3 and T_4	P[b]
Control	36	12.58 ± 5.78	43.23 ± 9.58	39.05 ± 5.28	5.15 ± 2.15	
TSH	36	16.34 ± 7.24	28.77 ± 2.96	39.09 ± 4.31	15.87 ± 6.33	< 0.005
Thyroid RNA	24	21.31 ± 5.99	37.73 ± 14.22	26.58 ± 9.65	14.40 ± 2.99	< 0.005
RNase[c]	12	22.64 ± 6.47	33.53 ± 7.23	37.51 ± 11.37	6.32 ± 1.58	> 0.2
Liver RNA	12	18.69 ± 4.51	45.10 ± 9.16	29.90 ± 13.21	6.32 ± 1.53	> 0.2

[a] Percent of radioactivity in particular area; mean ± S.D. MIT = monoiodotyrosine, DIT = diodotyrosi
[b] P value given for (T_3 and T_4).
[c] Treated with RNase.

3. Conclusion

The evidence from these experiments indicates that thyroid RNA mediates the TSH-stimulated iodine metabolism in the thyroid. This TSH-like effect of thyroid RNA is organ-specific and not due to contamination of RNA with TSH.

References

[1] A.M. Mansour and M.C. Niu, Proc. Natl. Acad. Sci. U.S. 53 (1965) 764.
[2] S.J. Segal, O.W. Davidson and K.Wada, Proc. Natl. Acad. Sci. U.S. 54 (1965) 782.
[3] M.M. Fencl and C.A. Villee, Endocrinology 88 (1971) 279.
[4] T. Fujii and C.A. Villee, Proc. Natl. Acad. Sci. U.S. 57 (1967) 1468.
[5] T. Fujii and C.A. Villee, Proc. Natl. Acad. Sci. U.S. 62 (1969) 836.
[6] K.S. Kirby, Biochem. J. 64 (1956) 405.
[7] T. Fujii and C.A. Villee, Endocrinology 82 (1968) 453.
[8] D.B. Villee, Science 158 (1967) 652.
[9] E.J. Wayne, Clinical aspects of iodine metabolism, 1st Ed. (Scientific Publications, Oxford, 1966) 233.

Niu and Segal (eds.). The role of RNA in reproduction and development
North-Holland Publ. Co., 1973

In vivo uptake of RNA and its function in the castrate uterus

M.C. NIU, L.C. NIU and S.F. YANG
Department of Biology, Temple University, Philadelphia, Pa. 19122, USA

In vivo uptake of ^{3}H-RNA was traced in uteri of spayed mice by autoradiography. Silver grains were found predominantly in the epithelial layer. Intracellularly, there were more grains in nuclei than cytoplasm. These ^{3}H-molecules were re-isolated from pooled uteri and compared with the ^{3}H-RNA used for intra-uterine injections by means of sucrose gradient centrifugation. The radioactivity peaks of these two samples had identical sedimentation coefficients. This finding coupled with the lack of RNase activity in the uterine homogenates showed that the RNA molecules in cells of uterine wall were not degraded.

The action of exogenous-uterine-RNA was to promote growth of the atrophic uterus with particular reference to the epithelial layer and to increase the activities of alkaline phosphatase and β-glucoronidase. This function of RNA was sensitive to boiling, RNase and actinomycin D treatments. The nuclear distribution of the exogenous-RNA and the actinomycin D inhibition of the RNA-stimulated biosynthesis indicate that the exogenous-RNA functions as a derepressor of the genome.

RNAs were isolated from organs rich in alkaline phosphatase, e.g. seminal vesicle and ovary. They were equally capable of elevating the activity of alkaline phosphatase. Liver is a poor source of alkaline phosphatase, and liver-RNA was unable to induce this enzyme, as did uterine-RNA. However, when liver-mRNA (poly A attached RNA) was used, synthesis of serum albumin was found in the epithelial cells of mouse uterus.

1. Introduction

Estrogen stimulates uterine synthesis of new RNA [1, 2] which in turn contributes to the maintenance of morphological and enzymatic entities of the cyclic uterus. Elimination of estrogen, as in the spayed animal, results in the degeneration of the uterus. RNAs were prepared separately from pooled uteri of calf [3] and of estrogen-treated adult rats [4]. They were injected and instilled respectively into the intrauterine lumen. The atrophic uteri thus treated returned to normalcy in a few days. Although these findings were confirmed subsequently by our own and other laboratories [5–11], the mechanism by which the RNA acted upon the target tissue is still obscure. For the study of RNA action in vivo, the first question to be resolved deals with the site of localization of the RNA molecules. Are they absorbed to cell surfaces or do they enter into the recipient cells? Are recipient cells scattered all through uterine tissues or are they restricted to particular areas, e.g., epithelial layer, circular muscle or longitudinal muscle? If RNA molecules are entering the cells, are they located primarily in cytoplasm or nucleus? Is input RNA incorporated directly, or is it broken down and resynthesized before incorporation? The first part of this communication is aimed at answering these and other related questions. The second part deals mainly with functional potentiality of the exogenous-RNA. While uterine-RNA restores the normalcy of the atrophic uterus, rat liver-RNA-treated uterine epithetial cells acquire the capacity to synthesize an antigen (albumin) that reacts with fluorescent rabbit anti-serum against rat albumin.

2. Materials and methods

2.1. Removal of ovary

Swiss-Webster mice, 25–30 g body weight, were used. Under light ether anesthesia, both ovaries were excised through a dorsal incision of the body wall. They were maintained in groups of five in cages. Twelve to fifteen days later, the right uterine horn was exposed for separate intra-uterine injections of ^{3}H- and cold-uterine-RNAs, ovary-RNA, seminal vesicle-RNA, boiled-RNA, RNase-digested-RNA, ^{3}H-liver RNA and liver-poly A attached RNA (10 O.D. in 0.05 ml of physiological saline). Injections of these materials were accomplished by the use of 1 ml plastic syringe with a 30-gauge hypodermic needle.

2.2. RNA isolation

Pooled uteri from mice and calves, seminal vesicles of mice, cow endometrium and calf ovaries were used for RNA preparation. The procedure was reported elsewhere [12, 13]. Unless otherwise stated, all RNA samples were stored in 2 vols of ice-cold 95% ethanol in a freezer (–20°C). Half of the uterine-RNA was used for direct methylation with ^{3}H-dimethylsulfate according to the procedure of Smith et al. [14]. Liver-^{3}H-RNA was isolated from pooled livers of rats, 250–300 g body weight. A group of 6–10 rats were starved for 24 hr. Four hours prior to sacrifice by guillotine, they received intraperitoneal injection of ^{3}H-uridine (Schwarz Biochemicals, 100 μCi per rat) and food. The livers were perfused and pooled for isolation of ^{3}H-RNA. The poly A attached RNA was isolated from rat-liver using the procedure of Rosenfeld et al. [15].

2.3. Autoradiography

^{3}H-RNA from calf-uteri and rat-liver were separately injected into the lumen of right uterine horns of spayed young mice (10 O.D. in 0.05 ml saline). The horn with calf-^{3}H-RNA was excised 1 hr later, washed with saline and then fixed in glutaraladehyde for 2 hr on ice, while those with rat-^{3}H-RNA were processed at intervals of 15, 30, 45, 60 and 120 min. After washing 3 times with glutaraldehyde buffer (pH 7.4) they were refixed individually in 2% osmium tetraoxide for 1 hr on ice. Embedding was made in Araldite. An automatic Blum-Porter microtome was used to cut 0.5 μ and 90–150 nm sections. Gold colored sections were chosen and mounted on slides for low power and grids for high resolution autoradiographs. The method for electron microscopic autoradiography was published elsewhere [16].

2.4. Sucrose gradient sedimentation

Uterine ^{3}H-RNA was separated with 1 M NaCl into high (H) and low (L) molecular fractions. Ten O.D. of L fraction in 0.05 ml of saline was injected into the lumen of the right uterine horn of spayed mice. Forty-five minutes later, the treated uteri were pooled and used for RNA isolation. The RNA thus obtained and the L-^{3}H-RNA were subjected to sucrose density gradient centrifugation. About 20 O.D. of RNA were layered on 5–20% sucrose gradient in Tris buffer (0.0.1 M Tris-HCl (pH 7.6)–0.1 M KCl-1 mM EDTA) and centrifuged in a Beckman S.W. 25.1 rotor at 25,000 rpm for 12 hr. Fractions were collected (60 drops each vial) through a hole at the bottom of the polyethylene tube. The absorbance at 260 mM was determined. Radioactivity was recorded by Packard Tri-Carb-Scintillation Spectrophotometer.

2.5. Enzyme determination

Three days after injection of saline and RNA, the uteri were excised and homogenized separately [3]. Cellular debris was removed by low speed centrifugation. The supernatant was used for estimation of enzyme activity. The p-nitrophenyl phosphate method [17] and the phenolphthalein glucuronide method [18] were employed respectively for the determination of alkaline phosphatase and β-glucuronidase.

The uteri treated with boiled uterine-RNA, uterine-RNA digests, ovary-RNA, and seminal vesicle-RNA were used for the determination of alkaline phosphatase activity. Actinomycin D was a gift from Merck Co., West Point, Pa. Boiled uterine-RNA was prepared by heating at 100°C for 20 min. The RNAse-digested-uterine-RNA was prepared by incubating uterine-RNA (200 O.D./ml) with pancreatic ribonuclease (200 μg, preboiled) in Tris buffer (0.0.1 M), pH 7.5 at 37°C for 30 min. The RNAse activity was neutralized by rabbit antiserum against ribonuclease.

The site of alkaline phosphatase activity was shown by histochemical staining technique. The excised uteri were fixed in dry-ice–acetone mixture (–78°C) and sectioned by the freeze substitution microtechnique [19]. The staining procedure was that of Gomeri [20].

2.6. Immunological identification of albumin

Twenty-four hours after separate injections of saline and poly A attached RNA from liver into one uterine horn of spayed mice, the uterine horns were removed and fixed in dry-ice–acetone mixture (–78°C). They were sectioned at 6 μ and transferred to clean cover slips for drying. Rabbit anti-bovine and anti-rat albumin sera were purchased from Cappel Laboratories, Doylestown, Pa. They were absorbed with mouse albumin from Pentax. To each section a drop of the absorbed anti-serum was applied and then kept in a moist chamber over night. After washing with saline and rinsed through a series of distilled water, they were examined with a Zeiss fluorescent microscope.

2.7. Estmation of ribonuclease activity in RNA, uterine homogenate and mouse serum

Serum was obtained from blood collected by puncturing the capillary of the eye. It was adjusted to a protein concentration of 20 mg per ml. Uteri were removed from normal and spayed mice, weighed to 0.01 mg and individually kept in a prechilled teflon-glass homogenizer with 1 ml of double distilled water. Final concentration was adjusted to 2% (20 mg/ml). Homogenization was accomplished by 20 strokes and cell breakage was checked microscopically. Ribonuclease activity was determined by the procedure of Kunitz [21], using a standard curve with pancreatic RNAse (Worthington, crystalized) and calf-uterine or rat-liver-RNA.

3. Results

3.1. The purity and heterogenity of RNAs

RNAs were prepared from pooled uteri, ovaries, seminal vesicles and livers. They possessed the following properties:

(1) UV absorption spectra were typical of nucleic acids with minimal and maximal absorption at 230 and 258 μm respectively.

(2) The amount of RNA calculated from O.D. was similar to that of ribose-P, thus showing that RNA is the principal UV absorbing material.

(3) DNA contamination [22] was negligible.

(4) Protein contamination [23] was about 1–2%.

(5) Ribonuclease activity [21] was undetectable.

(6) Sucrose gradient sedimentation yielded three peaks (28 S, 18 S, and 4 S).

(7) Specific activity of the methylated uterine-RNA was approximately 1200 cpm/μg and of the rat-liver RNA 30 cpm/μg.

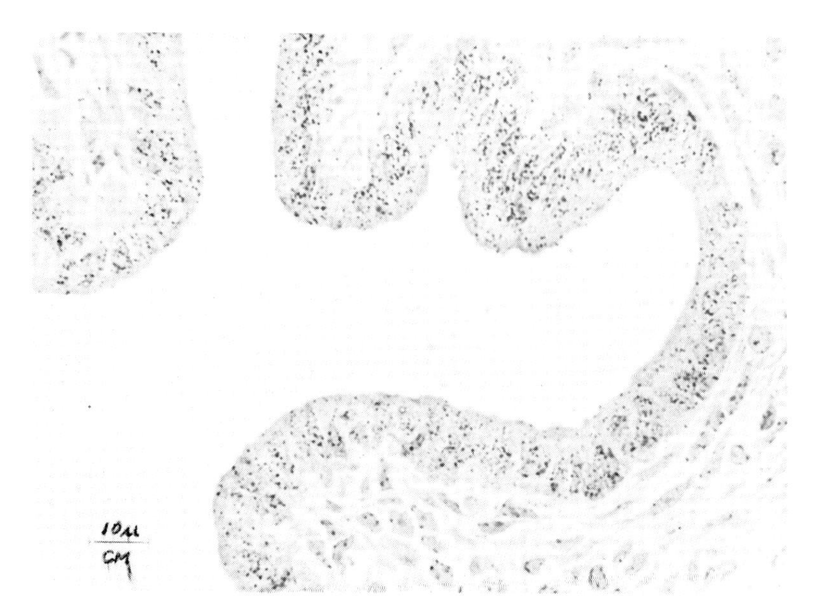

Fig. 1. Low magnification autoradiogram (×832) of uterine wall of the spayed mouse, treated with uterine-RNA, methylated with ^{3}H-dimethyl sulphate.

3.2. Uterine uptake of RNA as seen from autoradiograms

3.2.1. Uterine-^{3}H-RNA

3.2.1.1. Lower power autoradiograms. As reflected by the presence of silver grains in the autoradiograms, ^{3}H-RNA molecules were localized mainly in the epithelial layers (fig. 1). The exact intracellular localization can hardly be ascertained in this thick section.

3.2.1.2. High resolution autoradiograms. Studies on thin sections under electron microscope confirmed that there were more silver grains in the epithelium than in the stroma of endometrium (figs. 2 and 3).

3.2.2. Liver-^{3}H-RNA

During the early phases of our autoradiographic study on RNA uptake, ^{3}H-RNA was isolated from rats with ^{3}H-uridine given previously. On account of the low specific activity in the isolated ^{3}H-RNA and the small amount of label taken up by the target cells, high resolution autoradiography was chosen because the magnified silver grains and artifacts can readily be distinguished. The ^{3}H-RNA-treated-uterine horns were excised at intervals and autoradiograms were prepared from them. Our search for silver grains was focussed on three tissues from the lumen outward:

Fig. 2. High resolution autoradiogram (×1660) of uterine epithelium of spayed mouse, treated as for fig. 1.

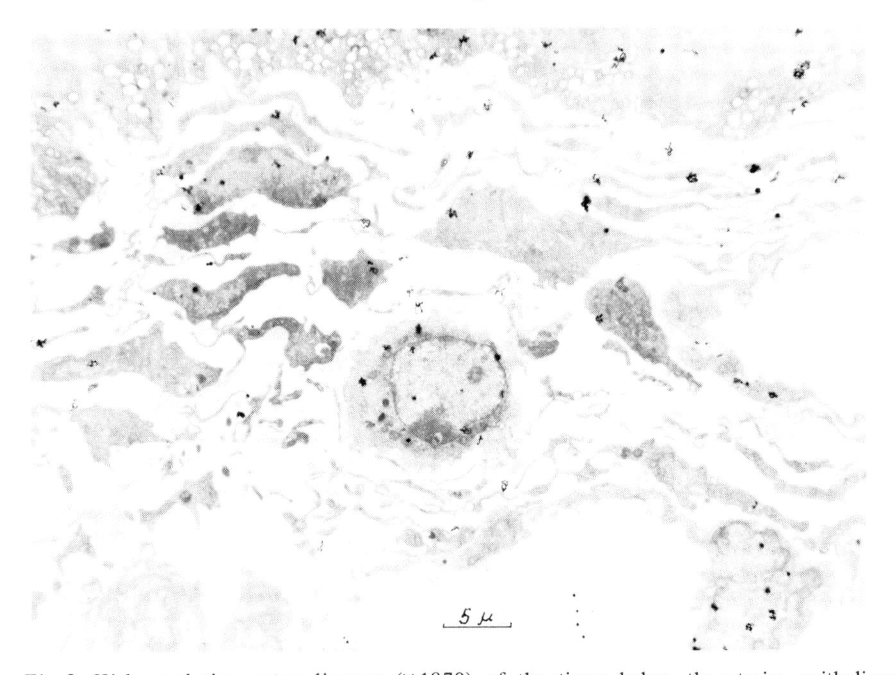

Fig. 3. High resolution autoradiogram (×1870) of the tissue below the uterine epithelium (fig. 2), treated as for fig. 1.

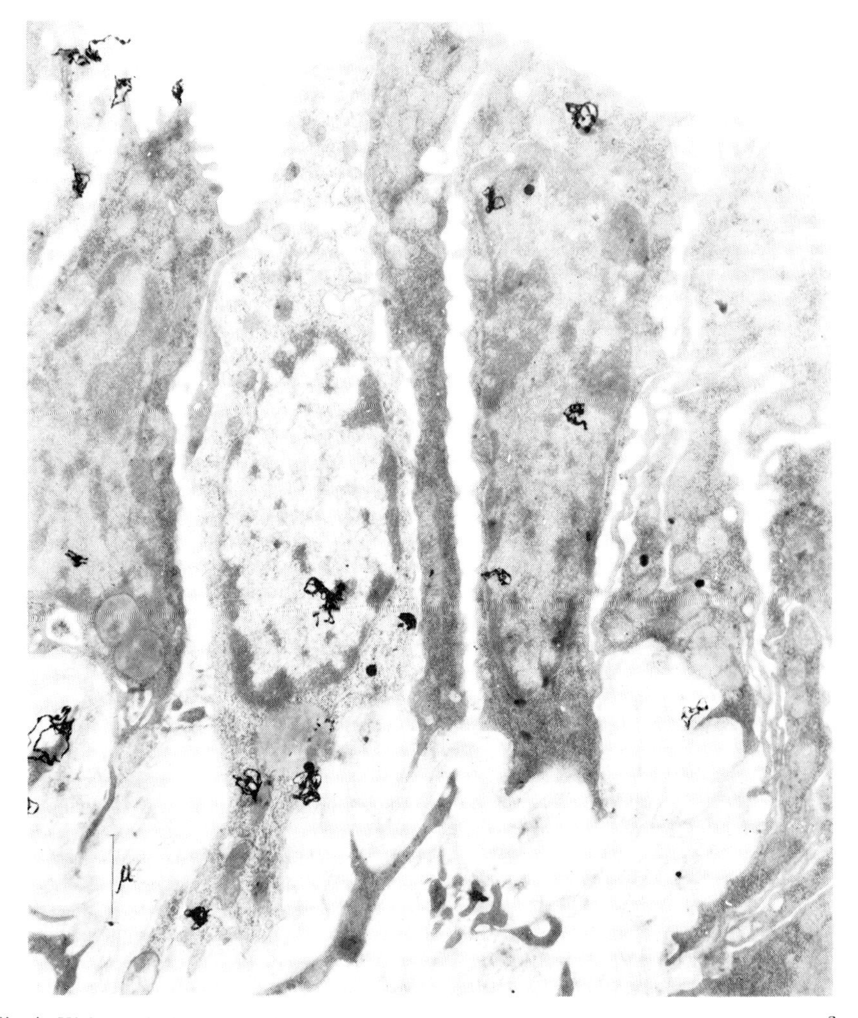

Fig. 4. High resolution autoradiogram (×17,137) of uterine epithelium of spayed mouse. ^{3}H-liver-RNA was isolated from pooled livers of rats. The rats were injected intraperitoneally with ^{3}H-uridine (0.1 mCi) 16 hr earlier.

epithelial layer (fig. 4), circular (fig. 5) and longitudinal (fig. 6) muscle layers. Intracellular distribution of silver grains in the 3 uterine tissues is summarized in table 1. Two points from the table are of significance: (A) the average of silver grains in epithelial cell, circular and longitudinal muscle cells is 2+, 1+ and 1+ respectively and (B) the ratio between nuclear and cytoplasmic grains in the 3 kinds of cells is 1.5+,1 and below 1 respectively. This pattern of RNA distribution can easily be explained by pervasion of the injected RNA solution instead of by circulation. If circulation was responsible for the distribution of RNA molecules, silver grains

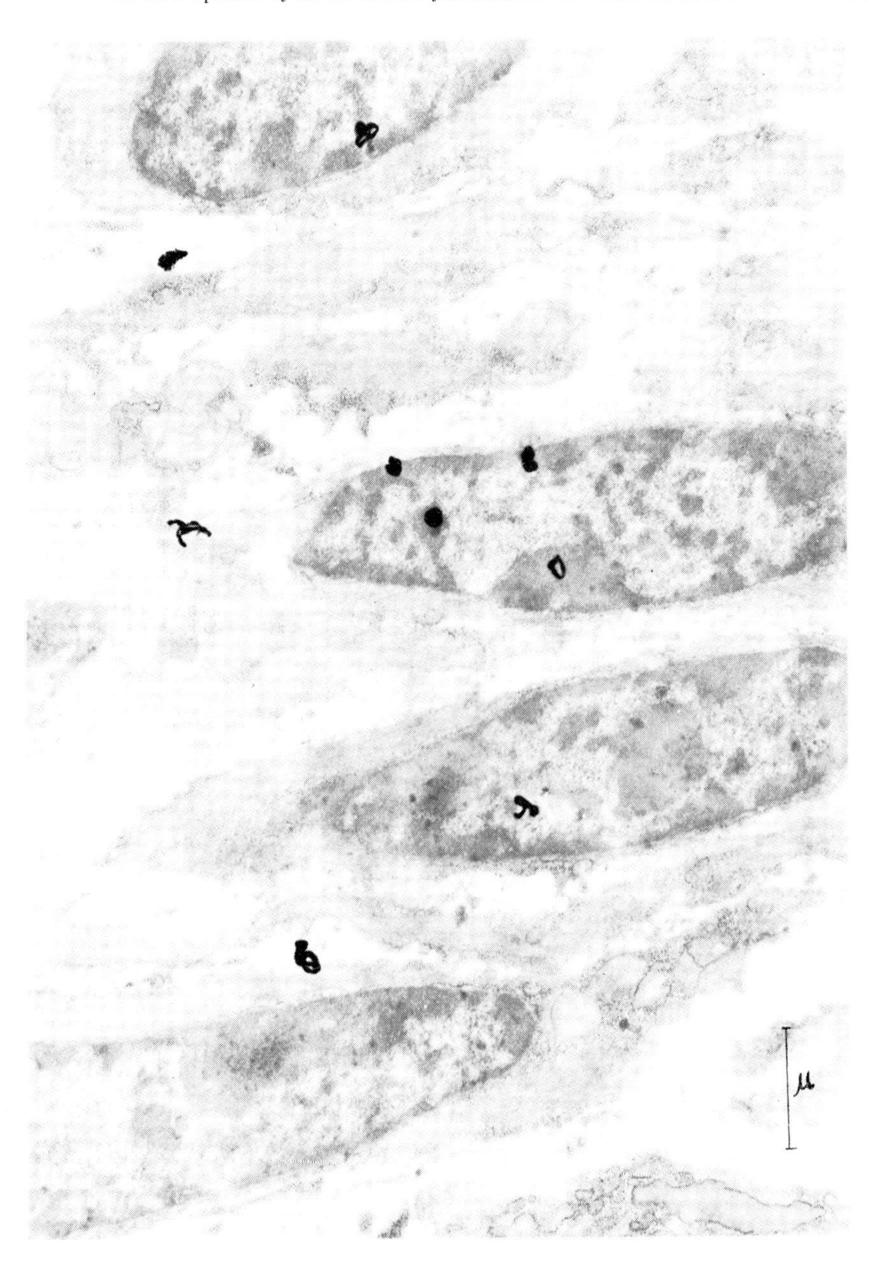

Fig. 5. High resolution autoradiogram (× 19,699) of circular muscle layer from the same cross section as fig. 4.

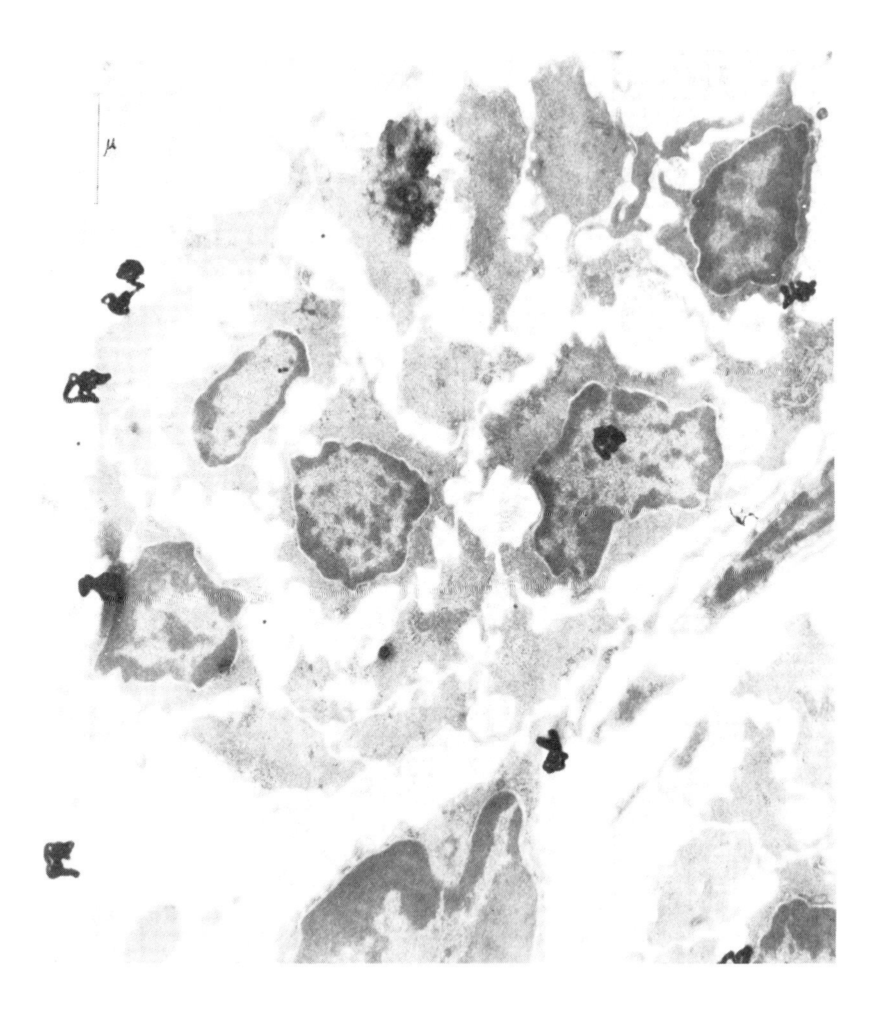

Fig. 6. High resolution autoradiogram (×14,775) of longitudinal muscle from same cross section as fig. 4.

would be more in and around the capillary wall, endothelial cells and/or blood cells. A search for capillaries was carried out. A cross section of the capillary is depicted by fig. 7 on which silver grains are no more than other cells.

The manner by which macromolecular RNA enters the epithelial cells requires further study. As a working hypothesis, pinocytosis should be involved. Therefore, careful study of the fine structure was made of the free border of the epithelial cells. Tiny vesicles were seldom seen complicated by the fact that most epithelial cells were ciliated and electron dense (fig. 4). Only a limited number of electron light cells bore no cilia. In the latter cells, membrane associated vesicles have not yet been observed.

TABLE 1

Distribution of silver grains in castrate uterus (mouse) treated with rat liver ^{3}H-RNA

Minutes after RNA application	Number of cells counted	Total number of silver grains in		
		Cytoplasm (*C*)	Nucleus (*N*)	*N*/*C*
		Epithelial cells		
15	88	85	78	0.88
30	123	130	158	1.22
41	340	258	440	1.71
60	108	60	150	2.50
120	149	101	163	1.61
Total	754	604	989	1.64
		Circular muscle cells		
15	124	114	86	0.75
30	66	115	124	1.08
45	313	153	245	1.60
60	326	258	194	0.75
120	234	134	130	0.97
Total	1063	774	779	1.05
		Longitudinal muscle cells		
15	230	225	120	0.53
30	80	65	60	0.91
45	315	200	275	1.38
120	325	275	140	0.51
Total	950	765	595	0.78

3.3. Sucrose gradient sedimentation

Fig. 8 depicts the profile of RNA sedimentation in 5–20% sucrose. The radioactivity peak in the ^{3}H-RNA used for injection coincides with that of the ^{3}H-macromolecules extracted from the ^{3}H-RNA treated uteri. These 2 peaks differ only in magnitude. If the exogenous RNA break down, thc ^{3}H-labelled products could be used by uterine tissue to resynthesize ^{3}H-RNA. The newly synthesized RNA should sediment at different peaks because most new RNAs are of ribosome and messenger types. Therefore, the ^{3}H-RNA extracted from the ^{3}H-RNA treated uteri was apparently not degraded.

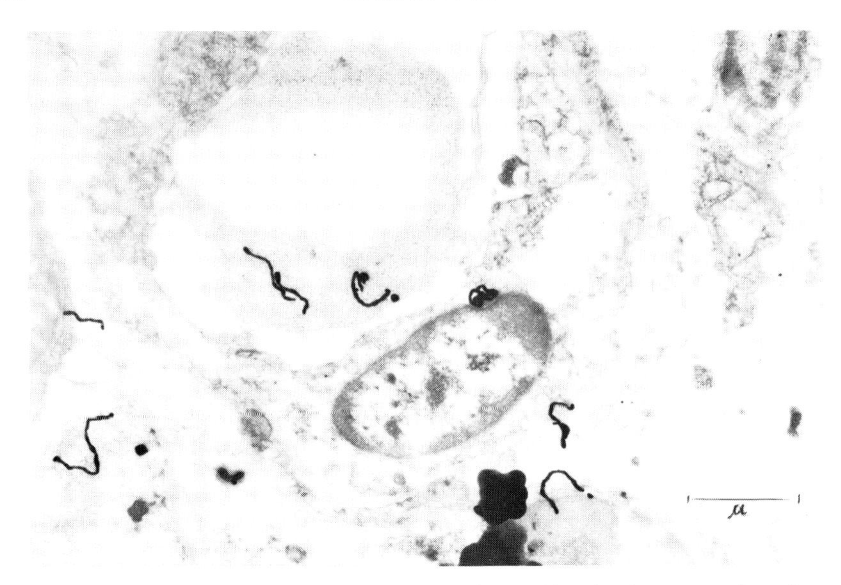

Fig. 7. High resolution autoradiogram (× 16,416) of a capillary in the stroma of uterine wall from the same cross section as fig. 4.

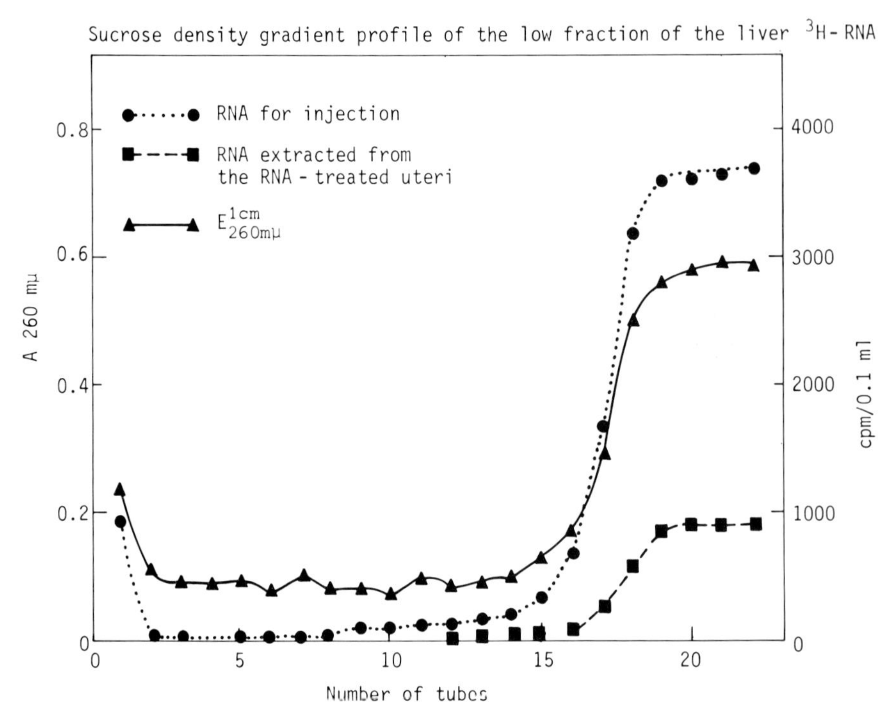

Fig. 8. Profile of the sucrose density sedimentation of the ^{3}H-uterine RNA and the ^{3}-macromolecules (^{3}H-RNA) isolated from the ^{3}H-uterine RNA treated uteri. The radioactivity peaks of the two samples differ only in quantity.

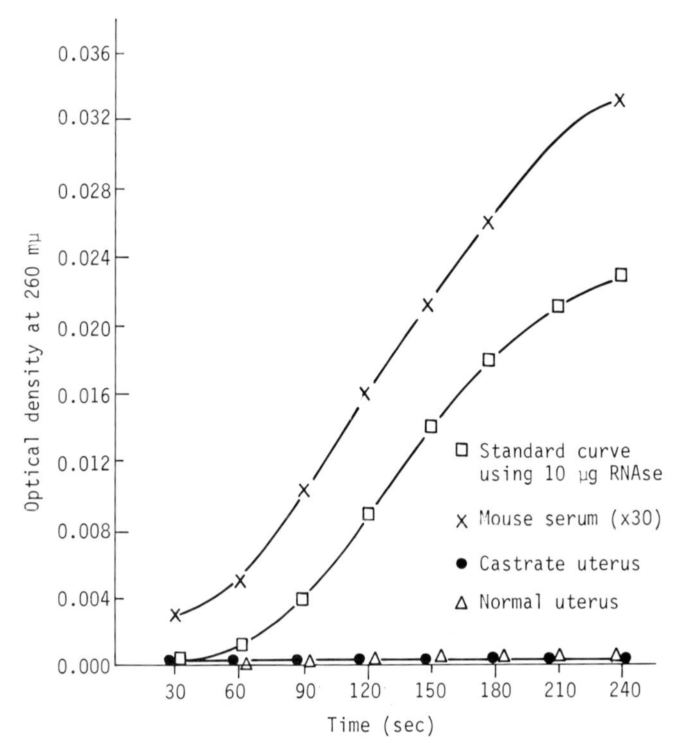

Fig. 9. Determination of ribonuclease activity in mouse serum, uterine homogenate of spayed mice and the RNA used in experiments of this paper.

3.4. Ribonuclease activity of mouse serum and homogenate

Crystallized pancreatic ribonuclease from Worthington was used as standard. The ribonuclease activity of mouse serum (30 X dilution) was found higher than the activity of 10 ng of pancreatic ribonuclease (fig. 9). Ribonuclease activity in the uterine homogenate and the RNAs used in this report was not detectable.

3.5. RNA-induced synthesis of enzymes and protein

The level of enzyme activity in uterus depends upon its physiological state. Removal of ovaries results in a significant reduction. RNA from cyclic uterus or from estrogen-treated uterus was capable of elevating the activity of alkaline phosphatase (fig. 10) and β-glucuronidase (fig. 11) in the uterus of the spayed mouse. The activity of uterine-RNA was sensitive to denaturation by boiling and treatment with pancreatic RNase but, most importantly, specific because liver-RNA prepared from the same animal by the same procedure was unable to raise the enzyme level (table 2). Repeated washing with ethanol and ether would eliminate the hormone,

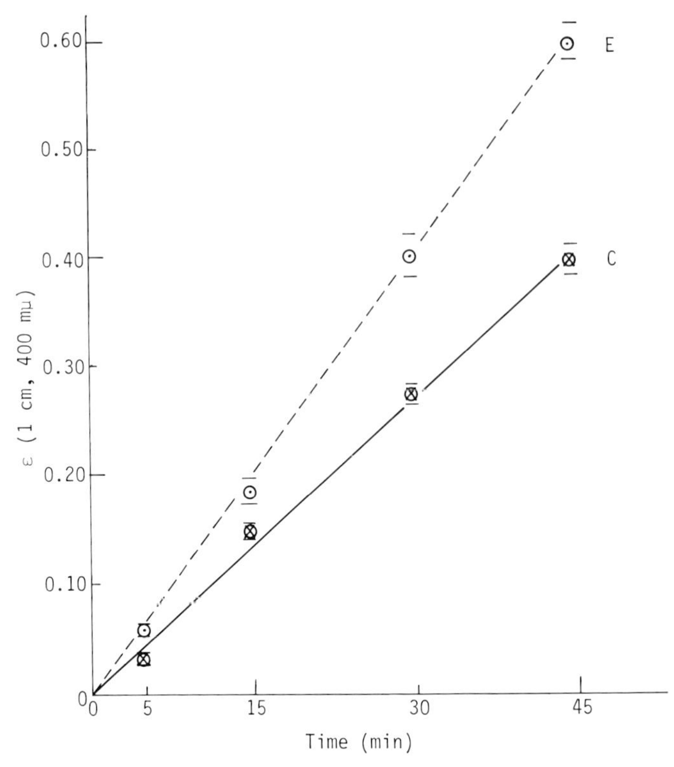

Fig. 10. The activity of alkaline phosphatase in the control and uterine-RNA-treated uteri of the spayed mice.

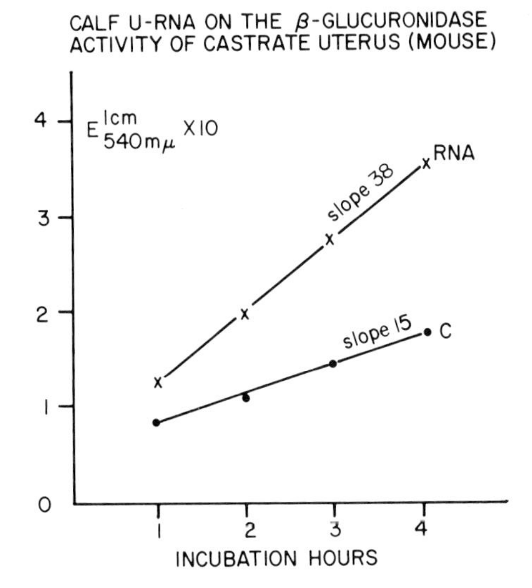

Fig. 11. The activity of β-glucuronidase in the control and uterine-RNA-treated uteri of the spayed mice.

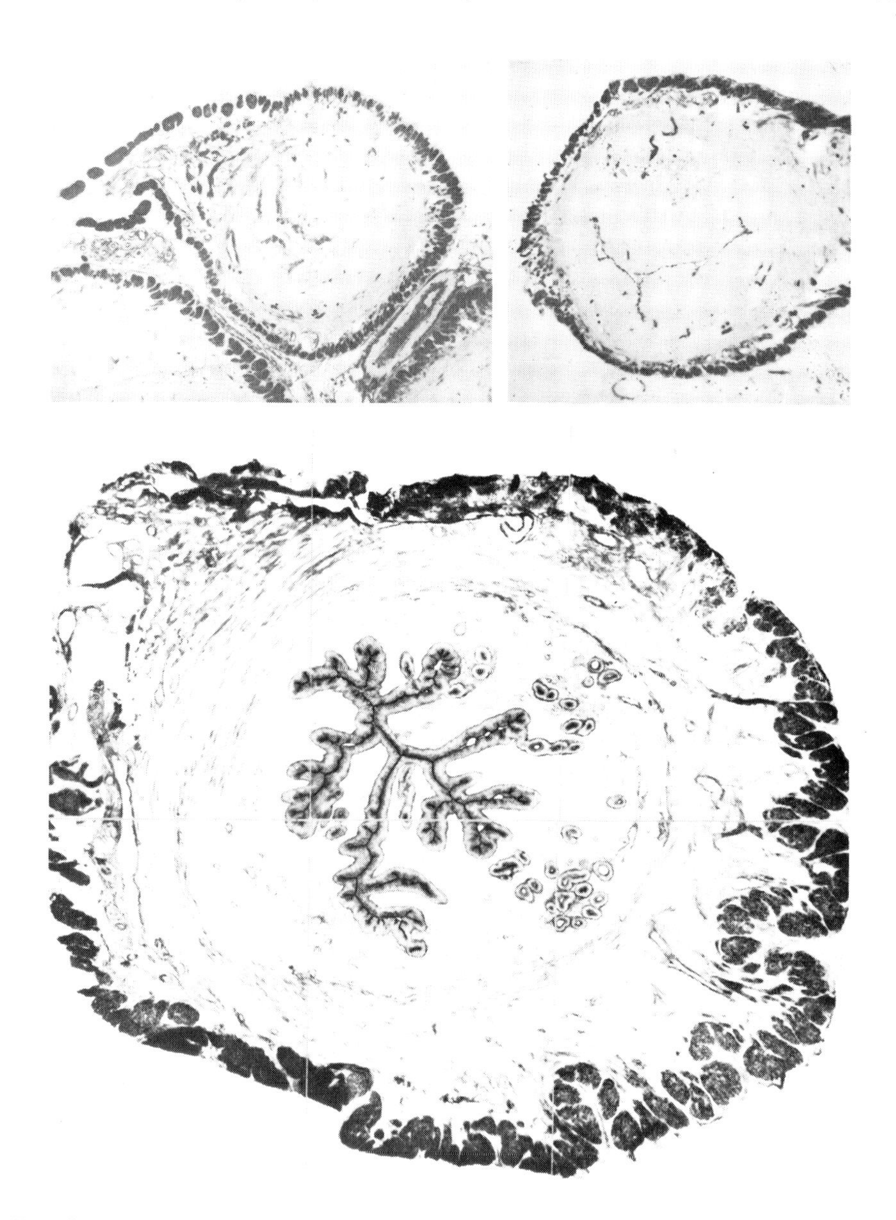

Fig. 12. Histochemical demonstration of the presence of alkaline phosphatase in the liver-RNA- (top left), the boiled uterine-RNA- (top right) and uterine-RNA-treated uteri of spayed mice (× 36).

TABLE 2

The effect of different RNA on the activity of alkaline phosphatase in homogenate of ovariectomized mouse uterus

Treatment	No. of animals	Body wt (g)	Uterus wt (g)	Enzyme activity [a]	Sigma units
Control	55	31.52	23.19	221 ± 27.0	3.2
Liver RNA	37	28.70	21.67	267 ± 15.5	4.0
Uterine RNA (U-RNA)	61	29.01	22.46	454 ± 15.5	6.7
Boar siminal vesicle	5	31.52	19.25	374 ± 15.8	5.4
RNase treated U-RNA	5	29.80	25.30	269 ± 39.1	3.7
Boiled U-RNA	6	28.40	18.00	180 ± 33.1	3.2

[a] The enzyme activity is expressed by changes in optical density (E^{cm}_{400nm}) × 10^3. The number of each column are mean ± standard error of mean.

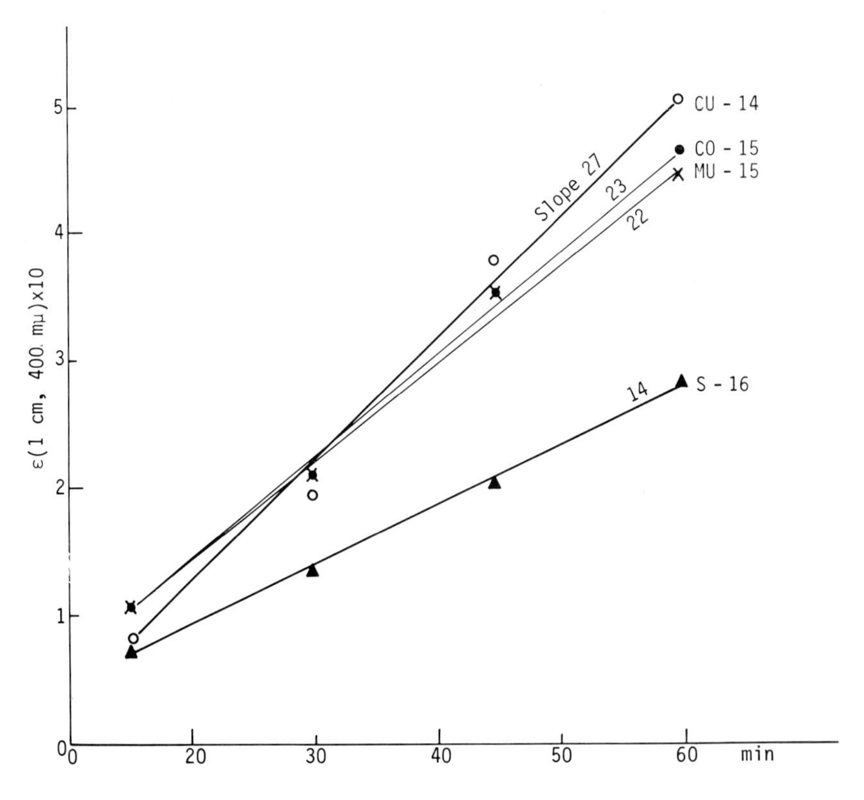

Fig. 13. The activity of alkaline phosphatase in the control, mouse-uterine-RNA-, calf-ovary-RNA-, and calf-uterine-RNA-treated uteri of spayed mice.

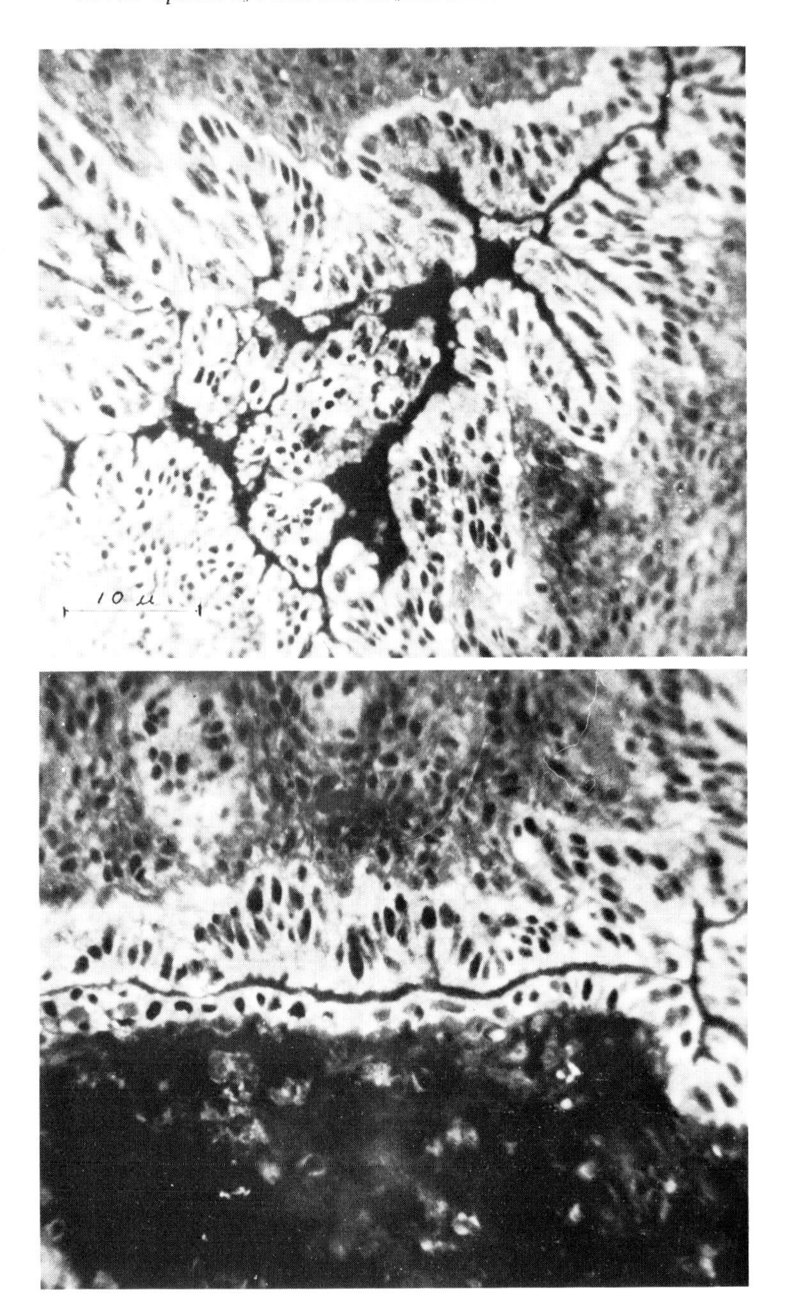

Fig. 14. (A) and (B) Photomicrographs (×360) of the rat-uterine-mRNA-treated uterine epithelium of spayed mouse showing cytoplasmic fluorescence after incubation with fluorescent rabbit anti-serum against rat albumin. The fluorescence indicates the presence of albumin in the cytoplasm of epithelial cells. No fluorescence was found in the epithelium incubated with fluorescent anti-serum against bovine albumin.

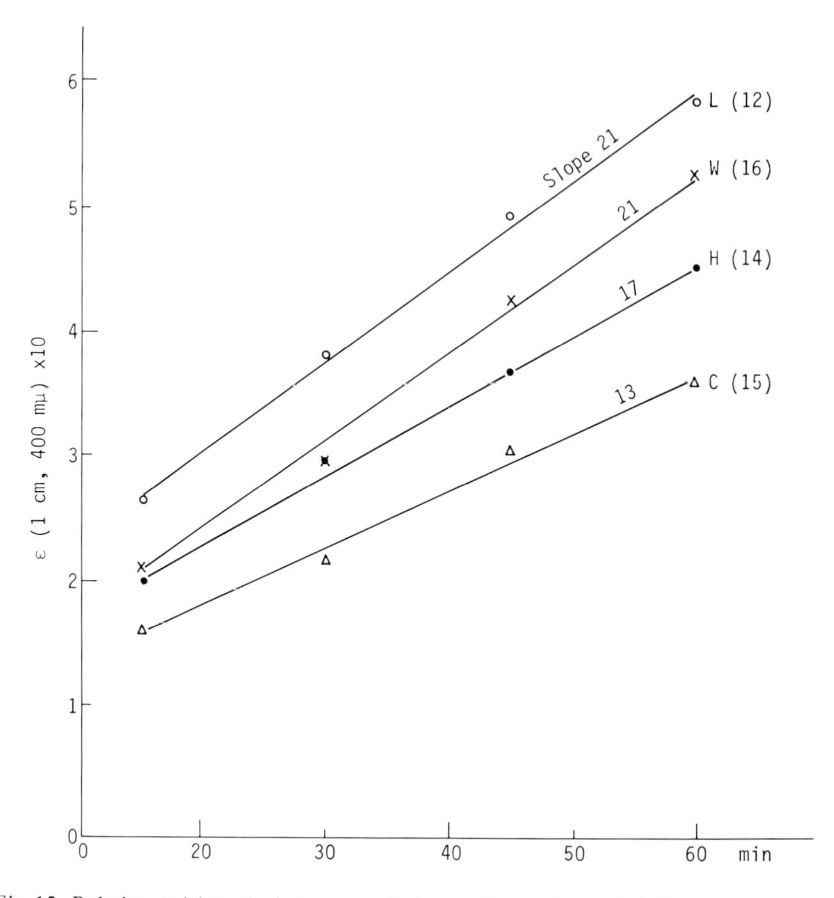

Fig. 15. Relative activity of alkaline phosphatase in the control, whole-liver-RNA and L-liver-RNA treated uteri of spayed mice.

e.g., estrogen, but did not reduce the effect of uterine-RNA. The uterus of 18 spayed mice was used for histochemical staining of alkaline phosphatase. Fourteen of them showed the presence of the enzyme in the epithelial layer (fig. 12). None was found in the boiled uterine-RNA treated, RNase-digested uterine-RNA treated, liver-RNA treated and control series.

The effect of uterine-RNA on enzyme activity was shared by RNAs from other organs which are rich in alkaline phosphatase, e.g., ovary (fig. 13) and seminal vesicle (table 2). Liver is a poor source of the enzyme, and liver-RNA was incapable of increasing the level of enzymatic activity, nor did it alter the cellular size of the epithelium. In view of the finding that the poly-A-attached-RNA of liver programmed the synthesis of serum albumin in mouse ascites cells, attempts were made to see whether or not the liver-mRNA-treated-uterus could synthesize albumin.

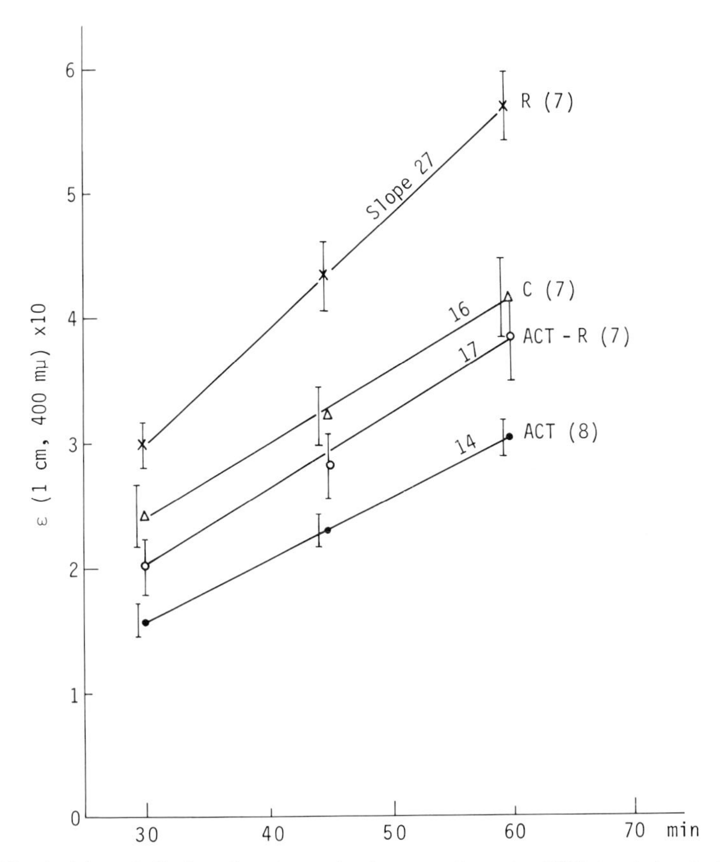

Fig. 16. The activity of alkaline phosphatase in the control, uterine-RNA-treated-, uterine-RNA and actinomycin D-treated- and actinomycin D-treated uteri of spayed mice.

Intracellular demonstration of albumin was achieved by staining frozen sections with fluorescent antiserum against albumin. Preliminary experiment has shown that albumin was found only in the cytoplasm of epithelial cells (fig. 14A and B) and none in the untreated (control) epithelium.

The function of whole uterine RNA, and its high and low molecular fractions was compared. The low fraction was most active (fig. 15). The uterine-RNA stimulated effect was abolished by treatment with actinomycin D (fig. 16).

4. Discussion

The heterogeneous RNA used in this study was contaminated with 1–2% protein. This RNA and tissue homogenate of the uteri of spayed and normal mice possessed of no detectable ribonuclease activity. It seems therefore that the RNA used to treat the uterus could not have been degraded before reaching its site of action. The strategy to substantiate this point is 2-fold: (A) showing the intracellular distribution of ^{3}H-labelled macromolecules in uterine wall and (B) proving the equivalence between the ^{3}H-RNA treated uteri and the ^{3}H-RNA used for injection. To the former, we traced the whereabouts of ^{3}H-RNA molecules by low and high resolution autoradiography. There were more silver grains in the epithelial layer than in other uterine tissues (figs. 1 and 2) and practically all silver grains in the epithelium were intracellular. The number of silver grains appearing in the nucleus varied with time and reached a maximum within an hour (table 1). The ratio between nuclear and cytoplasmic silver grains was greater than 2. This would show the genotropic nature of the exogenous RNA [25]. The evidence presented in this paper supports the conclusion that exogenous RNA is incorporated in uterine cells as such, rather than after degradation and resynthesis [16]. Enzymatic breakdown is unlikely since uterus has insignificant ribonuclease activity. Furthermore, labelled RNA used for injection and that recovered from treated uteri had identical sedimentation coefficients on sucrose gradient centrifugation.

The physiological function of intact RNA in uterine wall has been a subject of active study in recent years. The earliest report of this kind dealt with the uterine-RNA promoted increase of alkaline phosphatase activity in the atrophic uterus [3]. The sepcificity of this enzyme induction and kinetics has been established [5, 6]. Alkaline phosphatase was found mainly in the epithelial layer [26]. In view of the size increase in the RNA treated uterus (2–3 times), however, the effect of endometrial RNA had to spread all over the uterine tissue. Both β-glucuronidase [27] and glucose-6-phospate dehydrogenase [28] also were responsive to treatment with uterine RNA. The results of these in vivo studies lend direct support to the view that liver-RNA induced in vitro synthesis of glucose-6-phosphatase [29–30]. Uterine-RNA mediated alteration of epithelial morphology was striking [8, 9, 31]. Exceptional efforts have been made to rule out the possibility that the RNA effect might be due to trace contamination with female sex hormone [4, 11].

The effect of uterine-RNA on enzyme activity was shared by RNAs from seminal vesicle [3] and ovary [5, 6] and other tissues [7] with high activities of alkaline phosphatase. The question raised in this connection concerns the organ- and the species difference of the enzyme. All we can say at the present time is that RNAs from these organs, irrespective of species: mouse or calf, stimulated the enzyme activity to almost the same degree. Qualitative differences remain to be examined. Liver is a poor source of alkaline phosphatase, and appears incapable of raising the enzyme activity or of changing epithelial morphology. However, the poly A-attached RNA (mRNA) from rat-liver was found to program albumin synthesis in the uterine epithelium. Apparently, this heteroplastic site was well equipped to carry on protein synthesis. Once mRNA entered the cells, it began to program the production of specific protein in a manner similar to those found in mouse ascites cells [13, 24, 32] and *Xenopus* oocytes [33, 34].

Acknowledgement

This work was supported in part by research grants from the National Foundation and the Population Council.

References

[1] A.B. DeAngelo and J. Gorski, Proc. Natl. Acad. Sci. U.S. 66 (1970) 693.
[2] D.N. Luck and T.H. Hamilton, Proc. Natl. Acad. Sci. U.S. 69 (1972) 157.
[3] A.M. Mansour and M.C. Niu, Proc. Natl. Acad. Sci. U.S. 53 (1965) 764.
[4] S.J. Segal, Develop. Biol., Suppl. I (1967) 264
[5] M.C. Niu and L.C. Niu, In: Proceedings of the Symposium on the Mutational Processes (Publishing House of Czechoslovak Academy of Sciences, Praha) 101–108.
[6] A.M. Mansour, Acta Endocrinol. 54 (1967) 541.
[7] A.M. Mansour, Acta Endocrinol. 57 (1968) 465.
[8] O. Unjhem, A. Attramadal and J. Solna, Endocrinology 58 (1968) 227.
[9] P. Galand, N. Dupont-Mairesse, F. Leroy and J. Chretien, In: Basic actions of sex steroids on target organs, Eds. Hubinont, Leroy and Galand (Karger, Basel), 227–239.
[10] M.N. Fenel and C.A. Villee, Endocrinology 38 (1971) 279.
[11] P. Galand and N. Dupont-Mairesse, Endocrinology 9 (1972) 936.
[12] M.C. Niu, Proc. Natl. Acad. Sci. U.S. 44 (1958) 1264.
[13] M.C. Niu, C.C. Cordova and L.C. Niu, Proc. Natl. Acad. Sci. U.S. 47 (1961) 1689.
[14] K.D. Smith, J.L. Armstrong and B.J. McCarthy, Biochem. Biophys. Acta 142 (1967) 323.
[15] G.C. Rosenfeld, J.P. Comstock, A.R. Means and B.W. O'Malley, Biochem. Biophys. Res. Commun. 47 (1972) 387.
[16] M.C. Niu, L.C. Niu and A. Guha, Proc. Soc. Exptl. Med. 128 (1968) 550.
[17] O.A. Bessey, O.H. Lowry and M.J. Brook, J. Biol. Chem. 164 (1946) 321.
[18] W.H. Fishman, B. Springer and R. Brunette, J. Biol. Chem. 173 (1948) 449.
[19] J.P. Chang and S.H. Hori, J. Histochem. Cytochem. 9 (1961) 292.
[20] C. Gomeri, Proc. Soc. Exptl. Biol. Med. 42 (1939) 23.
[21] M. Kunitz, J. Biol. Chem. 164 (1946) 563.
[22] R. Burton, Biochem. J. 62 (1956) 315.
[23] O.H. Lowry, A.L. Rosenbrough, H.J. Farr and R.J. Randall, J. Biol. Chem. 193 (1951) 265.
[24] C.J. Granzow, L.C. Niu, T. Lin and M.C. Niu, Am. Zool. 11 (1971) 684.
[25] M.C. Niu, R. Cerroni and P. Piver, Am. Zool. 2 (4) (1962) 225.
[26] S.F. Yang and C.Y. Hsu, Proc. Soc. Exptl. Biol. Med. 133 (1970) 485.
[27] M.C. Niu, L.C. Niu and A.M. Mansour, Federation Proc. 24, No. 2, Part I (1965) 600.
[28] C. Villee, this symposium, page 167.
[29] M.C. Niu, C.C. Cordova, L.C. Niu and C.L. Radbill, Proc. Natl. Acad. Sci. U.S. 48(1962) 1964.
[30] M.C. Niu, Science 148 (1965) 513.
[31] S.J. Segal, O.W. Davidson and K. Wada, Proc. Natl. Acad. Sci. U.S. 54 (1965) 782.
[32] E. Zimmerman, M. Zoller and F. Turba, Biochem. Z. 339 (1963) 53.
[33] J.B. Gurdon, C.D. Lane, H.R. Woodland and G. Marbaix, Nature 233 (1971) 177.
[34] C.D. Lane, G. Marbaix and J.B. Gurdon, J. Mol. Biol. 61 (1971) 73.

Niu and Segal (eds.). The role of RNA in reproduction and development
North-Holland Publ. Co., 1973

Injection of messenger RNA into living cells and its application to the study of gene action in *Xenopus laevis*

J.S. KNOWLAND and J.B. GURDON
Medical Research Council, Laboratory of Molecular Biology, Hills Road, Cambridge, England

and

R.A. LASKEY
Imperial Cancer Research Fund, Lincoln's Inn Fields, London WC2A 3PX, England

In this paper we briefly discuss the control of gene activity in embryonic cells of *Xenopus laevis* at three main levels: DNA replication, transcription of DNA into RNA, and translation of RNA into protein. We concentrate on the last of these processes, and our main purpose is to show how RNA can be injected directly into normal living cells of *Xenopus laevis*. We discuss the particular advantages of this technique as a means of transferring messenger RNA from one cell to another, and for studying the translation of messenger RNA in a normal cellular environment. We illustrate this principle by describing the translation of rabbit haemoglobin mRNA in *Xenopus* oocytes, and by showing how *Xenopus* eggs can be used as a sensitive micro-assay for the presence of collagen message activity in a complex mixture of RNA's.

1. Introduction

To a biochemist, the answer to the problem of how to understand and account for differentiation often lies in the words "differential gene activity". He would consider his contribution essentially complete if he could explain how different genes come to be expressed in different cell types that are derived originally from a single cell, the zygote. Although a process as complex as differentiation may involve more than sequential activation or repression of particular genes as the formation of specialized cell types proceeds, adjustments in gene activity must play a major part in development.

It is probably true that the majority of such gene activity is expressed as the production of characteristic proteins, and two main possibilities could explain how different proteins come to be made in different cells. Specialized cells could differ from each other either in their content of genetic information or in the expression of it. In the first case, different cell types would come to contain characteristic sets of genes, implying that the pattern of genetic activity is ultimately determined at the level of DNA replication. In the second, every cell would contain a copy of every gene, but the expression of the genes would be adjustable, with transcription or translation being perhaps the most obvious processes on which controls might operate. The replication of genes, the transcription of DNA into RNA, and the translation of RNA into protein emerge as fundamental aspects of differentiation, and in this article we briefly discuss their operation in embryonic cells of *Xenopus laevis,* the South African clawed toad. In particular, we show how RNA can be injected directly into normal living cells, and discuss the unique advantages of this technique as a means of transferring mRNA from one cell to another, and for studying the translation of mRNA in a normal cellular environment.

2. Nuclear equivalence during cell differentiation

The first explanation of cell specialization mentioned above suggests that different cell types contain different sets of genes, and that their characteristic function is a consequence of their particular genetic content. This possibility can be tested by assessing the genetic content of specialized cells. Existing biochemical methods are not powerful enough for this purpose, but a considerable body of biological evidence shows that cell specialization does not result from irreversible losses of genetic information inherited from the zygote. The most convincing evidence comes from experiments whose general principle is to construct a normal adult organism using a specialized cell as the only source of genetic information. Whole plants have been grown from single tobacco and carrot cells [1, 2]. Normal adults of *Xenopus laevis* have been obtained after replacing the nucleus of an unfertilized egg with a nucleus from a differentiated cell [3–5]. Although considerable manipulation is required to achieve these spectacular results, apparently because it is not easy to reactivate quiescent genes in specialized cells, such experiments show that the genes of the zygote that function during differentiation are retained in the adult cells derived from the zygote. Many genes are, no doubt, permanently inactivated during normal differentiation, but the inactivation is, in principle, reversible. Cases exist, however, in which parts of the genome are lost during differentiation. Examples are to be found chiefly in insects, where heterochromatic regions of chromosomes may sometimes be eliminated during cell division [6]. Such changes are typically associated with inability to divide further, and were first described for the rather gross case of whole chromosome elimination in *Ascaris* [7]. It seems more likely that irreversible elimination of certain portions of the DNA is the consequence rather than the cause of cell differentiation, and that events leading up to this stage, which are undoubtedly an essential part of differentiation, do not involve qualitative changes in the genome, but do involve differential gene activity.

The synthesis of specific proteins certainly accompanies differentiation, and a reasonable working hypothesis is that most, if not all, of the genetic activity that directs differentiation should be expressed as synthesis of specific proteins. In protein synthesis, the two most obvious processes on which controls might act are transcription and translation. It seems likely that these events may be well separated in time, particularly in developing systems. Differentiation in *Acetabularia,* involving the formation of a characteristic cap, can still occur months after the nucleus whose genes determine the form and biochemical properties of the cap has been removed [8]. Enucleated amphibian eggs retain the information that directs their division into several hundred cells [9]. In sea-urchins, normal embryonic development is refractory to the inhibitor of RNA synthesis, actinomycin-D, until the gastrula stage [10]. Observations such as these imply the existence of long-lived RNA templates and are often interpreted in terms of translational control [11].

It is therefore necessary to study the control of transcription and translation independently. The investigation of transcriptional control is simplified by studying a type of RNA that is synthesized in large quantities, and in the following section we describe some experiments on the production of one such RNA species, namely ribosomal RNA.

3. Mutations affecting ribosomal RNA synthesis in Xenopus laevis

In studying the control of transcription, the chief problem is to identify a particular species of RNA. One outstanding example is the ribosomal RNA of *Xenopus laevis.* It is easy to label and prepare this RNA, and consequently there is a better description of the course of its synthesis during both oogenesis and embryogenesis than of any other RNA species [12, 13]. Furthermore, because the ribosomal RNA genes of *Xenopus* are repeated many times in the genome [15], they can participate easily in RNA–DNA hybridization. Most important of all, the ribosomal RNA genes can be isolated, and a mutant in which they are eliminated is available [14, 15].

In the first reported example of the fractionation of a eukaryotic genome, Wallace and Birnstiel [15] showed that *Xenopus laevis* DNA contains a high density fraction that occupies a characteristic position after equilibrium centrifugation in a caesium chloride density gradient and which contains all copies of the sequences complementary to ribosomal RNA. In the mutant which has one nucleolar organizer (1–*nu* or +/*o nu*) and one nucleolus per nucleus [14], the number of rRNA genes was half the number present in the wild-type, which has two nucleolar organizers (2–*nu* or +/+ *nu*) and either one or two nucleoli per nucleus. In the mutant which lacks both nucleolar organizers (*o*/*o nu*) and has no normal nucleoli [14, 16], constructed by crossing two +/*o nu* heterozygotes, more than 95% of the rRNA genes were deleted, agreeing with the earlier observation that the *o*/*o nu* does not synthesize rRNA [17]. Intermediate situations should be more useful for analyzing the control of transcription. Their existence was originally suggested from

TABLE 1

Relative rates of rRNA synthesis in nucleolar mutants of *Xenopus laevis* with different amounts of rDNA [a]

	Nucleolar genotype				
	$+/+$	$+/p^{l-1}$	$+/o$	p^{l-1}/p^{l-2}	p^{l-1}/o
Relative amount of rDNA (%)	100	60	45	35	23
Relative rate of rRNA synthesis (%)	100	100	100	50	25

[a] Values are taken from ref. [21].

cytological work in which mutants of *Xenopus laevis* which have one normal nucleolar organizer (+ *nu*) and one that forms only a partial nucleolus (*partial nu* or *p nu*) were isolated [18]. By crossing such animals (+*p nu*) with a +/*p nu* animal, embryos containing only a *p nu* organizer were obtained. Some *p nu* organizers (*p lethal* or p^l) cannot support development beyond tadpole stage, when such embryos (p^l/o *nu*) die, although living a little longer than anucleolate mutants (*o/o nu*), which also die as tadpoles. They synthesize rRNA at approximately 25% of the normal rate [19] and contain 25% of the diploid number of ribosomal RNA genes [20]. The available data on the rates of rRNA synthesis in the various nucleolar mutants of *Xenopus laevis* are summarized in table 1. It is clear that the rate of rRNA synthesis is not directly proportional to the number of rRNA genes. Thus, if a second p^l *nu* organizer is introduced into a p^l/o *nu* mutant by crossing two $+/p^l$*nu* heterozygotes, the increase in rRNA synthesis is greater than can be accounted for by the number of rRNA genes added with the second p^l *nu* organizer [21]. This suggests the existence of controlling sequences within the rRNA gene system, in agreement with the earlier finding [17] that the rate of rRNA synthesis in the +/*o nu* is the same as in the +/+ *nu*.

Although the ribosomal gene system of *Xenopus laevis* provides an attractive model for studying the control of transcription in development, it is not certain that the control systems for ribosomal RNA are shared by messenger RNA's. Ribosomal RNA is found in all cell types, and there is no evidence that it contributes to cell specificity. Its synthesis appears to be correlated with cell growth rather than with cell differentiation. These features may distinguish the signals that set the synthesis of rRNA in motion from those that dictate which structural genes are to be transcribed, but an understanding of how a genetic activity common to all cell types is regulated should help in the search for the means whereby differential gene activity is controlled.

4. *The use of* Xenopus *oocytes and eggs as an* in vivo *messenger translation system*

4.1. The principle of injecting mRNA into cells

In the study of differential gene activity, translation is somewhat more amenable to experimental analysis than transcription because it is very much easier to separate different proteins from each other than it is to fractionate messenger RNA. Nevertheless, the part played by translational control in development is uncertain, partly because very little is known about the kinds of proteins synthesized during embryonic development, but also because very few ways exist of experimentally modifying the normal course of translation within cells exist. One of the main difficulties is to develop a system, resembling a normal cell as closely as possible, in which the function of molecules that participate in translation can be studied. A possible solution to the problem is to inject normal cellular components into intact cells, and this approach is outlined here, with reference to the use of oocytes and eggs of *Xenopus laevis* to translate injected messenger RNA [22]. The injection technique is especially valuable as a means of inserting RNA into intact cells, because it eliminates the possibility that the RNA is altered or degraded during entry. This possibility may apply if RNA enters a cell by other, less direct means. With the exception of the special case of virus infection, no other method offers this particular advantage. The main attraction of an *in vivo* assay system so constructed is that translation takes place in a natural environment, so that normal controls affecting translation are likely to operate. It should be possible, for example, to test for factors that restrict or enhance the rate of translation of a particular mRNA, and to discover whether all mRNA's are translated at the same rate in a certain kind of cytoplasm. On the other hand, the high background of protein synthesis by the host cell is likely to make the sustem unsuitable for analysis of the finer details of the translation process itself.

4.2. Translation of mRNA injected into cells

An embryonic cell seems most likely to provide a cytoplasmic environment suitable for the translation of injected messengers, because restrictions on translation, if they exist, are less likely to be present in a non-specialized cell. It is also important that the translation product of the injected mRNA should be easily distinguishable from those made by the host cell. These principles were applied in the first successful application of the method, in which 9 S RNA from rabbit reticulocytes was found to direct the synthesis of rabbit globin chains in *Xenopus* oocytes (fig. 1 and refs. [22, 23, 24]). Two features of this experiment are especially noteworthy. One is that only rabbit globin chains are made, there being no evidence that *Xenopus* globin chains are synthesized. This shows that intact mRNA molecules that enter normal embryonic cells do not in some way activate the corresponding genes of the host cell; rather they are themselves translated. The second point is that translation of injected mRNA does not compete with endogenous protein synthesis, so that the overall rate of protein synthesis can be increased substantially by injecting exogenous mRNA (fig. 2). This shows that the translation systems in the cells used are not fully occupied [25] and suggests that the supply of mRNA to the cytoplasm is of prime importance in determining the rate of protein synthesis. It seems unlikely that such a convincing demonstration of spare translational capacity in intact cells could be achieved using a cell-free system. The oocyte system has been used to translate a wide variety of mRNA's [22–29], and it now seems true that any mRNA from a vertebrate can be translated in the *Xenopus* oocyte. These results suggest that any message, once offered to the cytoplasm, is automatically translated, and that the choice of which structural genes are to be expressed is not made at the level of translation but at some preceding level. It of course remains possible that this conclusion applies only to embryonic cells, and that differentiated cells possess translational control mechanisms that are absent in the oocyte. This could be tested by injecting one kind of specialized cell with mRNA characteristic of another. For example, reticulocyte mRNA, which directs the synthesis of globin, might be injected into a nerve cell, which does not make globin. The chief difficulty in performing such an experiment is to find a specialized cell that is as amenable to micro-injection as is the *Xenopus* oocyte.

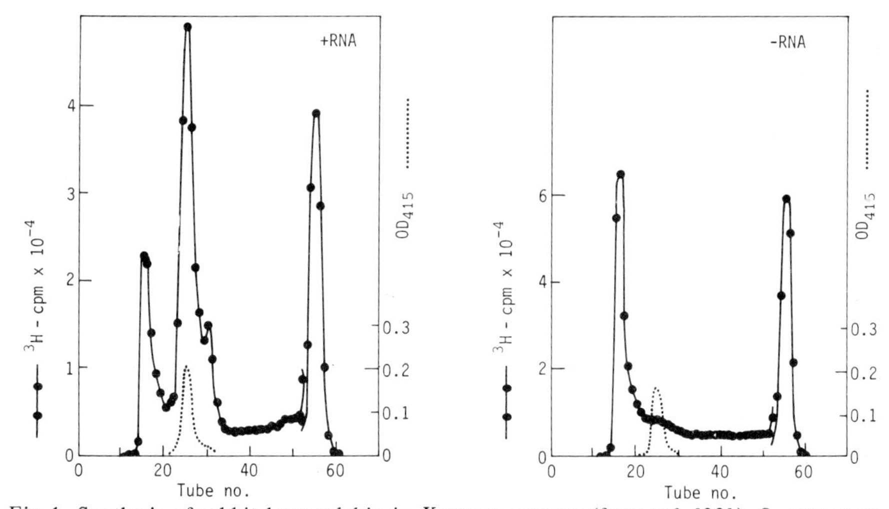

Fig. 1. Synthesis of rabbit haemoglobin in *Xenopus* oocytes (from ref. [23]). Oocytes were injected with 9 S RNA from rabbit reticulocytes and incubated for 6 hr in ^{3}H-histidine. They were then homogenised with carrier haemoglobin, and the supernatant was passed down a G-100 Sephadex column running in Tris-glycine buffer, pH 8.9. The peak resulting from injection of RNA has been identified as globin by various criteria [23].

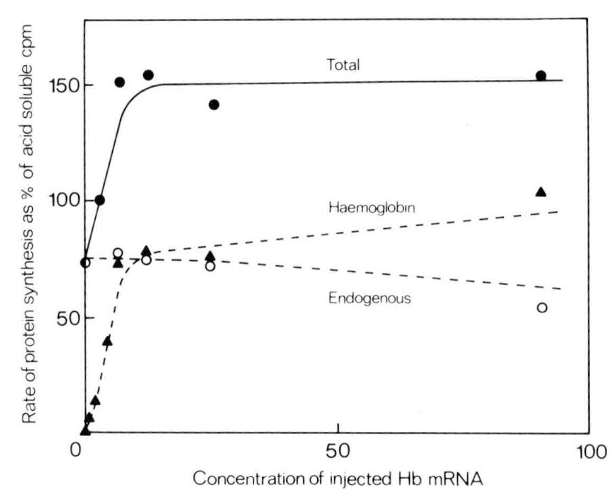

Fig. 2. Spare translational capacity of *Xenopus* oocytes. Oocytes were injected with increasing concentrations of globin mRNA(ng/oocyte), and the proportions of endogenous and exogenous protein synthesis estimated by Sephadex chromotagraphy (fig. 1). This figure is derived mainly from data in table 1 of ref. [25].

4.3. Stability of mRNA injected into oocytes

Because *Xenopus* oocytes kept in a simple saline solution continue to synthesize proteins in vitro for several days, the stability of mRNA injected into the oocyte can be estimated. Thus, if oocytes are injected with rabbit globin mRNA, and radioactive amino-acids added to the incubation medium only after several days have elapsed, rabbit globin synthesis is still observed. Experiments of this kind have shown that rabbit globin mRNA can be translated at 70% of the initial rate after 13 days of incubation inside the *Xenopus* oocyte, and it may well be that globin mRNA has an infinite life in oocyte cytoplasm (Gurdon, Lingrel and Marbaix, submitted).

4.4. Injection of RNA into cells to detect messenger RNA for particular proteins

Another aspect of the *Xenopus* oocyte translation system concerns its use as a sensitive micro-assay for mRNA. The high efficiency of the oocyte assay system, which translates reticulocyte 9 S RNA at about the same rate as the reticulocyte itself, means that only minute amounts of RNA are needed for an effect to be detectable. The system should therefore be useful for detecting individual messengers in a complex mixture; and its sensitivity will depend only upon the sensitivity of the assay for the translation product of the RNA in question. A test of this principle requires RNA from a cell which makes a variety of proteins, one of which is distinctive and easily recognized. The collagen-producing mouse fibroblast 3T6, in

TABLE 2
Hydroxyproline : proline ratios in oocytes injecyed with 3T6 RNA

Injection	Acid-insoluble Hypro : Pro
RNA from light polysomes (80–190 S)	0.93
RNA from heavy polysomes (190–300 S)	2.23
Saline; oocytes still attached to connective tissue	0.40
Saline; oocytes dissected free from connective tissue before analysis	0.05

After injection, oocytes were incubated in saline containing 1 mCi/ml of (5-^{3}H)proline. Batches of 10 oocytes were homogenised in 2 ml of cold 5% TCA, and the precipitate was collected and thoroughly washed with cold 5% TCA by suction filtration through GF 83 filters. After drying, each filter was treated with 2 ml of 6 M HCl for 18–24 hr at 108°C. HCl was removed from the hydrolysate using AG1108 resin (Bio-Rad), and residual salt was removed according to Drèze et al. [34]. Proline and hydroxyproline, which were shown by tests using radioactive amino-acids to be quantitatively recovered by this procedure, were then separated by chromatography in butanol : acetic acid : water (12 : 3 : 5) on Whatman 3 MM paper. The region of the chromatogram spanning hydroxyproline and proline was sliced into 0.5 cm strips. Each strip was moistened with 0.01 M HCl, covered with scintillant and counted. As little as 0.002% of the proline radioactivity is found in the hydroxyproline region, setting a lower limit of about 0.005% on the hypro : pro ratio than can be detected using this method.

which collagen production accounts for approximately 10% of total protein synthesis, is a convenient cell for this purpose because collagen synthesis is easily determined from the amount of proline converted to hydroxyproline in the acid-insoluble material. When RNA from 3T6 cells is injected into *Xenopus* oocytes, collagen synthesis can be detected in this way (table 2). Unfortunately, collagen synthesis is also detectable in oocytes injected with saline alone (table 2), and this remains true even if the oocytes are free of associated ovarian tissue, although the level is reduced. The sensitivity of the assay used here is high enough to reveal this background of collagen synthesis, which may arise either from the oocyte itself or from the follicle cells that surround it. The background can, however, be eliminated by using eggs, in which collagen synthesis is undetectable. The use of eggs is to be preferred because of the possibility that the results from oocytes simply reflect a non-specific response to the wounding caused by injection.

If 3T6 polysomal RNA is injected into *Xenopus* eggs, collagen synthesis is stimulated (fig. 3). The analytical method used is sensitive enough to detect collagen synthesis is at a level of about 0.01% of total protein synthesis. If unhydroxylated collagen labelled with (5-^{3}H) proline [30] is injected into *Xenopus* eggs, then after incubation for 2 hr a hypro : pro ratio of 0.74 is obtained. This is close to the value of 0.80 characteristic of fully hydroxylated 3T6 collagen [31], and shows that in the intact egg the conversion of proline to hydroxyproline is not limited by the hydroxylating enzyme, in agreement with the finding that protocollagen hydroxylase can be extracted from unfertilized eggs [32].

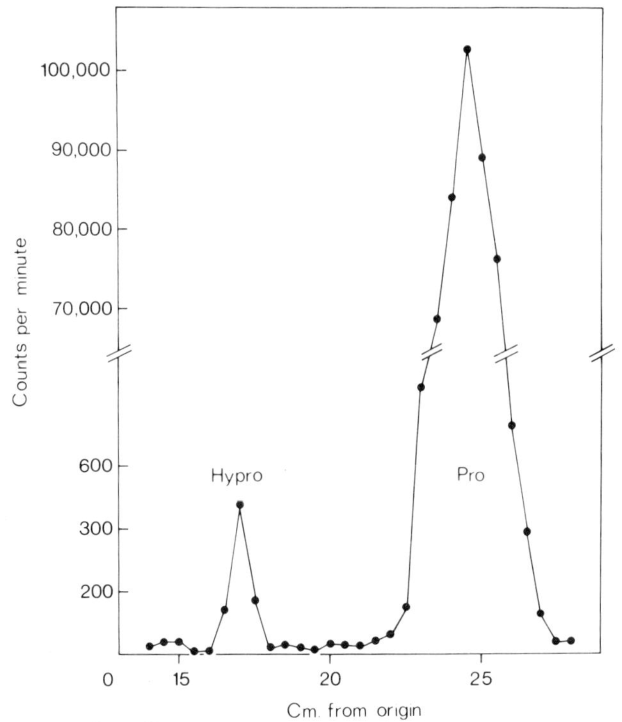

Fig. 3. Collagen synthesis in *Xenopus* eggs following injection of 3T6 RNA. RNA from heavy polysomes (table 2) was injected with (5-^{3}H)proline (10 mCi/ml) into *Xenopus* eggs. After incubation for 4 hr, the jelly coats were removed and the eggs processed as described in table 2. Hydroxyproline : proline ratio – 0.12%, showing that collagen synthesis represents about 0.04% of total protein synthesis [33]. No collagen synthesis can be detected in eggs injected with saline alone.

5. *Conclusions and discussion*

In this section we list and briefly discuss the most important conclusions to emerge from the injection of messenger RNA into living cells.

5.1. Injection of RNA into cells

The micro-injection technique has unique advantages as a method of inserting RNA into cells. The RNA enters the cell without being degraded; the amount of RNA that enters can be accurately measured, and it is easy to ensure that all cells receive the same amount of RNA. The technique is, of course, only applicable to cells that are large enough to inject without undue difficulty, and resilient enough to withstand the operation. Oocytes and eggs of *Xenopus laevis* meet both these criteria, and the conclusions below are based on experiments using these cells.

5.2. Translation of messenger RNA injected into cells

As shown in sect. 4, rabbit globin mRNA injected into *Xenopus* oocytes is faithfully translated, there being no evidence that the corresponding host cell genes are activated. This demonstrates that the immediate function of mRNA that enters cells is to participate in translation, and also shows that mRNA from both a foreign cell type (a reticulocyte) and a foreign species (a rabbit) can be translated inside a normal, living cell (a *Xenopus* oocyte). No mRNA-specific factors are needed to effect translation or, if factors are required, they already exist in the oocyte. These conclusions apply to intact cells and are drawn from experiments using intact cells. Experiments based on cell-free systems could not yield such reliable information about normal cells because specificity might be lost during the preparation of a cell-free system.

5.3. Translational capacity of living cells

The synthesis of globin in oocytes that is supported by injected mRNA does not compete with endogenous protein synthesis; rather it occurs independently (fig. 2). Oocytes at least, therefore, have unused translational capacity, and this has not so far been shown for any other kind of cell. The existence of spare capacity means that certain competition experiments are possible; for example one could test whether a given mRNA is preferentially translated when others are injected with it. Again, cell-free systems would be unsuitable for answering such questions about intact cells.

5.4. Stability of mRNA in cells

Globin messenger RNA injected into *Xenopus* oocytes is remarkably stable, but it is not yet clear whether stability is a property of this particular mRNA, or whether the oocyte confers stability on any injected mRNA. Elucidation of this question requires many more purified mRNA preparations.

5.5. Efficiency of the oocyte translation system

Rabbit globin mRNA is translated inside *Xenopus* oocytes almost as efficiently as it is under normal conditions inside rabbit reticulocytes [22]. This high efficiency makes the oocyte system very attractive, especially when only very small amounts of RNA are available. It also places the general conclusions drawn from message injection experiments on a secure basis, for it is most unlikely that such high translational efficiency could be achieved if injected mRNA were handled in a fundamentally different way from internally supplied mRNA.

Acknowledgement

We are grateful to the Medical Research Council and the Imperial Cancer Research Fund for financial support.

References

[1] A.C. Braun, Proc. Natl. Acad. Sci. U.S. 45 (1959) 932.
[2] F.C. Steward, Proc. Roy. Soc. B175 (1970) 1.
[3] J.B. Gurdon, J. Embryol. Exptl. Morphol. 10 (1962) 622.
[4] J.B. Gurdon, Quart. Rev. Biol. 38 (1963) 54.
[5] J.B. Gurdon and R.A. Laskey, J. Embryol. Exptl. Morphol. 24 (1970) 227.
[6] J. Schultz, Brookhaven Symp. Biol. No. 18 (1965) 116.
[7] T. Boveri, In: Festschrift zum 60 - Geburtstage, Richard Hertwig, Vol. 3 (Jena, G. Fischer, 1910) 131–214.
[8] J. Hammerling, Ann. Rev. Plant. Physiol. 14 (1963) 65.
[9] R. Briggs and T.J. King, In: The cell, Vol. 1, Eds. J. Brachet and A.E. Mirsky (Academic Press, New York and London, 1959) 537–617.
[10] P.R. Gross, Curr. Topics Devel. Biol. 2 (1967) 1.
[11] H. Harris, Nucleus and cytoplasm (Clarendon Press, Oxford, 1968).
[12] D.D. Brown, Curr. Topics Devel. Biol. 2 (1967) 47.
[13] J.B. Gurdon, In: Heritage from Mendel, Ed. R.A. Brink (The University of Wisconsin Press, Madison, 1967) 203–241.
[14] T.R. Elsdale, M. Fischberg and S. Smith, Exptl. Cell Res. 14 (1958) 642.
[15] H. Wallace and M.L. Birnstiel, Biochem. Biophys. Acta 114 (1966) 296.
[16] E.D. Hay and J. B. Gurdon, J. Cell Sci. 2 (1967) 151.
[17] D.D. Brown and J.B. Gurdon, Proc. Natl. Acad. Sci. U.S. 51 (1964) 139.
[18] L. Miller and J.B. Gurdon, Nature 227 (1970) 1108.
[19] J. Knowland and L. Miller, J. Mol. Biol. 53 (1970) 321.
[20] L. Miller and J. Knowland, J. Mol. Biol. 53 (1970) 329.
[21] L. Miller and J. Knowland, Biochem. Genet. 6 (1972) 65.
[22] J.B. Gurdon, C.D. Lane, H.R. Woodiand and G. Marbaix, Nature 233 (1971) 177.
[23] C.D. Lane, G. Marbaix and J.B. Gurdon, J. Mol. Biol. 61 (1971) 73.
[24] G. Marbaix and C.D. Lane, J. Mol. Biol. 67 (1972) 517.
[25] V.A. Moar, J.B. Gurdon, C.D. Lane and G. Marbaix, J. Mol. Biol. 61 (1971) 93.
[26] A.J.M. Berns, M. van Kraaikamp, H. Bloemendal and C.D. Lane, Proc. Natl. Acad. Sci. U.S. 69 (1972) 1606.
[27] C.D. Lane and J. Knowland, to be published in: The biochemistry of animal development, Vol. 3, Ed. R. Weber.
[28] J. Knowland and R.A. Laskey, unpublished.
[29] R.A. Laskey, J.B. Gurdon and L.V. Crawford, Proc. Natl. Acad. Sci. U.S. 69 (1972) 3665.
[30] B. Goldberg and H. Green, Proc. Natl. Acad. Sci. U.S. 59 (1968) 1110.
[31] K.A. Piez, E.A. Eigner and M.S. Lewis, Biochemistry 2 (1963) 58.
[32] H. Green, B. Goldberg, M. Schwartz and D.D. Brown, Devel. Biol. 18 (1968) 391.
[33] H. Green and B. Goldberg, Nature, 204 (1964) 347.
[34] A. Drèze, S. Moore and E.J. Bigwood, Analyt. Chim. Acta 11 (1954) 554.

Session Four

Transfer Of Tissue Specificity

Chairman

SHELDON J. SEGAL

Population Council,
Rockefeller University,
New York, New York

Chairman's introduction

Can RNA be transcribed in one animal, extracted from the host tissues, put through a series of chemical purifications, exposed to tissues of another animal, pass into the cytoplasm via the apical cell membrane – and then elicit a biological response? Under these circumstances why is the exogenous RNA not hydrolyzed by the ubiquitous RNase of the host tissue or extracellular fluid?

These were typical of the questions posed, quite understandably, when the first reports of Niu and his colleagues were published in 1958. Fifteen years later the body of evidence has grown. Now, the questions are concerned with whether the activity reflects direct translation of the RNA or whether the biological effect is mediated through the nucleus. When it was first demonstrated that estrogen-stimulated uterine RNA itself exhibits biological activity characteristic of estrogen, the suspicion of estrogen contamination was strong. The burden of proof was on the few investigators who had made and reported their findings. A series of carefully executed and controlled investigations resulted, and now contamination is no longer a viable interpretation.

Concurrently, investigation of the mechanism of action of steroid hormones has advanced rapidly. The role of the gonadal hormone in eliciting quantitative and qualitative changes in RNA synthesis has emerged as a key event, believed by many investigators to be primary to subsequent biological changes brought about by the hormone. The evidence has emerged from studied of the rate of RNA synthesis in response to the hormone, the inhibition of hormone action by the use of specific metabolic inhibitors that interfere with RNA synthesis and, more recently, the in vitro effects of hormones on RNA or protein synthesis in cell free systems.

Valuable as they are, these lines of evidence are indirect. They are supported strongly, however, by the direct evidence first provided by the methodology introduced by Niu and now used by many investigators. The application of this methodology goes far beyond the study of hormone action and has proven useful in studying embryogenesis, tumorogenesis and immunological phenomena. Advances in these areas of investigation, based on the study of biological responses to RNA are included in this section of the symposium.

In his "Introduction to experimental physiology", Claude Bernard made the observation that: "It often happens that a good method gives more service to science than the highest theoretical speculation."

The extent to which science has been served by those contributing the methodology for the extraction, purification and testing of biological potential of RNA is now beginning to be appreciated.

Sheldon J. Segal

Niu and Segal (eds.). The role of RNA in reproduction and development
North-Holland Publ. Co., 1973

The role of macrophage RNA in the immune response*

Marvin FISHMAN
The Public Health Research Institute of The City of New York, Inc., New York, N.Y. 10016, USA

Peritoneal exudate cells that have been incubated in vitro with bacteriophages yield ribonucleic acid (RNA) fractions which induce the formation of specific phage neutralizing antibodies by lymphoid cells in organ or cell cultures. Two active fractions can be distinguished; one is a complex of phage antigen and preexisting RNA and the other is informational RNA (i-RNA) synthesized after exposure of the cells to antigen. In cultures of lymph node fragments the former evokes IgG antibody and the latter gives rise to IgM antibody that has the allotypic specificity of the donor of the peritoneal exudate cells.

While both classes of RNA originate in macrophages of the peritoneal exudate, each can be shown to be formed by a separate subpopulation, and both subpopulations constitute minorities among the heterogeneous macrophages. Besides the synthetic functions ascribed to certain cells of the "adherent population", an additional role appears to be that of effective presentation of the product to receptive lymphocytes.

It is suggested that under in vivo conditions the most likely role of the RNA-antigen complex is that of providing a metabolically stable form of antigen which insures continued stimulation of antibody formation and immunological memory. The minimal function of i-RNA may be that of contributing to the rapid initial antibody response by lateral spread of information. Possible replication of such RNA, indicated by the studies of others, suggests a broader role for it with implications for its eventual use in the treatment of certain immunodeficiencies.

1. Introduction

The initiation or transfer of specific immune responses from antigenically stimulated cells or tissues to the cells of "normal" recipients through extracts rich in ribonucleic acid or polynucleotides has been under intensive study for many years. The transfer of both humoral [1] and of cellular [2, 3] immunity has been demonstrated both in vivo and in vitro. The nature of the active RNA fractions, their origin and mode of action, and their functional roles are beginning to come into focus and the more recent relevant experimental findings will be presented in this paper.

Biologically active RNA has been obtained from cells of immune as well as nonimmune animals. The latter cells were stimulated in vitro with specific antigens

* This work was supported in part by grant GB-29018X from the National Science Foundation.

before extracting the RNA. In most of the experiments spleen cells and peritoneal exudate cells were employed as the source of RNA [1] although other cell types, such as lymphocytes [4] and liver parenchyma cells [5], have also been used by some investigators. The peritoneal exudate cells, consisting mostly of macrophages (80–90%), were the prime source of immunogenic RNA studied in this laboratory. Because these cells were not considered to be immunocompetent, their RNA would have to be transferred to other cell types to bring about antibody formation. The capacity of the RNA to initiate the immune response was assayed in vitro by adding a minimum of 100 μg to lymph node organ cultures [6, 7] or to spleen cell cultures [8]. Antibody activity was measured in the tissue culture fluids by neutralization [9] or hemagglutination [10]. In lymph node organ cultures the RNA initiated a biphasic antibody response in which the first peak was attained 4–6 days after start of the culture and the second at 10–14 days. Two types of RNA were involved in eliciting the immune responses and have been designated as super-antigen or RNA antigen complex, and as informational RNA (i-RNA), respectively.

The basic characteristic of these active fractions, as revealed by biochemical and cellular studies, are described in the following sections of the paper.

2. RNA-antigen complexes

RNA-antigen complexes, first described by Garvey and Campbell [5], were originally isolated from the livers of rabbits that had been immunized months before by a single injection of antigen. Subsequent work from their laboratory [11] reported the presence of immunogenic complexes in the serum of animals as early as 5 hr after a primary exposure to antigen. These complexes were identified as polyribonucleoproteins.

Biologically active RNA-antigen complexes have also been obtained from nonimmune cells that had been incubated with specific antigens in vitro [12, 13]. One source of such nonimmune cells has been the peritoneal exudate cells from rabbits or rats. When such cells were incubated with a particulate antigen, bacteriophage T_2, and the mixture contained at least 100 phage particles per macrophage, an RNA preparation was obtained that initiated an immune response in lymph node organ cultures. Several methods were used to detect the presence of antigen associated with the RNA. These included radioactive tracing of the antigen in the RNA [14], separation of the RNA-antigen complexes by precipitation with specific antisera, or inhibiting the immunogenicity of the RNA by suppressing it with antibody specific for the phage antigen or by inactivating the complex with the proteolytic enzyme, pronase [15]. The amount of phage antigen associated with the RNA was calculated to be about 10^{-5} μg nitrogen.

It was evident that the immunogenicity of these RNA-antigen complexes depended on the RNA moiety as well as the antigen. Inactivation of these complexes was accomplished with RNAse, freed of DNAse activity. A high concentration of

RNAse had to be used (enzyme : substrate ratio = 1 : 15) to destroy immunogenicity. This relative resistance to degradation may be a result of the presence of bound phage protein in the RNA.

Treatment of the RNA preparation with trace amounts of antiphage antibody (quenching), or digestion by pronase eliminated only the activity responsible for initiation of the late immune response in lymph node organ cultures which was shown to be solely IgG immunoglobulin, leaving intact the activity responsible for the emergence of the early IgM antibody response. Further support for the view that the RNA-antigen complex caused synthesis of the late IgG antibody response in lymph node cultures was obtained by the demonstration that purified RNA-antigen complexes, obtained by specific precipitation with antiphage sera, evoked this type of response exclusively, and that chromatography on methylated bovine serum albumin–Kieselguhr [16] yielded RNA-antigen which acted as did the product obtained by specific precipitation. In addition, studies on the cell-free synthesis of RNA-antigen complexes [17] added further support to this concept.

Estimates of molecular size of the RNA–antigen complex were obtained by analytical centrifugation, acrylamide gel electrophoresis and by methylated albumin–Kieselguhr chromatography (MAK). With the aid of MAK columns the RNA was resolved into 3 fractions with S values of 4 S, 16 S and 23–28 S (MAK I, II and III respectively). All 3 fractions were assayed for their capacity to initiate the late IgG antibody response in organ culture. Only MAK fraction III was found to be responsible for the late 7 S antibody. The molecular weight of the RNA in this fraction (1×10^6) was calculated from the S value according to the equation reported by Spirin [18]. The relatively large molecular size of this RNA differed from that reported by Gottlieb and coworkers [19, 20]. They have shown that the RNA complexed to antigen (synthetic amino acid polymer), was of low molecular size (4–5 S) and of a unique species. Current joint experiments are being conducted with Gottlieb to resolve these differences.

The question as to whether the RNA complexed to antigen was newly synthesized RNA or RNA present in the cell prior to antigen contact was answered by applying RNA inhibitors to the incubation mixture of peritoneal exudate cells and phage. One such inhibitor used for this purpose was actinomycin D. The RNA preparation obtained from antigen-stimulated peritoneal cells in the presence of actinomycin D was fully able to initiate the formation of the late IgG response in lymph node cultures. These results, therefore, indicated that this RNA existed in the peritoneal exudate cells before their exposure to antigen.

A summation of the characteristics of the RNA–antigen complex is given in table 1. Briefly, it was concluded that upon incubation of peritoneal exudate cells with bacteriophage, the phagocytized phage was solubilized and that some of the products were bound to preexisting large molecular weight RNA forming an immunogenic RNA-antigen complex. This complex, when placed in lymph node organ cultures, would initiate the formation of IgG antibody directed against the phage.

An attempt was made to synthesize the RNA-antigen complex in a cell-free

TABLE 1

Characteristics of the immunogenic RNA–antigen complex

Presence of antigen	
Radioactive tracer	+
Precipitation with specific antiserum	+
Specific quenching with antiserum	+
Inactivation with pronase	+
Minimum ratio of macrophage to phage needed to form complex	1 : 100
Antibody response in LN fragments stimulated by complex	
Peak response	days 10–14
Immunoglobulin	IgG
Molecular size, 23S	1×10^6
Inactivation with RNAse (substrate:enzyme)	15 : 1
Effect of actinomycin D on its formation	–
Amount of antigen associated with complex	10^{-5} μg protein
Ability to transfer allotype marker (messenger activity)	–

system in order to elucidate its mode of formation. This approach seemed particularly important since other laboratories had reported that such RNA–antigen complexes were artifactual [21]. Roelants et al. [22] described the formation of complexes between antigens and RNA, of a mammalian or bacterial origin, with unpredictable biological activity. The formation of nonimmunogenic RNA–antigen complexes had also been observed in this laboratory. It was shown that, when RNA from nonstimulated peritoneal exudate cells was incubated with radioactive solubilized phage, complexes were formed and these could be isolated from the MAK III fraction. However, they did not possess any immunogenic activity when added to the tissue culture systems. When the same mixture of RNA and solubilized antigen was incubated in the presence of cell sap preparations, the RNA–antigen complex was capable of initiating an immune response in vitro. This response, as

TABLE 2

Formation of immunogenic RNA–antigen complexes in cell-free system. Requirement for cell-sap components

Tissue culture fluids days	Incubation of RNA with solubilized T_2 phage Per cent neutralization [a]	
	Presence of cell sap	Absence of cell sap
4	3	0
7	20	2
10	32	3
14	48	0

[a] Tissue culture fluids were diluted 8-fold; neutralization assay performed in the presence of rat serum complement [9].

Lymph node cultures incubated with 100 μg of RNA reextracted from the above incubation mixtures.

seen in table 2, was indistinguishable from the late response observed with the RNA–antigen complexes formed when peritoneal cells were incubated with intact phage in vitro. The apparently essential cell sap factor or factors have been found in the 100,000 *g* supernatant of organ homogenates of liver, peritoneal exudate cells and spleen of nonimmune rabbits [17]. The cell sap components were protein in nature and thermolabile. Preparations of RNA that were successfully used in the formation of active RNA–antigen complexes were from spleen, peritoneal exudate and liver, while RNA from brain and kidney were found to be unsuitable, suggesting that the source of suitable RNA are immunocompetent cells. The synthesis of RNA–T_2 complexes was pH and temperature dependent which, along with other observations, would favor the hypothesis that the cell sap component was an enzyme. In contrast to the results reported by Roelants et al. [22], yeast RNA and HeLa cell RNA did not yield an immunogenic complex when incubated with solubilized antigen and cell sap components. Additional experimentation with the cell sap factors are in progress.

3. *Informational RNA*

The other type of immunogenic RNA (i-RNA) differs from the RNA–antigen complex in several ways and appears to induce antibody formation by functioning as a messenger [7]. The evidence to prove the absence of antigen in this RNA fraction rests on several criteria. Attempts to detect antigen in this RNA by radioactive tracer experiments, or by precipitation with specific antiphage serum, both led to negative results. Pronase digestion failed to destroy its activity and quenching with antiphage sera also was unsuccessful. These results clearly distinguished this RNA from the RNA–antigen complex. Additionally, the RNA free of antigen was highly sensitive to the action of RNAse and was readily inactivated by small amounts of enzyme (enzyme : substrate = 1.75).

The antibodies engendered by this RNA appeared 4–6 days after the start of the culture and was shown to be IgM immunoglobulin. There was no evidence for the synthesis of IgG antibody, in response to the antigen-free RNA.

Separation of this RNA fraction was attained by MAK column chromatography which yielded the activity in the second elution peak, whereas the antigen–RNA complex was retained at this salt concentration. These observations and data from analytical centrifugation suggested an S value for i-RNA of 10–12. A single-stranded RNA with an S value of 10 would have a molecular weight of about 2×10^5. It could also be calculated that an RNA of this size could theoretically code for a protein of about 27,000 mol. wt. This could account for the synthesis of a complete light chain of immunoglobulin but only for half of the heavy chain. As will be discussed later, the i-RNA transferred information for both light and heavy chain synthesis.

Evidence for the messenger nature of this immunogenic RNA was reinforced by experimental data that showed rapid incorporation of radioactive uridine into this

RNA, and inhibition of formation of such RNA by an RNA inhibitor, actinomycin D. These observations supported the view that i-RNA was newly synthesized by the peritoneal exudate cells after their exposure to antigenic stimulation.

The most direct evidence for the messenger nature of the RNA in question is the finding that it directs the synthesis of a new protein. This was shown for the immunogenic i-RNA by its ability to cause synthesis of antibodies in the cells of a recipient which possessed the allotypic markers of the donor [4, 7]. Allotypic markers in rabbits are expressed in both the light and heavy chains of immunoglobulins and show no evidence of crossover in matings within a homozygous group of rabbits. An RNA preparation, prepared from antigenically stimulated peritoneal exudate cells of a rabbit of allotype a_1a_1/b_4b_4 (donor), was added to a culture of lymph node fragments obtained from a recipient rabbit of allotype a_3a_3/b_5b_5. The immumoglobulins formed in culture were assayed for anti-T_2 activity and for their allotypic specificity. Two methods were used to determine allotype specificity; one the inhibition of antiphage neutralizing activity with specific anti-allotype sera and the second the direct measurement for the foreign allotype sera by means of the hemagglatination inhibition. By both procedures it was shown that the IgM immunoglobulin synthesized by the lymph node organ cultures had the allotypic markers of the macrophage donor (a_1a_1/b_4b_4), whereas the subsequently produced IgG antibody was of allotype a_3a_3/b_5b_5. These results clearly illustrated the messenger activity of i-RNA in coding for both the light and heavy chain allotypes.

Since, as previously stated, this i-RNA could theoretically code for a protein of 27,000 mol. wt, a single hit phenomenon could not account for these allotype transfer results. It was necessary to postulate, therefore, that multiple hits must occur to account for the observation that the RNA coded for both chains. This requirement could easily have been met in the tissue culture experiments since at least 100 μg of RNA were routinely added to the estimated 10^7 lymphoid cells

TABLE 3
Characteristics of the immunogenic i-RNA

Presence of antigen:	
Radioactive tracer	–
Precipitation with specific antiserum	–
Quenching with specific antiserum	–
Inactivation with pronase	–
Minimum ratio of macrophage to phage needed to form i-RNA	1000 : 1
Antibody response in LN fragment culture stimulated with i-RNA:	
Peak response	days 4–6
Immunoglobulin	IgM
Molecular size, 10–12 S	2×10^5
Inactivation with RNAse (substrate:enzyme)	75 : 1
Effect of actinomycin D on its synthesis	I
Ability to transfer allotype marker (messenger activity)	+

contained in the lymph node fragments. A summary of the properties of i-RNA is shown in table 3.

In vitro tests for messenger activity were not attempted because it seemed that neither the total RNA nor the partially purified MAK fraction II would contain the specific RNA in sufficiently pure form to make this approach worthwhile. Gel electrophoresis of the MAK II fraction showed, for example, that it contained at least 7 species of RNA. Further data, to be discussed below, suggested that concentration and purification of the cells responsible for synthesis of this RNA might be feasible and thus provide a more suitable source of RNA for such studies.

4. Cellular involvement in antibody formation initiated by immunogenic RNAs

In considering cellular activities relevant to the formation and the transfer of the immunogenic RNA fractions, it appears convenient to discuss the latter of these activities next.

Since cell suspensions provide better conditions for observations on cellular interactions than do organ fragments, suspensions of spleen cells were substituted for the lymph node fragments used in the earlier experiments. Such suspensions, in which the immunogenic RNA fractions elicited specific IgM and IgG antibodies, could be fractionated into adherent and nonadherent populations by the procedure of Mosier [23]. It was found that cultures of the macrophage-rich adherent cells or lymphocyte-rich nonadherent cells were unable to respond to immunogenic RNA. However, it was observed that when the adherent cells were incubated with RNA for an hour, under tissue culture conditions, and the nonadherent cells were added to the culture after the excess RNA had been removed, antibody formation occurred. Thus it was apparent that the adherent cells were required in order to achieve successful stimulation by RNA. It had previously been shown that radioactive RNA appeared to be "fed" into cultured lymphocytes by the macrophage [24]. From these observations it would appear that one of the functions of the macrophage was the transfer of undegraded RNA into immunocompetent cells.

Regarding the cells in which immunogenic RNA-antigen complexes and i-RNA are formed, it will be recalled that these were among the adherent cells of the peritoneal exudate. It became evident from a series of experiments, that at least two types of macrophages from this source were participating in RNA synthesis. When the peritoneal exudate cells were incubated with a low input of phage (ratio of bacteriophage to macrophages was 1 : 1000), an RNA was formed that initiated only the early IgM (19 S) antibody in lymph node organ cultures. It thus appeared that only a small population of macrophages, capable of recognizing the antigen, was involved in the synthesis of i-RNA. In contrast, the RNA-antigen complex was obtained only when the peritoneal exudate cells were incubated with a large excess of antigen (macrophage : phage ratio 1 : 100).

More direct evidence that the peritoneal macrophages were heterogeneous in respect to their role in forming immunogenic RNA was obtained by physically separating subpopulations of macrophages by density gradient centrifugation. Peritoneal exudate cells placed on BSA [25] or Ficoll [26] gradients separated into 5 subpopulations. Only the lighter density macrophages, bands obtained at 5/8 and 8/11% interfaces (BSA), yielded both i-RNA and RNA–antigen complexes after appropriate antigenic stimulation. The heavier density macrophages did not participate in the formation of immunogenic RNA. Only 15% of the total peritoneal macrophages composed the active 5/8 and 8/11% bands. Cells from these populations were further divided into 8/9 and 9/11% bands in attempts to obtain more specific macrophage donors for i-RNA and RNA–antigen complexes. This separation was successful and strongly suggests that the two RNAs were formed in different macrophages. Differences among macrophages from various tissues were also observed. Alveolar macrophages were not able to form any immunogenic RNA when incubated with antigen [12].

5. *Conclusions*

Antibody formation in vitro in response to RNA from macrophages exposed to antigen has been demonstrated in a number of laboratories [1]. In discussing this phenomenon it is important to separate the events leading to antibody production by the RNA-antigen complex from that obtained with the informational i-RNA.

The formation of the RNA-antigen complex in vivo would require the uptake of antigen by the macrophage, solubilization of the antigen, and coupling of solubilized and possibly fragmented antigen to preexisting RNA. Evidence to support an enzymatic requirement in the complex formation has been obtained in studies using a cell-free system. Once formed, the complex could become exteriorized and could stimulate immunocompetent lymphoid cells attracted to the macrophage. The cells in which these complexes are formed belong to a subpopulation of macrophages in which the dominant cell type is indistinguishable from the typical macrophage but appears to possess only moderate phagocytic activity. Another, and possibly more important function of these or similar cells may be the long term preservation of antigen in the form of the metabolically stable complex. Long term retention of antigen in macrophages [27] as well as other cells [5] in the form of RNA-antigen complexes has been reported in vivo.

The informational RNA, which was also extracted from antigen stimulated macrophages, was capable of initiating an IgM response in both lymphoid organ and cell cultures. These immunoglobulins possessed the allotypic marker for both the light and heavy chain globulins of the macrophage donor rabbit. These macrophages may, in a manner similar to lymphocytes, be stimulated by antigen. The formation of antibody by macrophages has been reported by Noltenius and Chahin [28]. In our experiments a distinct population of peritoneal exudate cells has been isolated by density gradients which formed i-RNA when incubated with antigen. These cells found in the light density regions of BSA gradients (8–11% interface) consisted of typical macrophages, and another mononuclear cell type. The latter cells had some of the properties of macrophages, such as adherence and phagocytosis, but were morphologically different from typical macrophages or lymphocytes. Definite experiments have not been done to establish which cell type formed the immunogenic RNA, however, it is possible that the atypical cell represents a macrophage of lymphoid origin which has been described by Howard [29].

The cellular transfer of immunogenic RNA, from macrophages to lymphocytes, has not been observed in vivo. Several investigators, however, have reported that material could be transferred from macrophages to lymphocytes through cytoplasmic bridges [30–32]. Both types of immunogenic RNAs could be transferred in this manner, especially the informational RNA which would have to express itself intracellularly.

The role of macrophage i-RNA in antibody formation in vivo could be visualized as occurring at the initial phase of IgM induction where the number of antibody-forming cells would require a doubling time of 4 hr. However, the actual cell-generation time for lymphocytes is 8 hr. To explain this discrepancy a lateral transfer of the immune response with RNA to nonproliferating lymphocytes might be envisioned.

Informational RNA could also conceivably be used clinically to introduce a specific immune response in individuals lacking this capacity. Bioengineering with i-RNA is possible since it has been shown that this RNA can be injected into animals and still convey its antibody-forming message [4]. Successful application of this type of therapy for certain immunodeficiency diseases would depend on the presence of an RNA replicating system in the recipient cell to insure a continuous production of this RNA. Recent reports have demonstrated that normal lymphocytes do contain RNA-dependent DNA polymerase which is different from the reverse transcriptase isolated from oncogenic viruses [34, 35]. Therapy has been attempted with the transfer factor [36], a polynucleotide fraction isolated from lymphocytes of individuals showing delayed hypersensitivity to certain antigens. Nonsensitive recipients of this transfer factor have acquired the delayed hypersensitivity pattern of the donor. Experiments of this kind with i-RNA should be forthcoming in the near future.

Acknowledgment

The author acknowledges and expresses appreciation to Dr. F.L. Adler in the preparation of the manuscript.

References

[1] M. Fishman, Ann. Rev. Microb. 23 (1969) 199.
[2] R.E. Jureziz, D.E. Thor and S. Dray, J. Immunol. 101 (1968) 823.
[3] R.E. Paque and S. Dray, J. Immunol. 105 (1970) 1334.
[4] C. Bell and S. Dray, J. Immunol. 103 (1969) 1196.
[5] J.S. Garvey and D.H. Campbell, The retention of S^{35}-labeled bovine serum albumin in normal and immunized rabbit liver tissue. J. Exptl. Med. 105 (1957) 361.
[6] M. Fishman, J.J. van Rood and F.L. Adler, In: Molecular and cellular basis of antibody formation, Prague Symp. Ed. J. Sterzl, (Publ. House, Czechoslovakia Acad. Sciences, Prague, 1965) 491–501.
[7] F.L. Adler, M. Fishman and S. Dray, J. Immunol. 97 (1966) 554.
[8] H.K. Meiss and M. Fishman, J. Immunol. 108 (1972) 1172.
[9] F.L. Adler, W.S. Walker and M. Fishman, Virology 46 (1971) 797.
[10] W.S. Walker, M. Fishman and F.L. Adler, J. Immunol. 107 (1971) 953.
[11] L. Yuan, J.S. Garvey and D.H. Campbell, Immunochemistry 7 (1970) 601.
[12] M. Fishman and F.L. Adler, In: Mononuclear phagocytes, Ed. R. van Furth (Blackwell Scientific Publ., England, 1970) 581–594.
[13] B.A. Askonas and J.M. Rhodes, Nature 205 (1965) 470.
[14] H.B. Herscowitz and P. Stelos, J. Immunol. 105 (1970) 779.
[15] M. Fishman and F.L. Adler, Cold Spring Harb. Symp. Quant. Biol. 32 (1967) 343.
[16] J.D. Mandell and A.D. Hershey, Anal. Biochem. 1 (1960) 66.
[17] M. Fishman and F.L. Adler, The formation of immunogenic RNA-antigen complexes in a cell-free system. Cell. Immunol. (1973) in press.
[18] A.S. Spirin, In: Progress in nucleic acid research, Vol. I, Eds. J.N. Davidson and W.E. Cohn (Academic Press, New York, 1963) 301–345.
[19] A.A. Gottlieb, Biochemistry 8 (1969) 2111.
[20] A.A. Gottlieb and R.H. Schwartz, Cell. Immunol. 5 (1972) 341.
[21] G.E. Roelants and J.W. Goodman, Biochemistry 7 (1968) 1432.
[22] G.E. Roelants, J.W. Goodman and H.O. McDevitt, J. Immunol. 106 (1971) 1222.
[23] D.E. Mosier, Science 158 (1967) 1573.
[24] M. Fishman, R.A. Hammerstrom and V.P. Bond, Nature 198 (1963) 549.
[25] S.G. Rice, Federation Proc. 31 (abst.) (1972) 780.
[26] W.S. Walker, Nature New Biol. 229 (1971) 211.
[27] R.S. Speirs and E.E. Speirs, J. Immunol. 92 (1964) 540.
[28] H. Noltenius and M. Chahin, Z. Immunitätsforsch. 139 (1970) 312.
[29] J.G. Howard, In: Mononuclear phagocytes, Ed. R. van Furth (Blackwell Scientific Publ., London 1970) 178–199.
[30] H.R.P. Miller and S. Avrameas, Nature New Biol. 229 (1971) 184.
[31] J.A. Clarke, A.J. Salsbury and D.A. Willoughby, Nature 227 (1970) 69.
[32] D. Sulitzeanu, R. Kleinman, D. Benezra and I. Gery, Nature 229 (1971) 254.
[33] R.W. Dutton and R.I. Mishell, Cold Spring Harb. Symp. Quant. Biol. 32 (1967) 407.
[34] H.M. Temin, J. Nat. Cancer Inst. 46 (1971) III.
[35] S.N. Borrow, R.G. Smith, Z.M.S. Reit and R.C. Gallo, Proc. Natl. Acad. Sci. U.S. 69 (1972) 3228.
[36] A.S. Levin, L.E. Spitler, D.P. Stites and H.H. Fudenberg, Proc. Natl. Acad. Sci. U.S. 67 (1970) 821.

Niu and Segal (eds.), The role of RNA in reproduction and development
North-Holland Publ.Co., 1973

Studies on biological potentiality of testis-RNA
I. Induction of axial structures in whole and excised chick blastoderms

H. LEE and M.C. NIU
Departments of Biology, Rutgers University, Camden, N.J., USA, and Temple University, Philadelphia, Pa., USA

The biological potentiality of three types of calf testis-RNA, i.e., whole RNA (wRNA), poly A-attached RNA (mRNA) and filtrate of wRNA (fRNA), was investigated using chick blastoderms at the definitive streak stage. Both wRNA and mRNA induced the formation of a structure resembling a secondary axis in whole blastoderms and nodal pieces of blastoderms (NPs). Post-nodal pieces of blastoderms (PNPs) treated with chick Ringer's solution did not undergo differentiation, but those treated with wRNA or mRNA acquired the capacity to differentiate. The most striking feature of the differentiation was the emergence of a structure resembling a defective early chick embryo in the median axis. mRNA was most potent in all three series of experiments. The specificity of mRNA action was shown by the fact that neither fRNA nor native heart-mRNA was capable of inducing the formation of a "secondary embryo". Furthermore, the effect was abolished by heat denaturation, digestion with pancreatic ribonuclease, and treatment with actinomycin D.

1. Introduction

Functional studies of RNA have shown that it induced the formation of specific structure and/or specific protein (enzyme). The type of differentiation initiated by RNA varied with the donor tissue of RNA [1–9]. Thus, for example, brain-RNA induced in the majority of cases the formation of neural tissue. Kidney- and heart-RNAs seldom induced neural differentiation, but could often transform some undifferentiated tissue into tubular and cardiac tissues, respectively. All of the RNAs previously used were prepared from somatic tissues. From the developmental point of view, somatic tissues are derived from the union of germ cells. The question is what effect may the RNA from gonad or germ cells have in developing undifferentiated tissue? Would it function differently from soma-RNAs? As a working hypothesis, the authors prefer to think that testis-RNA differs functionally from soma-RNAs. Firstly, it may invoke the development of a secondary axis, thus paving the way for epigenetic formation of various organs and tissues. Secondly, it may transform some embryonic tissue into gonad or primordial germ cells. As the first part of a serial study on the biological potentiality of testis-RNA, we have explored the first possibility. In this report, evidence for the capacity of testis-RNA to induce the formation of a structure resembling a secondary axis or embryo in whole and excised chick blastoderms is presented.

2. Materials and methods

2.1. Culture procedure

2.1.1. Whole blastoderms

Fertile hen's eggs were incubated at 37.5°C for 17–19 hr to obtain embryos at the definitive streak stage (stage 4 of Hamburger and Hamilton [10]). Embryos were grown in vitro for 19–21 hr using Spratt's [11] technique.

2.1.2. Nodal pieces (NPs) and post-nodal pieces (PNPs)

Stage 4 blastoderms were isolated from the yolk, trimmed off the area opaca, and transected 0.6 mm posterior to the primitive pit in the manner shown in fig. 1. The anterior pieces (NPs) were grown in vitro for about 24 hr using Spratt's [11] technique. The posterior pieces (PNPs) were placed ventral side upward on a piece of vitelline membrane laid on an agar medium [12] and incubated for 4 days. The PNPs were subcultured every other day. Protocol was made prior to each subculturing. A Ringer-agar plus yolk-albumen extract medium [13] was used to culture all whole and excised blastoderms. This will be referred to as the basic medium.

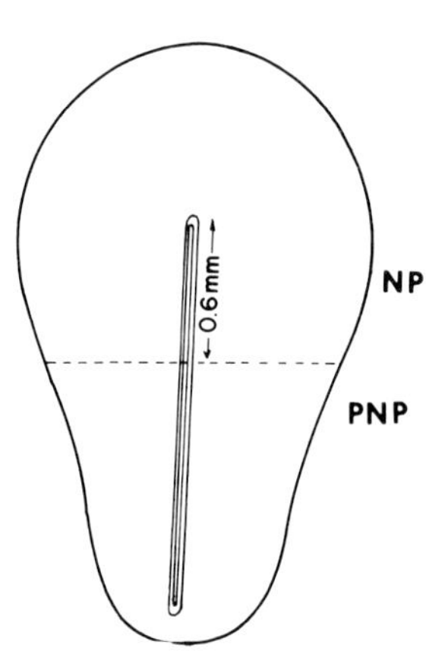

Fig. 1. A schematic diagram of the area pellucida of stage 4 chick blastoderm. The dotted line indicated the level of the cut. The tip of the primitive pit was considered as 0.0 mm and cut was made with reference to it. NP, nodal piece; PNP, post-nodal piece.

2.2. Grafting experiments

All of the grafts used were obtained from the node area (size = 0.3 × 0.3 mm) of stage 3+ blastoderms [10]. The grafts consisted mainly of the mesoderm and the underlying endoderm. Each graft was inserted between the epiblast and hypoblast of a host embryo at stage 4 according to the technique described by Grabowski [14]. All host embryos were grown for about 24 hr on the basic medium.

2.3. Preparation of RNA

2.3.1. Whole testis-RNA (wRNA)

Calf testes were excised immediately after slaughtering and cooled in ice-cold isotonic solution (0.25 M sucrose in 0.001 M $MgCl_2$). Unless otherwise specified, all preparations were carried out in the cold room at 4°C. Connective tissue and fat were removed from the testes. Groups of 100 g were chopped up with scissors into fine pieces and blended in a Waring blender with 6 vol of buffer solution (0.075 M NaCl, 0.025 M EDTA, 0.5% SDS, and 0.01 M Tris-HCl, pH 8.0). The homogenate was filtered through 2 layers of cheese cloth. The filtrate was mixed with an equal volume of phenol saturated with the above buffer solution and blended at 60 V for 5 min. The mixture was centrifuged at 1800–2000 rpm for 30 min and the top aqueous layer (supernatant I) saved. To the interphase and phenol layers, 0.5 vol of saline was added and blended at 30–35 V for 5 min. The top aqueous layer was combined with the supernatant I and once again extracted with 0.5 vol of buffered phenol as before. To the top aqueous layer, 2 vol of 95% ethanol and K-acetate (up to 2%) were added. This was kept in a freezer (–20°C) for 20 min or overnight. Nucleic acid was collected by centrifugation at 2000 rpm for 10–20 min and washed with ethanol twice. The precipitate was redissolved in 0.9% NaCl and centrifuged at 18000 rpm for 20 min to remove insoluble material. The supernatant was washed 3–5 times with ether to remove phenol. Ether was removed under negative pressure. Nucleic acid was precipitated, dissolved in buffered saline (pH 7.5), and then treated with DNase (Worthington, RNase free). The enzyme was denatured by chloroform containing 1% isoamyl alcohol. RNA was reprecipitated by 2 vol of 95% ethanol. The precipitate was dissolved in a small volume of saline. RNA concentration was estimated with an UV spectrophotometer at 260 nm and then adjusted to 150–200 optical density (O.D.) per ml. This RNA solution was kept in a freezer (–20°C) after addition of 2 vol of ethanol.

2.3.2. Poly A-attached RNA (mRNA) and filtrate RNA (fRNA)

These two RNAs were prepared by a procedure recently described by Rosenfeld et al. [15]. mRNA was the fraction of wRNA that adsorbed to the nitrocellulose membrane (millipore filter, pore size = 0.45 μm). fRNA was the fraction of wRNA that passed through the filter. The latter contained all species of testis-RNA except mRNA. However, trace contaminant of mRNA was not avoidable.

2.3.3. Heart mRNA

Heart mRNA was prepared in the same manner as testis-mRNA. An UV spectrophotometric examination of testis-RNA solution revealed that it was typical of nucleic acids. Tests with the Lowry procedure [16] showed that it contained about 1% of protein. The Burton reaction [17] showed that mRNA had practically no DNA contamination.

2.3.4. RNase-digested mRNA

mRNA and crystallized RNase (preboiled at 100°C for 30 min) were dissolved separately in chick Ringer's solution. They were so combined that the mixture contained 60 O.D. units of RNA and 60 μg of RNase per ml. The mixture was incubated at 37°C for 30 min. RNase activity was neutralized by rabbit antiserum against RNase.

2.3.5. Denaturation of mRNA

mRNA (60 O.D./ml) was boiled at 100°C for 30 min immediately before use. The turbidity, if any, was removed by centrifugation.

2.3.6. Testis-mRNA and actinomycin D

A stock solution of actinomycin D (Sigma Chemical Co.) was made up in 70% ethanol at a concentration of 100 μg per ml and stored at 4°C. Before use, this stock solution was diluted with a stock solution of testis-mRNA and chick Ringer's solution to make up a working solution containing 0.02 μg of actinomycin D and 60 O.D. units of mRNA per ml. The presence of ethanol, at the concentration used, was not found to affect normal chick development.

2.4. Methods of treatment

2.4.1. Whole blastoderms

Eight to 10 isolated whole blastoderms were placed in 10 ml of chick Ringer's solution with or without test agent(s). They were incubated at room temperature (21°–23°C) for 2 hr. After this treatment, they were placed on the basic medium and incubated for 19–21 hr at 37.5°C.

2.4.2. PNPs and NPs

Two PNPs (or NPs) were placed on the basic medium. One drop of chick Ringer's solution with or without test agent(s) was applied directly onto each PNP (or NP). Culture dishes were kept at room temperature for 2 hr. Any excess solution was carefully removed. The explants were cultured for 4 days (or 24 hr).

2.5. Histological preparations

All explants were fixed in Bouin's fluid. A few were stained with Delafield's hematoxylin, and kept as whole mounts, while the rest were embedded in paraffin, sectioned at 6 μm, and stained with Delafield's hematoxylin and eosin.

3. Results

3.1. Whole blastoderms

3.1.1. Control series

A total of 48 chick blastoderms were cultured for 19–21 hr. Forty-two (88%) developed to stages 9–11 (fig. 2), 3 (6%) were abnormal, and 3 (6%) died.

3.1.2. Experimental series

3.1.2.1. mRNA. The biological potentiality of testis-mRNA was tested on the development of explanted chick blastoderms. The results are summarized in table 1. It can be seen that the concentration of mRNA used in this study interfered with the normal development of explanted chick embryos. Abnormal embryos were those

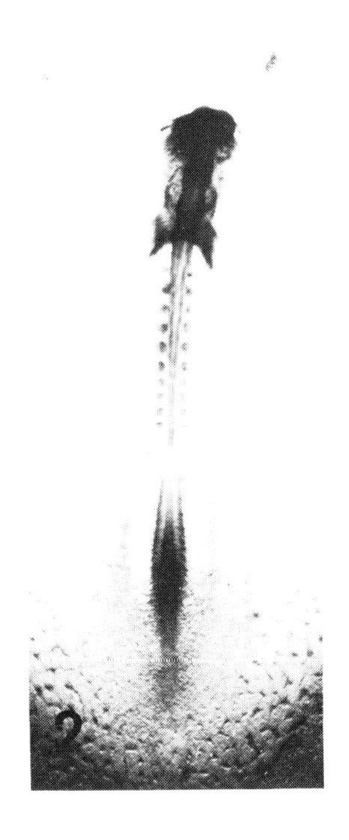

Fig. 2. Control embryo explanted at stage 4 and cultured for 19 hr on Spratt's agar medium. × 13.5.

TABLE 1
Effect of mRNA on the development of explanted stage 4 chick embryos

Concentration of RNA (O.D./ml)	No. of embryos	% of embryos			Embryos with secondary axis		
		Dead	Ab-normal	Normal	No.	% of total	% of surviving
0	48	6	6	88	0	0	0
20	28	3	36	61	8	29	30
40	45	4	76	20	21	47	49
60	62	3	86	11	48	77	80
80	41	22	78	0	27	66	84
100	24	87	13	0	3	13	100

possessing one or more discernable abnormalities. The common features of these embryos were malformed brain, short embryonic axis, less numerous and/or poorly defined somites, and edema on both sides of the body axis (fig. 3). These deleterious effects were observed at a concentration of 80 O.D./ml. That was the series with significantly increased mortality (see column 3, table 1). On the other hand,

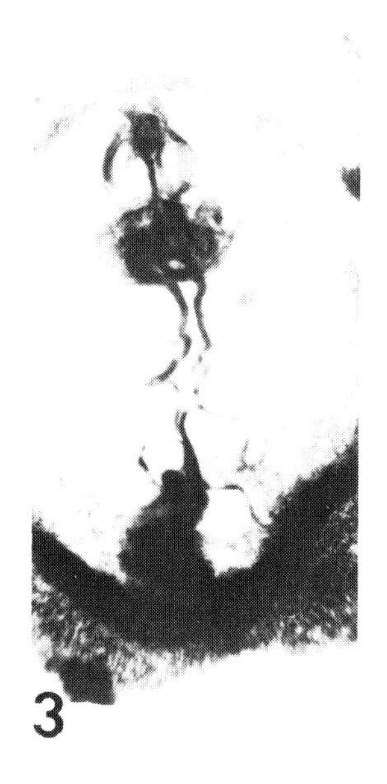

Fig. 3. Stage 4 embryo treated with 80 O.D./ml of mRNA and cultured for 21 hr. Note the greatly malformed brain region, poorly defined somites, and edema on both sides of the primary body axis. "Secondary axis" (head-like structure) is clearly visible at the posterior end. × 13.5.

Fig. 4. Stage 4 embryo treated with 60 0.D./ml of mRNA and cultured for 21 hr. Note the presence of "secondary axis" at the posterior end. × 13.5.

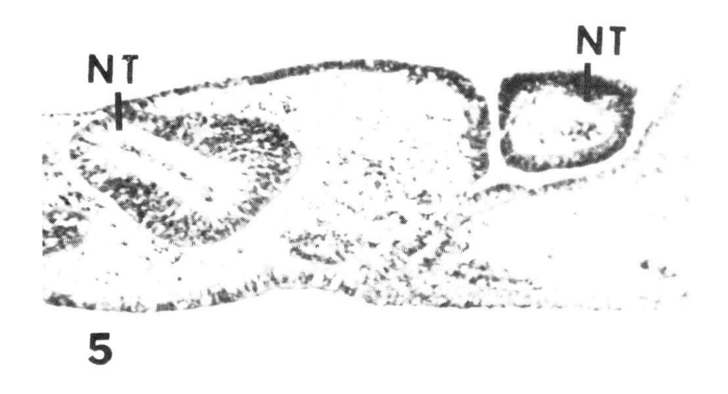

Fig. 5. Cross section through "secondary axis" of stage 4 embryo treated with 60 O.D./ml of mRNA and cultured for 19 hr. Note induced neural tubes (NT). × 160.

TABLE 2
Effects of wRNA on the development of explanted stage 4 chick embryos

Concentration of RNA (O.D./ml)	No. of embryos	% of embryos			Embryos with secondary axis		
		Dead	Ab-normal	Normal	No.	% of total	% of surviving
0	48	6	6	88	0	0	0
20	14	0	29	71	1	7	7
40	14	0	64	36	2	14	14
60	18	6	72	22	11	61	65
80	28	28	61	11	13	46	65
100	14	71	29	0	4	29	100

there was practically no difference in mortality at 60 O.D./ml and lower concentrations as compared to the controls, but the maximal frequency of "secondary axis" (or head-like structure) occurred in the 60 O.D. series, especially at the posterior end of the embryo (fig. 4).

Serial sections through the "secondary axis" showed in all cases the presence of well defined neural tissue in the form of neural tube or neural epithelium (fig. 5). Somite-like mesoderm and notochord were also found, but the frequency of their appearance was lower than that of the neural tissue.

3.1.2.2. wRNA. The effect of wRNA was similarly analyzed. Table 2 depicts the concentration effect on the formation of "secondary axis". It can be seen that the

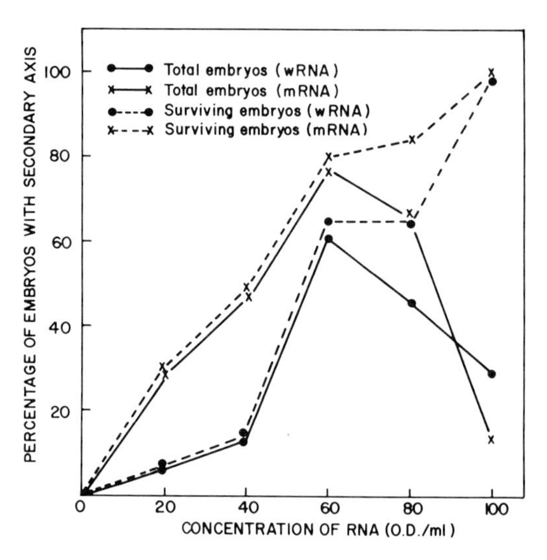

Fig. 6. The concentration effect of wRNA and mRNA on the formation of "secondary axis" in whole blastoderms.

TABLE 3
Effect of RNAs on the development of explanted stage 4 chick embryos

Type of RNA (60 O.D./ml)	No. of embryos	% of embryos		Embryos with secondary axis	
		Abnormal	Normal	No.	% of total
Controls	48	6	88	0	0
mRNA	62	53	11	48	77
wRNA	18	72	22	11	61
fRNA	38	68	24	2	5
RNase-digested mRNA	32	9	77	3	9
Boiled mRNA	32	9	81	0	0
mRNA plus actinomycin D (0.02 μg/ml)	28	39	39	0	0
Heart-mRNA	52	79	14	0	0

results of this series differ quantitatively, but not qualitatively, from those of mRNA-treated embryos (fig. 6). Apparently 60 O.D. per ml of testis-RNA is the optimal concentration for developmental studies.

3.1.2.3. fRNA. This is the fraction of wRNA with mRNA selectively removed. Out of the 38 blastoderms explanted, 9 were normal, 26 were abnormal, and 2 developed "secondary axis" (Table 3).

3.1.2.4. RNase-digested mRNA. Twenty-six out of the 34 blastoderms developed normally and 3 had "secondary axis" at the posterior end. The latter was probably due to the presence of undigested mRNA.

3.1.2.5. Boiled mRNA. Twenty-six of the 32 blastoderms developed normally, 4 had poorly developed brain and short body axis, and 2 died. "Secondary axis" was not seen in this series.

3.1.2.6. Testis-mRNA and actinomycin D. Of the 28 blastoderms, 4 showed little or no development and 2 were severely malformed. Eleven embryos had the syndrome of actinomycin D inhibition: poorly developed brain, few somites, and incomplete closure of neural folds. The other 11 were perfectly normal. The "secondary axis" was not seen in embryos of this series (table 3).

3.1.2.7. Heart-mRNA. This mRNA was prepared by the same procedure as that used in the isolation of testis-mRNA. Of the 52 blastoderms, 2 showed little or no development, 2 were severely malformed, and 48 were alive at the time of fixation. Of the surviving embryos, 7 were normal and others had two or more abnormalities: poorly developed brain (100%), short embryonic axis (94%), and less numer-

7

Fig. 7. Control NP cultured for 24 hr showing a normal embryo. × 30.

ous somites (93%). However, heart development was normal in 93% of the cases. The absence of "secondary axis" formation in this series illustrates the functional uniqueness of testis-mRNA.

3.2. Nodal pieces (NPs)

3.2.1. Control series

Of the 42 NPs, 3 died, 5 were abnormal, and 34 normal as described by Spratt

TABLE 4
Effect of RNAs on the development of NPs of stage 4 chick embryos

Type of RNA (60 O.D./ml)	No. of NPs	% of NPs		Embryos with secondary axis	
		Abnormal	Normal	No.	% of total
Controls	42	12	81	0	0
mRNA	54	83	17	42	78
wRNA	46	70	26	29	63
RNase-digested mRNA	24	29	50	2	8
Boiled mRNA	24	29	63	0	0
mRNA plus actinomycin D (0.02 μg/ml)	24	50	33	0	0
Heart-mRNA	28	71	18	0	0

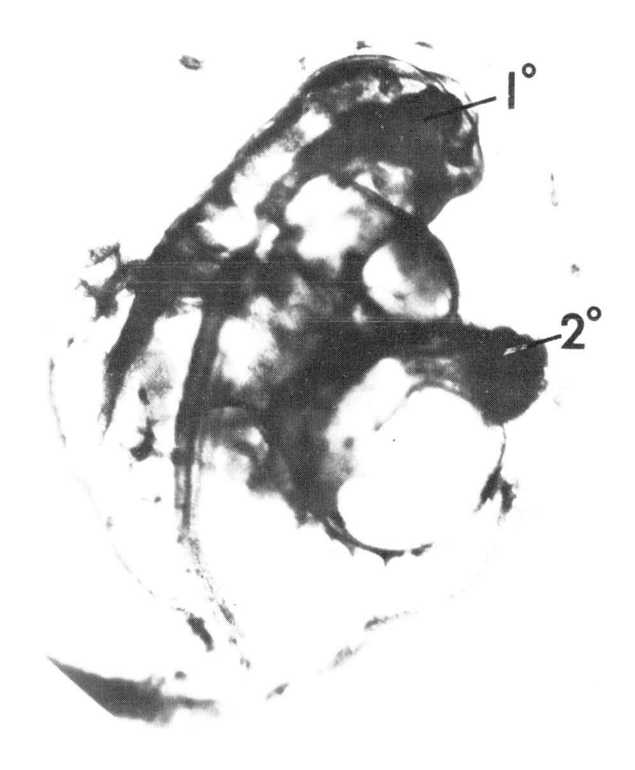

8

Fig. 8. A double headed embryo from the testis-mRNA-treated NP, cultured for 24 hr. × 25. 1°, Primary embryo; 2°, secondary embryo.

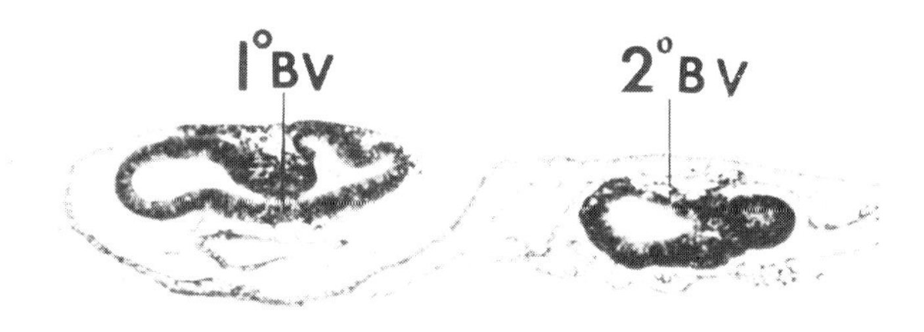

9

Fig. 9. Cross section of a specimen similar to that shown in fig. 8. × 50. 1°BV, primary brain versicle; 2° BV, secondary brain vesicle.

[18]. In all cases, the posterior end developed a projected "tail" consisting of neural tube, notochord, and somites (fig. 7). No "secondary axis" was seen in embryos of this series.

3.2.2. Experimental series

A total of 200 NPs were used. They were divided into 6 sets of experiments. The results are summarized in table 4. As can be seen, the pattern of the respective series is comparable to that obtained from whole embryos (compare table 4 with table 3). Both wRNA and mRNA series were featured by the development of secondary embryo (fig. 8). A cross section of the induced brain is shown in fig. 9.

The constituents of secondary embryo induced by mRNA were compared with the structure induced by Hensen's node. The latter is known to induce the formation of secondary embryo. Analysis of 15 mRNA-induced and 25 Hensen's node-induced secondary embryos showed that testis-mRNA and Hensen's node were equally potent: both induced neural formation, but to a lesser degree, notochord and somites.

3.3. Post-nodal pieces (PNPs)

Our first exploratory study of testis-RNA in development was performed on PNPs. The concentration used was 80 O.D. per ml instead of 60. Testis-RNA-treated PNPs showed, at low magnification, the most complicated yet recognizable structures, e.g., the thickened primitive streak and pulsatile tissue on 4th day of incubation. Therefore, a total of 256 PNPs from 8 experimental sets were fixed after 4 days of cultivation. One hundred forty-nine PNPs were serially sectioned and categorically analyzed. The results are summarized in table 5.

TABLE 5
Effect of RNAs on the development of PNPs of stage 4 chick embryos

Type of RNA (80 O.D./ml)	No. of PNPs	% of PNPs with the following structures				
		Neuroids	Notochord	Somite mesoderm	Nephric tubules	Pulsatile tissue
Controls	18	6	0	0	6	0
mRNA	33	100	64	36	55	82
wRNA	31	90	10	7	52	29
fRNA	23	52	0	0	44	17
Boiled wRNA	8	25	0	0	0	0
Boiled mRNA	8	38	0	0	0	0
RNase-digested mRNA	12	8	0	0	17	0
mRNA plus actinomycin D (0.02 μg/ml)	16	13	0	0	0	0

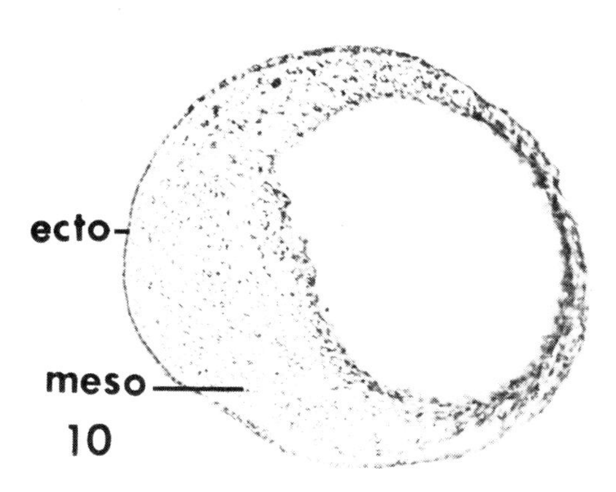

Fig. 10. Cross section of control PNP cultured for 4 days showing undifferentiated ectoderm (ecto) and mesoderm (meso). × 13.

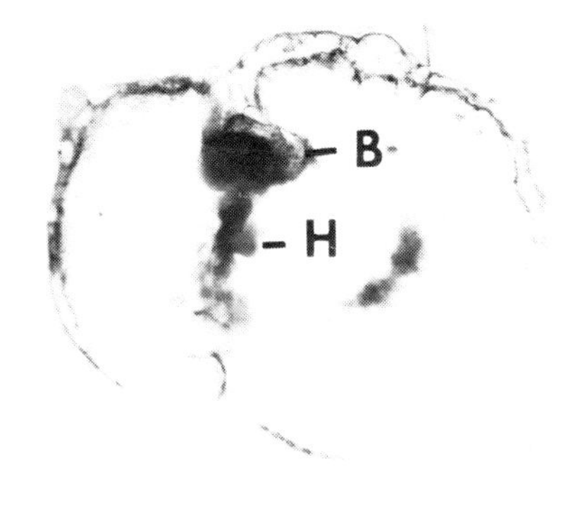

Fig. 11. mRNA-treated PNP cultured for 4 days showing a structure resembling a defective early chick embryo in the median axis. × 13. B, brain; H, beating heart.

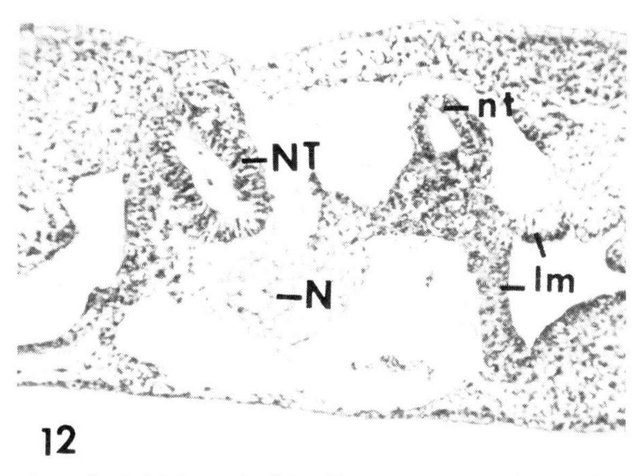

Fig. 12. Cross section of mRNA-treated PNP cultured for 4 days showing induced neural tube (NT), notochord (N), nephric tubules (nt), and split lateral mesoderm (lm). × 72.

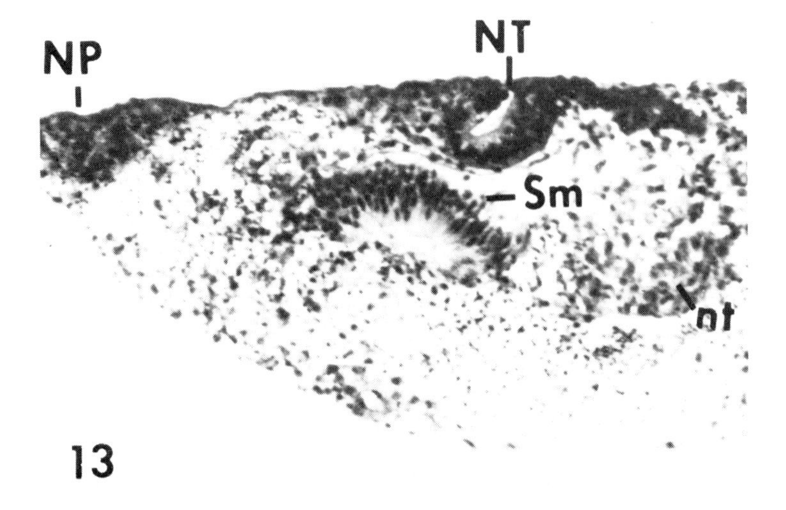

Fig. 13. Cross section of mRNA-treated PNP cultured for 4 days showing neural tube (NT), neural plate (NP), somite mesoderm (Sm), and nephric tubules (nt). × 90.

3.3.1. Control series

During 4 days of cultivation, such structures as thickened primitive streak and pulsatile tissue were not seen. Erythrocytes appeared as red patches on the 3rd day of incubation. Of the 18 PNPs studied histologically, only one showed signs of differentiation. A typical cross section of the PNP is shown in fig. 10. The differentiated one had neural epithelium and nephric tubules.

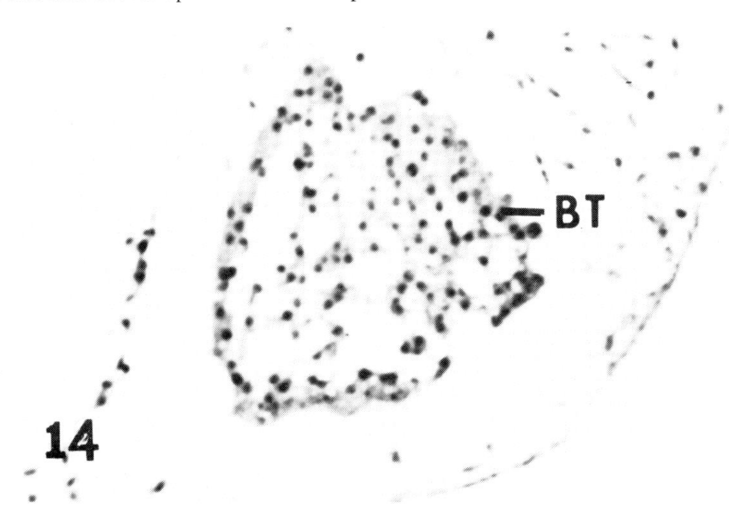

Fig. 14. Cross section of mRNA-treated PNP cultured for 4 days showing loosely arranged beating tissue (BT). × 90.

3.3.2. Experimental series

This series was featured by the development of a structure resembling a defective early chick embryo (fig. 11). Histological examination of serial sections revealed the presence of neural tube, notochord, somites and nephric tubules (figs. 12, 13). These structures were arranged in a pattern similar to that of those structures produced by abnormal inductor [19]. The results summarized in table 5 are in every aspect comparable to those of tables 3 and 4. As can be seen on the 4th row from the top of table 5, all 33 mRNA-treated PNPs had neural differentiation which was accompanied by notochord (64%), somite mesoderm (36%), nephric tubules (55%), and beating tissue (82%). Fig. 14 illustrates a cross section of another defective embryo with a patch of beating tissue far away from the median axis.

4. Discussion

In this study, 88% of stage 4 chick blastoderms cultured for 19–21 hr on agar medium developed normally as shown by early workers [11, 20, 21, 22]. If they were treated with exogenous testis-RNA, their developmental potentiality depended upon the RNA concentration used. The rate of accidental death due to handling was on the same level as that of the controls within a concentration of 60 O.D./ml (tables 1 and 2, column 3). If the concentration was 80 O.D./ml or above, the mortality increased abruptly. This increase was taken as resulting from some toxic effect of exogenous-RNA. The adverse effect was also shown by the fact that RNA, at concentrations of 80 O.D./ml or higher, abolished the development of normal embryos and simultaneously decreased the percentage of abnormal embryos. Therefore, within the optimal concentration of 60 O.D./ml, both wRNA and mRNA had not only affected the development of primary embryos but also stimulated the formation of a structure resembling a secondary axis, especially at the posterior end (fig. 4). Beyond the optimal concentration the frequency of secondary axis formation in surviving embryos was proportional to the RNA concentration (up to 100 O.D./ml) (fig. 6). In general, whole blastoderms responded to mRNA in a manner similar to those treated with heart-RNA. The only difference was that heart-RNA promoted heart development as previously shown by Niu and Mulherkar [7], while testis-mRNA induced a "secondary axis" and possibly stimulated germ cell formation. The study of the latter is presently underway.

In the series with NPs, mRNA induced the formation of what Spemann [23] called "duplicitas anterios" (fig. 8). Most of the controls developed into normal embryos as described by Spratt [19] and in no case did they show any sign of secondary embryo formation.

In the series with PNPs, those treated with chick Ringer's solution showed no appreciable differentiation. This is in agreement with Rawles [24], Spratt [25], and Niu and Deshpande [9]. PNPs treated with mRNA, however, acquired the capacity to undergo differentiation. The most striking feature of these PNPs was the presence of a structure resembling a defective early chick embryo in the median axis (fig. 11). PNPs had often been used for causal analysis of differentiation. The results of previous investigators are of two categories: (1) General effects – the treatment with small molecules (e.g., cysteine and glutathione) [26] or macromolecules (e.g., DNA) [27] resulted in the formation of some axial structures. These structures were scattered and did not appear in a pattern that would constitute an embryonic axis. (2) Specific effect – the treatment with RNAs from brain, kidney [5], liver [6], and heart [7, 9] resulted in the formation of tissue resembling the donor source of RNA. This aspect of the testis-mRNA function has yet to be analyzed.

Denaturation by boiling did not alter the molecular configuration of mRNA, but resulted in the loss of function. Digestion of mRNA with pancreatic RNase caused molecular breakdown and dysfunction as well. Both preparations were applied to whole blastoderms and NPs. The frequency of normal embryo formation in both series was comparable to that of the control series (tables 3 and 4). However, when the denatured mRNA was replaced by wRNA, mRNA, or heart mRNA, the frequency of normal embryo formation decreased significantly and conversely the abnormal embryo formation increased. This change could easily be explained by the dual function of RNA [7, 8]: wRNA or mRNA had interfered in various degrees with those cells involved in the development of normal embryos and simultaneously stimulated some responsive cells to become a secondary axis or embryo as described above.

Results of this study clearly show that germ cell (testis)-RNA differs functionally from soma-RNAs. This view is supported by a number of previous reports. It has been shown that germ cell-DNA and DNAs from nuclei of the zygote and early cleavage cells are multipotential. As the development of embryo proceeds, there is a sequential selective inhibition of the genome (DNA) which leads to the differentiation of specific tissue(s) or organ(s) [28]. Therefore, it appears that gene regulation is responsible for functional deviation of soma-DNAs from germ cell-DNA. Similarly, in *Ascaris* chromatin diminution is associated with the formation of somatic tissue and germ cells have no chromatin diminution [29]. Furthermore, Tobler et al. [30] showed that spermatid-DNA hybridized with spermatid-RNA efficiently but much less with soma-RNA. These observations clearly show the difference in the nucleotide sequence between the germ cell-RNA and soma-RNA. In eucaryotes the hybridization percentage of one soma-mRNA and DNA differs from another. With all soma-mRNAs taken together their hybridization with DNA may reach a frequency comparable to the hybrid between testis-mRNA and testis-DNA. Using hybridization percentage as a yard stick to scale the inductive capacity of various mRNAs, the germ cell-mRNA and DNA should score highest, and induced a secondary axis or embryo. The soma-mRNA and DNA that scored lowest would promote the formation of the least complex structures. In between the two extremes, the mRNA-induced structural complexity other than the kind of RNA-donor tissue is related to the percentage of its hybridization with DNA. The reason for using "least complex" instead of "unique type" of structures is because the mRNAs employed were heterogeneous. In other words, unless the mRNA is homogeneous, it would be nearly impossible to obtain the development of an unique tissue type.

In birds, Hensen's node is the organization center [31]. By analogy, it corresponds to the dorsal lip of the blastopore of the amphibian gastrula. Hensen's node has been grafted onto another embryo. The secondary embryo produced was very similar to that obtained from treatment with testis-mRNA. This finding raises an important question about the chemical nature of the inducing agent(s) in Hensen's node and the dorsal blastoporal lip. It has long been known that the chordamesodermal material is capable of forming not only mesodermal structures but also various epidermal and neural structures [32]. The chrodamesoderm, when cultured in Niu–Twitty solution [33], was found to release nucleoprotein. This diffusible substance initiated the differentiation of presumptive ectoderm into an outgrowth of melanophores, mesenchyme, and aggregate of nerve cells with radiating nerve fibers [33]. The active component of the nucleoprotein was found to be RNA [34]. In view of the variety of tissues derived from the chordamesoderm, the RNA released by various developing cells would be highly heterogeneous. Therefore, it would seem that either heterogeneous mRNA from combined somatic tissues or testis-mRNA could initiate the formation of a secondary embryo.

Acknowledgements

This work was supported in part by grants from the Rutgers University Research Council No. 07-2189 (H.L.), the Population Council of New York and the National Foundation (M.C.N.).

References

[1] M.C. Niu, Proc. Natl. Acad. Sci. U.S. 44 (1958) 1264.
[2] N.W. Hillman and M.C. Niu, Proc. Natl. Sci. U.S. 50 (1963) 486.
[3] J.M. Butros, J. Embryol. Exptl. Morphol. 13 (1965) 119.
[4] A.M. Mansour and M.C. Niu, Proc. Natl. Acad. Sci. U.S. 53 (1965) 764.
[5] S.Sanyal and M.C. Niu, Proc. Natl. Acad. Sci. U.S. 55 (1966) 743.
[6] M.C. Niu and A. Leikola, Biol. Bull. 135 (1968) 200.
[7] M.C. Niu and L. Mulherkar, J. Embryol. Exptl. Morphol. 24 (1970) 33.
[8] M.C. Niu and N. Sasake, Exptl. Cell Res. 64 (1971) 65.
[9] M.C. Niu and A.K. Deshpande, J. Embryol. Exptl. Morphol. 29 (1973) 485.
[10] V. Hamburger and H. Hamilton, J. morphol. 88 (1951) 49.
[11] N.T. Spratt, Jr., Science 106 (1947) 452.
[12] J.M. Butros, J. Exptl. Zool. 152 (1963) 57.
[13] N.T. Spratt, Jr. and H. Haas, J. Exptl. Zool. 144 (1960) 139.
[14] C.T. Grabowski, Am. J. Anat. 10 (1957) 101.
[15] G.C. Rosenfeld, J.P. Comstock, A.R. Means and B.V. O'Malley, Biochem. Biophys. Res. Comm. 47 (1972) 387.

[16] O.H. Lowry, N.J. Rosebrough, A.L. Farr and R.J. Randall, J. Biol. Chem. 193 (1951) 265.
[17] R. Burton, Biochem. J. 62 (1956) 315.
[18] N.T. Spratt, Jr., J. Exptl. Zool. 107 (1948) 39.
[19] L.Saxen and S. Toivonen, Primary embryonic induction (Academic Press, New York and London, 1962).
[20] E. Howard, J. Cell. Comp. Physiol. 41 (1953) 237.
[21] L.M. Duffey and J.D. Ebert, J. Embryol. Exptl. Morphol. 5 (1957) 324.
[22] R.C. Fraser, J. Exptl. Zool. 145 (1960) 169.
[23] H. Spemann, Embryonic development and induction (Yale University Press, New Haven, 1938).
[24] M.E. Rawles, J. Exptl. Zool. 72 (1936) 271.
[25] N.T. Spratt, Jr., J. Exptl. Zool. 120 (1952) 109.
[26] S.P. Chauhan and K.V. Rao, J. Embryol. Exptl. Morphol. 23 (1970) 71.
[27] J. Butros, J. Exptl. Zool. 143 (1960) 259.
[28] P. Gross, this symposium, page 4.
[29] T. Boveri, In: Festschrift zum sechzigsten Geburgstag Richard Hertwigs 3 (1910) 131.
[30] H. Tobler, K.D. Smith and H. Ursprung, Develp. Biol. 27 (1972) 190.
[31] C.H. Waddington, The epigenetics of birds (Cambridge University Press, Cambridge, 1952).
[32] H. Holtfreter and V. Hamburger, In: Analysis of development, Eds. B.H. Willier, P. Weiss and V. Hamburger (Saunders, Philadelphia, 1955) 230–296.
[33] M.C. Niu and V.C. Twitty, Proc. Natl. Acad. Sci. U.S. 39 (1953) 985.
[34] M.C. Niu, In: Cellular mechanisms in differentiation and growth, Ed. D. Rudnick (Princeton University Press, Princeton, 1956) 155–171.

Niu and Segal (eds.). The role of RNA in reproduction and development
North-Holland Publ. Co., 1973

Biological activity of RNA from estrogen-stimulated uterus

P. GALAND* and N. DUPONT

Biology group, Institut de Recherche Interdisciplinaire en Biologie Humaine et Nucléaire (LMN), Faculty of Medicine, Free University of Brussels, 115 Boulevard de Waterloo, 1000 Brussels, Belgium.

Numerous studies have shown that uterine RNA obtained from estradiol stimulated rats (U-RNA) reproduces the effects of estradiol on the uterine epithelium of spayed rats. The present work was aimed at clarifying whether estrogen contamination of the RNA preparations could be responsible for the hormone-like effects. U-RNA and vaginal RNA were prepared from rats treated with 10 μg of estradiol-17-β-6,7-^{3}H. The labelled RNA preparations were assayed for estrogen contamination and for their biological activity in the uterine horn of spayed rats. The following observations were made:

(1) U-RNA and vaginal RNA preparations were contaminated with tritium label. Assuming that this is all in the form of estradiol, U-RNA would contain 5×10^{-7} μg estradiol/100 μg RNA, and vaginal RNA 6×10^{-7} μg/100 μg RNA.

(2) In biological tests, U-RNA only was active in stimulating hypertrophy of the luminal epithelial cells. This activity was lost after treatment of the U-RNA with ribonuclease.

(3) Intraluminal administration of 10^{-6} μg estradiol into the uterine horn of spayed rats was without any effect on the morphological characteristics of endometrial epithelium; 10^{-5} μg estradiol stimulated epithelial hypertrophy.

It is concluded that estradiol contamination, the upper limits of which are fixed by our study, cannot account for the functional potency of U-RNA.

Experiments with (^{3}H)-methylated U-RNA indicate that a significant amount of intact U-RNA remains in the uterus of recipient animals, killed 4 to 20 hr after intraluminal instillation. This represents about 1% of the injected U-RNA.

1. Introduction

Growth of the atrophic uterus of spayed rats is under the dependence of estrogens. After estradiol injection, total RNA and protein contents, which decline in the uterus of ovariectomized rats, are restored to normal levels (see review in ref. [1]).

During the first 2 hr of estrogen action, incorporation of (^{32}P)-orthophosphate into RNA of all subcellular fractions increases in the uterus of immature rats [2]. Attempts to demonstrate that qualitative changes in the uterine RNA could me-

* Chercheur qualifié du Fonds National de la Recherche Scientifique (Belgium).

diate the uterotrophic action of estrogens in the hormone-deprived organ, were made as early as in 1965. At this time, there was no direct demonstration of an estrogen-induced synthesis of specific messenger RNAs. Experiments with actinomycin D however [3] and the fact that estrogen appeared to elicit marked changes in activity of uterine ribosomes [4] could be interpreted as favouring such a hypothesis.

Thus Segal et al. [5] showed that preparations of total uterine RNA from estrogen-stimulated rats, mimicked the hypertrophic action of estradiol, when administered intraluminally in the uterus of ovariectomized rats. Similar results were reported by Mansour and Niu [6] who used liver and uterine RNA from mature, cycling mice and observed that the latter only was able of inducing the spayed uterus to restore normal morphological appearance and alkaline phosphatase activity. Several analogous studies appeared thereafter in the literature, and seemed to confirm those earlier works [7–11].

However there was controversy, or at least a lack of clarity, about the possible role of estradiol contamination in biological activity of the uterine-RNA preparations. Such possibility was disclosed on the grounds that procedure of RNA isolation requires alcohol precipitation and washing with ether [5–7]. Direct measurement of estradiol contamination, using (^{3}H)-estradiol, seemed to confirm this conclusion. This, together with the fact that RNAase digestion of the uterine RNA preparation abolishes its biological activity, supports the conclusion that RNA itself is the active component. However, in apparent contrast to this, there were discordant reports on a decline [8] or complete loss [10] of RNA biological activity after washing with ether.

Therefore, we tried to reproduce some of the above described observations and to reevaluate the role of estradiol contamination in the biological activity of the uterine RNA. Attempts were also made to follow the fate of labelled uterine RNA after intraluminal injection in the uterus of spayed rats. Part of our results have been previously reported [12, 13].

2. *Materials and methods*

Adult Wistar strain rats ovariectomized 2–3 weeks before use were used in all experiments reported here. Efficiency of ovariectomy was assessed by repeated examination of vaginal smears.

2.1. RNA preparation

Eight hours before sacrifice, rats were injected s.c. with 17-β-estradiol benzoate (10 μg/animal). The uterine horns were trimmed free of adipose tissue and quickly frozen on dry ice. RNA preparation was according to “method B” of Segal et al. [5]. This involves homogenization in buffered sodium lauryl sulfate and extraction with phenol at room temperature and at 60°C. Washing with ether (Merck) and treatment with deoxyribonuclease were also included in the procedure. No detectable amount of DNA was demonstrated by the Burton-diphenylamine procedure [14] in the final purified RNA. Protein contamination represented 1% of the preparation. The preparation so obtained is termed U-RNA.

2.2. Ribonuclease digestion of U-RNA

The RNA preparation was dissolved in presence of 100 μg crystalline ribonuclease (Nutritional Biochemicals Corporation, Cleveland, Ohio)/ml of 0.01 M phosphate buffer (pH 7.2). The mixture was incubated for 1–2 hr at room temperature. The total mixture was used to compare its biological activity on the uterine horn, to that of intact U-RNA. In some experiments RNAse was eliminated by repeated washing with chloroform.

2.3. Measurement of contamination of U-RNA by (^{3}H)-estradiol

(6,7,^{3}H)-estradiol-17-β (New England Nuclear Corporation; S.A. 48 Ci/mmole) was given s.c. in pretreatment of RNA-donor rats, at doses of 10, 50 or 100 μg/animal, with a final radioactivity of about 100, 10 and 10 μCi/μg respectively. Treated animals were sacrificed 8 hr after injection as in other experiments.

Vaginal and uterine RNA were prepared as described above. The radioactivity bound to aliquots of 50–300 μg of purified RNA was measured in a Nuclear Chicago (Mark I) scintillation counter. Counting efficiency was assessed by the method of the 2 channels-counting ratio, thanks to external standard calibration. Efficiency (using Bray's solution) was equal to 25%.

2.4. RNA administration

Ovariectomized recipient animals were anaesthesized (Avertin–Bayer) and the uterine horns were exposed by a ventral incision. The solution was injected under ligature, directly into lumen, from the corpus towards the tubal end of the horns. The right horn received the test solution, consisting usually of 30–70 μg U-RNA (intact or enzymatically digested) dissolved in 20–30 μl saline. The left horn served as control, and thus received an equal volume of saline alone. Animals were sacrificed 16–18 hr after injection.

2.5. Labelling of U-RNA with (^{3}H)-dimethylsulfate

The technique described by Smith et al. [15] was followed. 1 mg U-RNA was incubated in 0.25 ml 0.1 M phosphate buffer pH 7.5 supplemented with 20 of (^{3}H)-dimethylsulfate (5.0 mCi/2.68 mg) purchased from New England Nuclear Corporation. (^{3}H)-dimethylsulfate was dissolved in ether and diluted with unlabelled dimethylsulfate so that the final concentration in the compound was ca. 4 μmole per μmole RNA-nucleotide. The mixture was incubated for 2 hr, at 25°C. RNA was then precipitated with 2 vols ethanol and washed repeatedly. In these conditions, about 4% of the U-RNA nucleotides were methylated. This level of methylation does not appear to modify significantly the RNA properties as shown by hybridization with DNA [15, 16] or by sedimentation and fusion characteristics [17]. Our methylated U-RNA, besides, was biologically active in the uterus of spayed rats.

2.6. Chromatographic analysis of U-RNA

The technique initially devised for fractionation of DNA on DEAE-cellulose paper pulp by Davila et al. [18] was employed. This makes use of Whatman DEAE-cellulose (DE81) homogenized, washed with distilled water and placed on small columns which may be submitted to centrifugation at 1000 rpm so as to accelerate elution. Eluants of increasing ionic strength or pH are used:

(1) Phosphate buffer 0.01 M pH 7.0.
(2) Phosphate buffer 0.01 M pH7.0 + 0.14 M NaCl.
(3) Phosphate buffer 0.01 M pH 7.0 + 0.5 M NaCl.
(4) Phosphate buffer 0.01 M pH 7.0 + 1.0 M NaCl.
(5) 0.2 M NH_3 + 2 M NaCl.
(6) NaOH, 1 M.

The nucleotide concentration of the different RNA fractions was estimated from the U.V. spectrum, recorded with a Cary recording spectrophotometer. Radioactivity was measured with a Nuclear liquid scintillation counter, by using the scintillating medium elaborated by Davila et al. [18].

3. Results

Estradiol-induced modification of the uterine luminal epithelium of spayed rats consist in transition of the epithelium from low, cuboidal to high columnar cells, with nuclei lined up in basal position. Microscopical observation of U-RNA treated horns revealed similar changes. We measured cell height in 100–300 epithelial cells taken at four different levels in the uterine horns, in treated horn and in contralateral horn, treated with saline. When increase in mean cell height was less than 10% over control value, the response was judged as negative. In fact, positive responses registered here below, represented a 20 to 60% increase over control. Table 1 shows representative results, obtained when comparing biological activity of either intact or RNAase-digested U-RNA. It can be seen that RNAase digestion abolishes the biological activity of either intact or RNA-ase-digested U-RNA. In some experiments, (as shown under "b" in table 1), the whole mixture was administered in view of the possible objection that elimination of the enzyme could also eliminate an active contaminant, freed from the digested U-RNA.

TABLE 1

Effect of intraluminal injection of U-RNA on endometrium of spayed rats

Treatment	Case	Response [a]	
		Positive	Negative
1. U-RNA 50–75	12	11	1
2. RNAase-digested U-RNA 50–75			
(a) RNAase eliminated (by chloroform)	5	0	5
(b) RNAase present	15	0	15

[a] Increase in mean cell height of luminal epithelium, at least 20% over control.

3.1. Measurement of radioactive contamination of U-RNA prepared from (^{3}H)-estradiol-treated rats

Hormonal treatment of U-RNA-donor rats were performed using (^{3}H)-estradiol at different concentrations and specific radioactivities. As shown in table 2, a significant amount of tritium contamination remained bound as well to U-RNA as to vaginal RNA, despite the fact that ether-washing was included in the routine procedure for isolating these RNAs. Further washing with ether of the purified RNA did not modify the contamination. Assuming that all counts are in estradiol molecules,

TABLE 2

Measurement of the radioactivity recovered with RNA from the uterus and vagina of (^{3}H)-estradiol-treated rats

Amount of (^{3}H)-estradiol injected/ uterine horn (μg)	S.A. cpm/μg $\times 10^{-7}$	Organ	Homogenate		RNA fraction [a]	
			Total counts (cpm)	Corresponding amount of (^{3}H)-estradiol [a] (μg $\times 10^3$)	Total counts in 100 μg RNA cpm	Corresponding amount of (^{3}H)-estradiol [a] (μg $\times 10^6$)
100	2.8	Uterus	94,000	3.3	316	11.0
		Vagina	95,000	3.4	–	–
50	3.4	Uterus	118,600	3.5	357	10.0
		Vagina	241,440	7.1	437	13.0
10	20.0	Uterus	251,450	1.2	97	0.5
		Vagina	95,200	0.7	110	0.6

[a] This calculation is based on the S.A. of the labelled estradiol (column 2), assuming that all counts correspond to estradiol molecules. This thus gives the upper limits of estradiol contamination.

[b] The uterine and vaginal RNA were prepared exactly as described under "Methods" (sect. 2) to obtain biologically active U-RNA.

TABLE 3
Biological activity of RNA from the vagina and uterus of spayed rats treated with 10 μg (^{3}H)-estradiol [a]

Treatment	Case	Response [b]	
		Positive	Negative
Uterine RNA (50 μg/30 λ)	9	7 [c]	2
Hydrolyzed uterine RNA (RNAase digestion)	9	0	9
Vaginal RNA (50 μg/30 λ)	4	0	4

[a] U-RNA and vaginal RNA tested are those from which aliquots were assayed for radioactive contamination (see table 2).
[b] See footnote to table 1.
[c] Mean increase in cell height: 41% over controls.

radioactive contamination was converted into μg estradiol, on the basis of specific radioactivities shown in the second column of table 2. Results of these calculations are given in the last column of the table. This shows that the highest amount of estradiol that could contaminate our U-RNA preparation in the treatment with 10 μg estradiol, represents, at maximum, 5×10^{-7} μg estradiol for 100 μg U-RNA.

The biological activity of this tritium-contaminated U-RNA (which was prepared as in all other experiments) was assessed. As shown in table 3, 50 μg of this U-RNA were active, and this activity was destroyed by RNAase treatment. Vaginal RNA, despite at least equal contamination was ineffective in provoking hypertrophy of the uterine luminal epithelium.

3.2. Effect of intraluminal injection of small doses of estradiol

Doses of estradiol corresponding to the highest amount of the hormone that could have been injected together with U-RNA (as indicated by results in table 2) were

TABLE 4
Effect of intraluminal injection of various doses of estradiol-17-β on endometrial epithelium hypertrophy in spayed rats

Estradiol dose (μg horn)	Case	Response		Epithelium height in % over control
		Positive	Negative	
0.3	3	3	0	+60%
10^{-4}	3	3	0	+30%
10^{-5}	9	3	0	+22%
10^{-6}	9	0	9	− 1%

Animals were sacrificed 18 hr after estradiol administration as in experiments with U-RNA.

tested for biological activity when administered intraluminally. Dillutions were made from stock solutions of ^{3}H-labelled estradiol. Counting of samples from the H 10^{-5} and 10^{-6} μg/30 λ doses, showed that they were correct (maximum error 5–7%). Treated animals were sacrificed 18 hr after estradiol administration, as in experiments with U-RNA. The results showed (see table 4) that while 10^{-5} μg estradiol is a biologically active dose, 10^{-6} μg is not effective in stimulating hypertrophy of the endometrial epithelium.

3.3. Fate of (^{3}H)-methyl-labelled U-RNA after intraluminal injection

Purified U-RNA was methylated in vitro, using (^{3}H)-dimethylsulfate. Fifty micrograms of labelled U-RNA, corresponding to a radioactivity of 10^{6} cpm were injected into the uterine lumen of spayed rats. Treated animals were killed after 2, 15, 30 and 60 min and after 4, 8 and 20 hr (3 animals at each time). The uteri were removed and washed with saline. They were then divided in three parts: one serving to measure acid-insoluble radioactivity, one being submitted to phenol-extraction of RNA, and the last one being fixed for histological examination. For each experimental point, pooled uterine fragments from the 3 injected animals were processed together. Measurements of acid-insoluble radioactivity are shown in table 5. Despite variability in the results, one may conclude that after time 2 min, the amount of radioactive material that was fixed by the uterus declined sharply, but remains significant even after 8 to 20 hr. It corresponded to 1% of the amount of (^{3}H)-U-RNA, administered, this however representing only an order of magnitude.

The distribution of radioactivity among the 6 fractions obtained by chromatography on DEAE-cellulose, of the uterine RNA from rats treated with (^{3}H)-methyl-U-RNA is shown in fig. 1. Methylation does not modify the chromatographic behaviour of U-RNA as shown by spectrophotometry of nucleotide contents of the 6 chromatographic fractions, before (fig. 1a) and after methylation (fig. 1b). Chroma-

TABLE 5

Acid-insoluble radioactivity recovered from the uterus of spayed rats at different times following intraluminal injection of 50 μg of (^{3}H)-methyl-U-RNA (10^{6} cpm).

Time after injection	Total radioactivity per uterine horn (cpm)	Percent of administered radioactivity
2 min	58,500	5.8
15 min	11,500	1.1
30 min	27,500	2.7
1 hr	9,000	0.9
4 hr	27,500	2.7
8 hr	7,000	0.7
20 hr	8,500	0.8

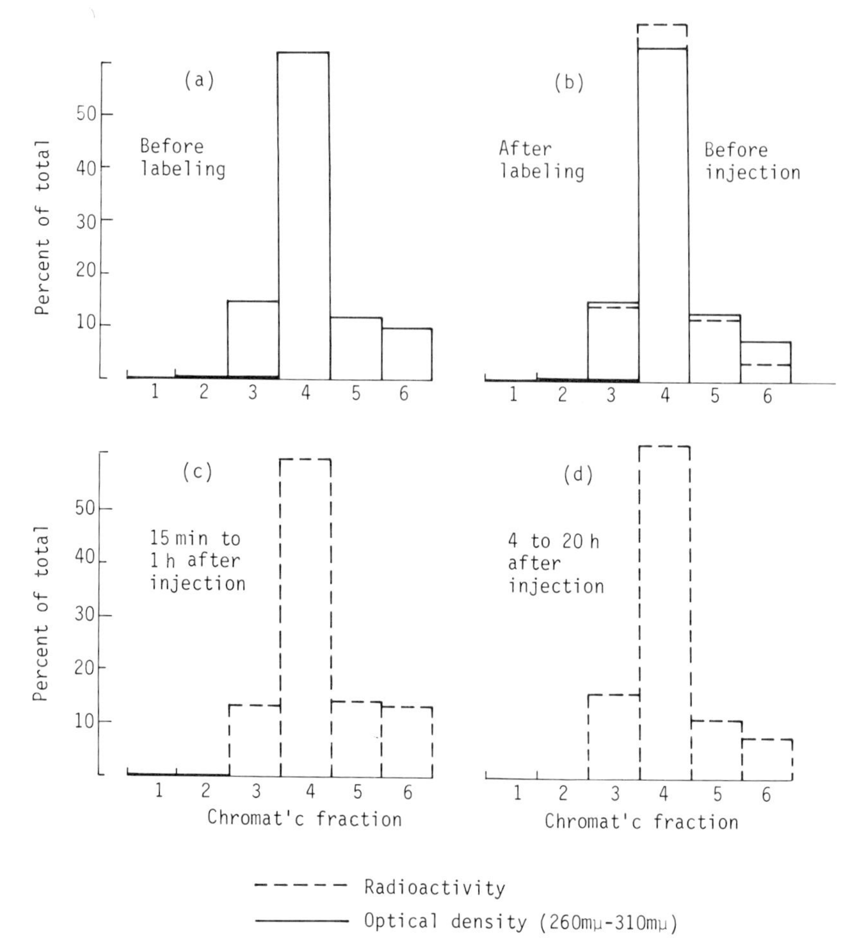

Fig. 1. Chromatographic profile of U-RNA labelled in vitro (on DEAE-cellulose) with (^{3}H)-methyl: distribution of radioactivity among the 6 fractions obtained by chromatography on DEAE cellulose in (^{3}H)-U-RNA before administration (B) to spayed rats and in (^{3}H)-U-RNA reextracted from the uterus of rats killed 15 min to 1 hr (C) or 4 to 20 hr (D) after intraluminal injection. Note also unmodified chromatographic pattern in (^{3}H)-methyl-U-RNA (B) as compared to unlabelled U-RNA (A).

tographic profile of the labelled RNA extracted from the uteri of spayed rats, that received intraluminal instillation of (^{3}H)-U-RNA, is identical to that exhibited by the latter before injection. This is shown in figs. 1c and 1d, corresponding to pooled RNA-extracts from animals killed 15 min to 1 hr, and 4 to 20 hr, respectively, after injection. Those results suggest that insoluble radioactivities measured hereabove, correspond to a fraction of U-RNA that remained bound in the treated horns, under a not grossly modified form, for as long as 4 to 20 hr following administration.

4. Discussion

Our results confirm previous reports by others [5–9] that it is possible to mimic estrogen action on the morphology of the endometrial epithelium by intrauterine administration of uterine RNA prepared from estradiol-treated animals. That undegraded RNA is the active component in the uterine RNA preparation is suggested, as in previous work [5–10], by the fact that ribonuclease treatment abolishes the biological activity of the preparation.

The results of the present work indicate that while there is almost certainly contamination of the U-RNA preparations with estradiol, the amount present cannot account for the observed biological effects.

Other authors have failed to detect this contamination. In the experiments of Segal et al. [5, 7], the S.A. of the injected (^{3}H)-estradiol was 10 times lower than in the present experiments, although the dose in μg was equivalent, so that contamination equal to that measured in our assays would have escaped detection in their work.

Fencl and Villee [19] could not detect any radioactivity in the uterine RNA preparation obtained from rats that were injected with (^{3}H)-estradiol, with SA similar to that of our labelled estradiol. This however could be due to the fact that they used a 100 times lower amount of estradiol (0.1 μg/animal).

As pointed out above, ether washing was included in the procedure of RNA preparation. Despite this, there remained some radioactivity bound to U-RNA that could not be eliminated by further ether washing of the purified RNA preparation. This treatment did not alter the biological activity of the preparation. Contradictory observation by others [8, 10], that either washing depresses U-RNA activity could be better explained by some denaturing effect on the active RNA [10]. This could depend on the batch of ether used.

The small amount of available RNA labelled with (^{3}H)-estradiol, did not allow us to investigate the nature of the labelled contaminant. Our calculation of estradiol contamination gives a maximum value because it is based on the premise that all the radioactivity is bound to estradiol, and not to less active metabolites.

It was important to establish whether such estrogen doses had biological activity in our experimental system. It was known [5] that intrauterine injection of 6×10^{-4} μg estradiol into ovariectomized rats provokes the increase in endometrial epithelium thickness which characterizes estrogen stimulation. A former report [20] suggested that 2×10^{-5} μg was the smallest quantity of estradiol which, when administered intraluminally, could produce these morphologic changes, together with modifications of the Golgi apparatus and the basal phospholipids.

Our results show accordingly that no changes in epithelial height are induced by intraluminal injection of 10^{-6} μg estradiol. This dose of estradiol is about 4 times higher than the highest amount which could have been present in our biological assays with tritium-contamined U-RNA (table 3). It is equal or inferior to the possible contamination of the U-RNA samples injected in our other assays (table 1).

Using very small amounts of U-RNA (0.75 μg/horn), Segal [7] observed blastocyst implantation in post-coitally ovariectomized rats, maintained with progesterone. In such conditions, implantation of the blastocysts (which remain viable) depends on the presence of small doses of estradiol. The minimal dose of 17-β-estradiol which, when applied in situ, is able to induce implantation is of the order of 3–5 ng [7, 11]. From our estimation of upper limits of estradiol contamination, it may be calculated that the highest possible estradiol contamination in these experiments [7] was of the order of 10^{-8}–10^{-7} μg/treated horn, thus 10^4–10^5 times less than the active dose of estradiol. Despite the evident limitations in comparing results from different laboratories, this may be taken as further indication that estradiol contamination is not the source of U-RNA activity. Other work on U-RNA activity [1, 2, 4–8] may not be examined from the same viewpoint, due to several differences in experimental schedules. One could imagine that small doses of estradiol, insufficient alone for stimulating uterine epithelium, may be more effective when given together with RNA molecules. That this could be the basis of U-RNA activity appears improbable since vaginal RNA was inactive (as already shown by Segal et al. [5, 7]) although it was contaminated by (^{3}H)-estradiol to, at least, the same degree as was U-RNA. This gives support to the conclusion that estradiol contamination may not account for the biological activity of the U-RNA preparation, and suggests that uterine cells are able to ingest RNA molecules.

Previous studies have shown that RNA molecules can enter ascite tumor cells, in vitro [20–24], and activate protein synthesis in those cells [23–25]. Attempts to demonstrate autoradiographically that uterine cells are actually endowed with a similar ability to take up intact RNA molecules, were unconclusive in our present study with (^{3}H)-methylated-U-RNA. However there are indications in this work for the possibility that a significant amount of unmodified U-RNA remains (bound ?) in the uterus up to 8–20 hr following intraluminal injection.

One may only speculate about the nature of the active RNA in the total uterine RNA preparation, and about its possible mode of action after intraluminal instillation in the uterus of ovariectomized rats. Several data must be taken into account when considering these questions:

(1) U-RNA is active at very low doses (50 μg in our study, but even as few as 0.75 μg in Segal's work on blastocyst nidation [7]). Besides, much of the injected U-RNA appears to be eliminated, 1–2% at most being present from 15 min to 20 hr.

(2) Active U-RNA may be isolated by different methods, whether they are aimed at preserving nuclear RNA (Segal's method B, also used in our laboratory) or not (Segal's method A, also used by Unjhem et al. [10]). These methods do not guarantee full protection of U-RNA from RNAase attack during isolation.

(3) Active U-RNAs have been extracted 4 hr [5, 7], 8 hr (our work) and at least 12 hr [10] after estradiol injection.

(4) Actinomycin D, administered together with U-RNA, "does not suppress completely" [7], or "depresses" [8] U-RNA activity. This however needs further evaluation.

(5) There are indications that all fractions of U-RNA are not equally active [8–10].

None of those facts excludes that U-RNA acts through messenger RNA activity, but they would also fit with an activating role on de novo synthesis of specific RNA. A combination of both direct and indirect action as already been suggested by Mansour [8].

The synthesis of all classes of RNA is accelerated by estrogen in the uterus of spayed rodents (see Hamilton for review [1]). New species of RNA molecules seem to be transcribed in the uterus of estradiol-stimulated, spayed does [26]. Besides, there is a marked change in the pattern of RNA molecules which are restricted to the nucleus: a larger proposition of molecules of very low molecular weight, and a decreased proportion of 20–45 S molecules are restricted to the nuclei in estrogen stimulated uterus [26]. It had already been suggested that the transport of RNA to the cytoplasm is sensible to regulation by estrogen, in the rat uterus, and that this was mediated through activation of the early, actinomycin D-sensitive, synthesis of RNA [1]. RNA molecules on the other hand, could conceivably be implicated in regulation of transcription [e.g., 27, 28].

Finally, one could suggest that administration of U-RNA amounts to furnish the recipient cells with the pattern of RNA molecules that characterizes the stimulated uterine cells. An equally reasonable hypothesis could be that some RNA(s) in the U-RNA preparation regulate(s) the synthesis and/or transport of new types of nuclear RNA, leading indirectly to the same final status, i.e., the appearance in the recipient cell of a new set of RNA, similar to that induced by estrogens. The results of the experiments with actinomycin D, cited above, [7, 8] tip the balance slightly in favor of the second hypothesis.

Acknowledgements

Work realized under Contract of the Ministère de la Politique Scientifique within the framework of the Association Euratom – University of Brussels – University of Pisa, and thanks to grant of the Caisse Générale d'Epargne et de Retraite (Fonds Cancer).

The authors thank Miss Ch. Borrey for the drawing of the figure and for the preparation of the manuscript.

References

[1] T.H. Hamilton, Science 161 (1968) 649.
[2] J. Gorski, W.D. Noteboom and J.A. Nicolette, J. Cell Comp. Physiol. 66 (1965) 91.

[3] H. Ui and G.C. Mueller, Proc. Natl. Acad. Sci. U.S. 50 (1963) 256.
[4] D.L. Greenman and F.T. Kenney, Arch. Biochem. Biophys. 107 (1964) 1.
[5] S.J. Segal, O.W. Davidson and K. Wada, Proc. Natl. Acad. Sci. U.S. 54 (1965) 782.
[6] A.M. Mansour and M.C. Niu, Proc. Natl. Acad. Sci. U.S. 53 (1965) 764.
[7] S.J. Segal, Develop. Biol. Suppl. 1 (1967) 264.
[8] A.M. Mansour, Acta Endocrinol. 54 (1967) 541.
[9] A.M. Mansour, Acta Endocrinol. 57 (1968) 465.
[10] O. Unjhem, A. Attramadal and J. Solna, Endocrinology 58 (1968) 227.
[11] A. Psychoyos, Arch. Anat. Microscop. 56 (1967) 616.
[12] P. Galand, N. Dupont-Mairesse, F. Leroy and J. Chrétien, In: Basic actions of sex steriods on target organs, Eds. P.O. Hubinont, F. Leroy and P. Galand (Karger, Basel, 1971) 227–239.
[13] P. Galand and N. Dupont-Mairesse, Endocrinology 90 (1972) 936.
[14] K. Burton, In: Methods of enzymology, Vol. 12, Eds. L. Grossman and K. Moldave (Academic Press, New York, 1968) 163.
[15] K.D. Smith, J.L. Armstrong and B.J. McCarthy,Biochem. Biophys. Acta 142 (1967) 323.
[16] M.R. Morrison, J. Paul and R. Williamson, Europ. J. Biochem. 27 (1972) 1.
[17] C. Bollack, G. Keith and J.P. Ebel, Bull. Soc. Chim. Biol. 47 (1965) 765.
[18] C. Davila, P. Charles and L. Ledoux, J. Chromat. 19 (1965) 382.
[19] M.M. Fencl and C.A. Villee, Endocrinology 38 (1971) 279.
[20] H. Elftman, Proc. Soc. Exptl. Biol. Med. 105 (1960) 19.
[21] P. Galand and L. Ledoux, Arch. Int. Physiol. Biochem. 69 (1961) 383.
[22] P. Galand, Arch. Int. Physiol. Biochem. 71 (1963) 816.
[23] P. Galand, Arch. Int. Physiol. Biochem. 72 (1964) 319.
[24] P. Galand, L. Rémy and L. Ledoux, Exptl. Cell Res. 43 (1966) 381.
[25] P. Galand and L. Ledoux, Exptl. Cell Res. 43 (1966) 391.
[26] P.B. Church and B.J. McCarthy, Biochem. Biophys. Acta 199 (1970) 103.
[27] R.J. Britten and E.H. Davidson, Science 165 (1969) 349.
[28] J.H. Frenster, Nature 206 (1965) 680.

Niu and Segal (eds.). The role of RNA in reproduction and development
North-Holland Publ. Co., 1973

Effects of exogenous polynucleotides on uterine enzymes*

Claude A. VILLEE
Department of Biological Chemistry and Laboratory of Human Reproduction and Reproductive Biology, Harvard Medical School, Boston, Mass. 02115, USA

Both synthetic polynucleotides and RNA extracted with cold phenol from the uteri of estrogen treated rats affect peptide synthesis when instilled into the uterus of an immature rat. The synthetic polynucleotides increase the incorporation into protein of the specific amino acid for which it codes (e.g. phenylalanine in response to polyuridylic acid) but not the incorporation of other amino acids. The addition of actinomycin D to the polynucleotide instilled into the uterine horn did not inhibit the increased incorporation of coded amino acid. Neither dibutyryl cyclic AMP nor theophyllin nor a combination of the two had any effect on the synthesis of either RNA or protein when instilled into the uterine lumen. The activity of glucose-6-phosphate dehydrogenase in the uterus is increased 45% 16 hr after the injection of estradiol. The instillation of RNA from the uteri of estradiol treated castrate adult females into one uterine horn led to a 37% increase in its glucose-6-phosphate dehydrogenase activity over the activity in the horn receiving RNA from the uteri of untreated castrates. RNA was separated by sucrose density gradients into 28 S, 18 S and 5 S RNA, and the separated species were instilled into the uteri of 21-day-old female rats. The glucose-6-phosphate dehydrogenase activity of the uterus was not altered by the instillation of 5 S RNA from estrogen treated rats, but the instillation of either 18 S or 28 S RNA from the uteri of estrogen treated rats increased uterine glucose-6-phosphate dehydrogenase activity 30% over that in the uterine horns receiving the comparable type of RNA from control uteri. The increased activity of glucose-6-phosphate dehydrogenase in the absence of any over-all increase in total protein would suggest that the RNA had stimulated the synthesis of some but not all proteins. These findings are consistent with the hypothesis that some of the RNA produced by the uterus in response to administered estradiol carries the code for the synthesis of specific enzymes such as glucose-6-phosphate dehydrogenase.

Cellular differentiation and the hormonal control of cellular processes may have much in common at the molecular level, for each may involve the turning on of a specific gene or set of genes to begin the production of one or more specific enzymes. One of the more popular theories regarding hormonal mechanisms suggests that the hormone may initiate the transcription of a specific portion of the genome by interacting either with the DNA or with a histone or some other nuclear protein. This theory, first formulated by Karlson [1] was based on an analysis of

* Aided by National Institute of Child Health and Human Development Grant No. HD 01232.

the control of molting in insects by ecdysone. A hormone may act by stimulating the read-out of a portion of the genome for a brief time, whereas differentiation appears to reflect the long-term activation of certain genes in certain cells. Each must involve some mechanism by which one portion of the genome, but not others, is selected for transcription.

The tissues that respond to a given hormone are characterized by the presence of a specific binding protein (a receptor) which reacts with that specific hormone. Perhaps the best known example of a receptor is the estrogen-binding protein found in the uterus and vagina [2,3], the anterior pituitary [4] and the anterior hypothalamus [5]. A comparable androgen binding protein has been described in prostatic nuclei [6,7], a progesterone binding protein in the chick oviduct [8], and cortisol binding proteins in the thymus [9] and retina [10]. It appears that there are receptor proteins for at least certain of the peptide hormones; receptors have been described in the testis for FSH [11], in the ovary for LH [12], and in the liver for insulin [13].

The finding that hormones may regulate the synthesis of specific proteins such as ovalbumin and avidin [14,15] and that they do this by some effect on the metabolism of ribonucleic acids, has strengthened the hypothesis that the hormone interacts with the genome and frees a certain part of the DNA so that it can be transcribed. By such an effect on the process of transcription, the hormone can mediate an increase in the specific messenger RNA(S) providing information for the synthesis of specific proteins.

Let us examine this concept step by step. The hypothesis predicts that we should be able to detect some sort of interaction of the steroid either with the DNA itself or with a nuclear protein such as histone or acidic protein. It predicts that the hormone should lead to a rapid increase in the synthesis of messenger RNA and perhaps of other kinds of RNA, and it would predict that this RNA, when isolated and reintroduced into a target cell, should lead to an increase in protein synthesis and to increased activity of those enzymes known to respond to the application of the steroid in vivo.

Interactions between steroids and nuclear proteins, specifically between testosterone and a histone present in the ventral prostate and between cortisol and a histone present in the liver, were shown by Sluyser [16]. Interactions between a variety of hormones and DNA prepared from the nuclei of human placental cells were measured by monitoring changes in the melting profile of the DNA [17]. Estrone, estradiol, cortisol, insulin, growth hormone and epinephrine all resulted in alterations in the melting profile which were interpreted as a weakening of the bonds between the DNA strands. If the DNA was denatured or sheared before the hormone was applied, the hormone was without effect on the melting profile. The synthetic polynucleotides, poly-d (AT) and poly-d (GC) were unaffected by the hormones. From these observations Goldberg and Atchley inferred that hormones may promote the separation of the complementary strands of specific segments of the double helix of DNA and thus initiate the transcription of specific genes.

TABLE 1

Effect of steroids on binding of actinomycin-D to chromatin from livers of adrenalectomized rats

Steroid added	^{3}H-AMD bound, cpm/μg DNA		
	Control	Steroid	Percent change
None	398	405	1.8
Cortisol, 30 μg	311	356	14.5
Cortisol, 30 μg	229	273	19.1
Corticosterone, 37 μg	528	622	17.8

Chromatin isolated from the livers of adrenalectomized rats was complexed with an excess (0.1 μM) of unlabeled actinomycin (AMD) to bind all available sites on the chromatin. The unbound AMD was removed by a Sephadex G-50 column. The chromatin was then reacted with ^{3}H-AMD with or without the steroid at 5°C for 18 hr. The unbound ^{3}H-AMD was then removed by passing the mixture through a second column of Sephadex G-50. The chromatin samples with and without steroid were passed through parallel columns of Sephadex.

It is generally held that actinomycin D complexes or binds specifically with the non-complexed, double-helical DNA of the chromatin [18]. To assay how much of the DNA is free, and presumably ready to be transcribed, in the control and in the steroid-treated state Mitchell [19] measured the binding of tritium labeled actinomycin D to chromatin. Using the procedure of Marushiga and Bonner [20] Mitchell prepared chromatin from the livers of normal and 7-day adrenalectomized male

TABLE 2

Effect of steroids on binding of actinomycin D to rat liver chromatin

Steroid added	DNA-bound AMD/free AMD
None	38 ± 2
Corticosterone, 1.9 μg	38 ± 2 [a]
Corticosterone, 9.4 μg	51 ± 3 [b]
Cortisol, 1.5 μg	43 ± 1 [c]
Cortisol, 30 μg	44 ± 2 [c]
Androstenediol, 10 μg	37 ± 1 [a]
None [d]	27 ± 1
Corticosterone [d], 7.5 μg	32 ± 1 [b]

To a solution of chromatin, 20 μg/4 ml reaction mixture, was added ^{3}H-AMD, 3.5×10^{-2} μM, and the mixture was incubated at 37°C for 10 min. It was then cooled rapidly and duplicate aliquots were placed in dialysis bags for equilibrium dialysis.

[a] Not significantly different from control.
[b] Significantly different, $p < 0.1\%$.
[c] Significantly different, $p < 5\%$.
[d] Chromatin from livers of normal rats; all other chromatin from livers of adrenalectomized rats.

rats. The chromatin was complexed with an excess of unlabeled actinomycin D to bind to all of the available sites in the DNA. The unbound actinomycin D was then removed by passage through a Sephadex G-50 column (table 1). The chromatin was then allowed to react for 18 hr at 5°C with tritium labeled actinomycin D in the presence or absence of the steroid. Unbound labeled actinomycin D was removed by passing the mixture through a second column. In other experiments the chromatin and labeled actinomycin D were incubated for 10 min at 37°C and then subjected to equilibrium dialysis at 6°C for three days (table 2). The amount of tritium-labeled actinomycin D bound to the DNA was then measured. Adding cortisol or corticosterone at 10^{-6} M led to increased binding of the labeled actinomycin D by the chromatin. This was interpreted as reflecting an increase in the amount of DNA that was free and available to bind with actinomycin D. Incubating chromatin with the biologically inactive steroid, androstenediol, was without effect on the subsequent binding of actinomycin D to the chromatin.

Other investigators [21,22] have shown that actinomycin binds to double helical DNA, but not to RNA, to polyribonucleotides, to DNA–RNA hybrids, to denatured or single-stranded DNA, or to the proteins of chromatin. Only double-helical, non-complexed DNA will bind actinomycin D at the concentrations of DNA (10^{-4} to 10^{-5} M) and actinomycin D (3×10^{-2} μM) used in these experiments. This effect appears to be specific for metabolically active steroids; they in some way promote the separation of the proteins from the DNA, thereby making the DNA available for binding.

Experiments in several laboratories [23, 24] have shown that steroid hormones increase the activity of RNA polymerase in the nuclei of their target cells. This is a different mechanism, one that should lead to an increased synthesis of all types or selected types of RNA, but not of specific messenger RNA's. Experiments in a number of laboratories including our own have shown that steroids increase the synthesis of all types of RNA as measured by the incorporation of labeled precursors such as cytidine or uridine into RNA. This was shown for the seminal vesicle in response to testosterone [25] or dihydrotestosterone [26] and in the rat uterus in response to estrogen [27].

The question of whether RNA produced in a target tissue in response to a hormone could be extracted and purified and then applied to another target tissue and mimick the effects of the hormone has been investigated by several laboratories. RNA from the uterus of estradiol treated mice instilled into the uterus of adult castrate mice led to an increase in uterine alkaline phosphatase activity [28], an effect also evident when estradiol itself is injected into the adult castrate mouse. Segal et al. [29] demonstrated that RNA extracted from the uterus of estradiol-treated rats, when perfused through the lumen of the uterus of an adult castrate rat would produce changes in endometrial cytology similar to those seen when estradiol was injected into the rat. RNA prepared from the uterus of castrate rats not injected with estradiol was without effect when instilled into the uterine lumen. Similar effects have been reported in other laboratories [30, 31]. Sherry and Nicoll

[32] showed that RNA extracted from the crop sac of pigeons treated with prolactin, when added to the crop sacs of control pigeons, would promote growth and protein synthesis in the sac.

Several earlier reports describe phenomena that may be related. Endometrial proliferation in ovariectomized rabbits was produced by the administration of crude, estrogen-free endometrial extracts [33]. Although these investigators did not identify the active agent it seems likely that it might be RNA. Endometrial cells were induced when necrotic uterine endometrium was implanted subcutaneously in rabbits [34]. This has been confirmed by the finding that endometrial cells are induced in the granulation tissue that forms around implants of homogenized necrotic endometrium in rats [35]. Liver RNA induces cells resembling liver and heart RNA induces cells resembling heart muscle when implanted on the chorioallantoic membrane of chick eggs [36].

RNA isolated from the seminal vesicle of androgen-treated rats, when instilled into the lumen of the seminal vesicle of a 3-week-old rat, led to increased weight and increased protein synthesis in the seminal vesicle [37] (table 3). RNA from any of the tissues of immature rats not treated with testosterone was without effect. Of several tissues from immature rats treated with testosterone, only the seminal vesicle yielded RNA that caused growth when instilled into the seminal vesicle of the test rat. Treating the RNA with ribonuclease or heating it to 100°C for 15 min before instillation destroyed the growth promoting effect. RNA extracted from the

TABLE 3

Effect of RNA from the tissues of adult male rats on the growth of the seminal vesicles of 3-week-old rats

	No. of animals	Body wt (g)	Seminal vesicle wt (mg)
Control	10	58.2 ± 1.3 [a]	8.2 ± 0.3 [a]
Forceps	4	51.6 ± 1.6	8.8 ± 0.2
Saline	11	63.7 ± 1.4	10.8 ± 0.6
Liver RNA	10	62.6 ± 2.7	13.7 ± 0.9 [b]
Liver RNA digested with RNase	10	53.3 ± 1.1	10.0 ± 0.1
Prostate RNA	12	61.8 ± 2.2	14.3 ± 0.9 [b]
Prostate RNA digested with RNase	9	54.7 ± 1.7	9.9 ± 0.3
Seminal vesicle RNA	16	60.0 ± 1.8	13.1 ± 0.5 [b]
Seminal vesicle RNA digested with RNase	11	53.1 ± 0.9	9.5 ± 0.8
Kidney RNA	8	54.3 ± 1.8	15.4 ± 1.2 [c]
Heated liver RNA	4	53.3 ± 1.3	10.1 ± 0.4
Heated prostate RNA	4	52.1 ± 2.2	9.9 ± 0.3
Heated seminal vesicle RNA	6	51.8 ± 2.1	8.8 ± 0.3

[a] Standard error of the mean.

[b] $p < 0.01$ (analysis of variance) that the difference between these values and the control value is due to chance.

[c] 72 hr after the injection of the RNA.

tissues of rats injected with testosterone ^{14}C contained no detectable ^{14}C. These experiments provide evidence that the increased protein synthesis results from RNA and not from any possible contamination of the RNA with testosterone. Examination of the seminal vesicle by light microscopy and by electron microscopy revealed evidence of hypertrophy following the instillation of RNA into the lumen of the gland [38]. The instillation of synthetic polyuridylic acid into the seminal vesicle did not lead to an increased weight of the gland or an increased amount of protein, but did lead to an increased incorporation of phenylalanine into protein.

To determine the nature of the RNA that is effective in stimulating protein synthesis in a seminal vesicle, RNA was fractionated on sucrose density gradients by centrifugation at 36,000 rpm for 5 hr [39]. This resulted in three distinct peaks corresponding to 4 S, 16–18 S, and 28 S. Each peak was then recovered and tested for its effect on protein synthesis when instilled into the seminal vesicle (table 4). The heaviest, 28 S RNA, and the lightest, 4 S RNA, had little or no stimulatory effect, but RNA with a sedimentation constant of 16–18 S showed a clear stimulation of protein synthesis when instilled into the seminal vesicle. The kind of RNA effective in stimulating protein synthesis and the kind of RNA that is rapidly labeled following injection of testosterone together with labeled cytidine into the animal appear in the same 18 S peak on density gradient centrifugation. These experiments established that RNA extracted from the seminal vesicle with phenol and sodium dodecyl sulfate can undergo partial purification and treatment with deoxyribonuclease and pronase, and then, when instilled into the seminal vesicle of a control rat, can produce the effects on growth and protein synthesis ordinarily seen in response to the injection of testosterone. They support the inference that RNA is involved in mediating the stimulatory effects of testosterone in the seminal vesicle.

TABLE 4
Effect of RNA fractions from the livers of adult male rats on the growth of the seminal vesicle in 3-week-old rats [a]

Source of RNA		No. of rats	Body weight (g)	Seminal vesicle weight (mg)	Increase (%)
Control liver	Saline	6	53.8 ± 2.4	11.6 ± 0.4	
	28 S	8	52.5 ± 1.4	12.6 ± 0.6	
	18 S	7	54.1 ± 1.3	14.0 ± 1.0 [b]	22
	5 S	6	54.0 ± 1.5	11.5 ± 0.6	
Deoxycholate-treated liver	Saline	10	55.5 ± 5.0	11.5 ± 0.6	
	28 S	11	54.1 ± 2.2	11.7 ± 0.6	
	18 S	10	56.6 ± 2.0	15.5 ± 0.7 [c]	32
	5 S	7	54.4 ± 1.5	12.9 ± 0.9	

[a] Values are expressed as mean ± standard error.
[b] $p < 0.025$ versus saline controls.
[c] $p < 0.001$ versus saline controls.

TABLE 5
Effect of uterine RNA from estradiol-treated immature rats on protein synthesis in the uterine horns of 3-week-old rats 24 hr after intrauterine application

Control uterine horn injected with	No. of of rats	Specific activity of proteins, cpm/mg		Geometric mean (±S.E.) of ratios of individual rats
		Mean of experimental horns	Mean of control horns	$\frac{\text{SA prot. exp}}{\text{SA prot. control}}$
Saline	11	923	723	1.34 ± 0.15 ($p < 0.05$)
Uterine RNA from untreated immature rats	8	1555	1196	1.28 ± 0.09 ($p < 0.02$)
Uterine RNA from E_2-treated rats, inactivated	5	1315	801	1.73 ± 0.35 ($p < 0.05$)
None	8	1794	1146	1.54 ± 0.16 ($p < 0.01$)

A comparable series of experiments has shown that RNA also mediates the effects of estradiol on protein synthesis in the rat uterus [40]. RNA was extracted with cold phenol and sodium dodecyl sulfate from the uteri of 3-week-old rats 12 hr after each rat had been injected with 5 μg of estradiol. The RNA was purified by treatments with deoxyribonuclease and pronase, washed with ethanol and ethanol-ether, and precipitated by the addition of ethanol and potassium acetate. Some 60 or 65 μg of RNA was instilled into one uterine horn of each 3-week-old rat. The other horn received an equivalent volume of a control solution, either saline, RNA treated with ribonuclease and boiled, or RNA extracted from the uteri of rats that had not been treated with estradiol (table 5). Two hours before the rats were killed, a mixture of ^{14}C-labeled amino acids was injected. After the rats were killed, the uteri were removed and homogenized, and the uterine proteins were precipitated with trichloracetic acid. They were washed, counted and measured, using the procedure of Lowry et al. [41] to estimate the amount of protein present. The instillation of RNA led to an increased protein synthesis in that horn of the uterus. Protein synthesis, as measured by the incorporation of labeled amino acids, was increased 34% when the experimental horn was compared with the control horn receiving saline, and 28% when compared with control horns receiving RNA from the uteri of untreated immature rats. In a third series of experiments, the control horn received RNA from estradiol treated rats, but the RNA was treated with ribonuclease before instillation. In these, the incorporation of amino acids into protein was 73% greater in the experimental horn than in the control horn. In subsequent experiments RNA from the liver of estrogen treated rats was also found to be effective in causing uterine growth when instilled into the uteri of the 3-week-old rats. In these experiments the control horn received either RNA from the livers of untreated immature rats, or RNA extracted from the livers of estradiol treated rats and inactivated with ribonuclease.

Uterine RNA was prepared from rats injected with 5 μCi of estradiol-6,7-^{3}H (specific activity 48 Ci/mmole). Although the initial tissue extracts contained a significant amount of radioactivity, the purified RNA preparations had less than 6 cpm per 1000 μg of RNA, which is less than the background level. From this we can calculate that the RNA instilled into a single uterine horn contained less than 2 $\times 10^{-9}$ μg of estradiol.

Recent studies by Tuohimaa et al. [42] confirmed our report that intraluminal instillations of RNA from uteri or liver of estrogen-stimulated rats can induce an increase in protein synthesis. They also found that the amount of contamination of the RNA with estradiol was far below the minimum effective dose for the measured response.

The relationship of the time of instillation of RNA and its effect on protein synthesis was investigated. Rats were killed at varying times after the RNA was instilled and protein synthesis was measured by a pulse of labeled amino acids injected 2 hr before the rats were killed. The effect of RNA on protein synthesis was maximal in 24 hr, and had decreased by 48 hr to a level that was not significant. Other 3-week-old rats were injected with estradiol, and in these the increased protein synthesis in the uterus was also maximal at 24 hr and had returned to normal values by 48 hr.

We were interested by the observation of Rysser [43] that polyamines increase the uptake of proteins by mammalian cells in tissue culture. In one series of experiments 1 μg of polyornithine was added to the RNA instilled into one horn of the uterus; the other horn received RNA without the polyornithine. There was no difference between the two horns in the subsequent rate of protein synthesis. This

TABLE 6
Effect of RNAs from different tissues of estradiol-treated immature rats on protein synthesis in the uterine horns of 3-week-old rats 24 hr after intraluminal application

	No. of rats	Specific activity of proteins, cpm/mg		Geometric mean (± S.E.) of ratios of individual rats
		Mean of experimental horns	Mean of control [a] horns	SA prot. exp. / SA prot. control
Uterine RNA	8	1555	1196	1.28 ± 0.09 ($p < 0.02$)
Liver RNA	7	1050	773	1.29 ± 0.13 ($p < 0.05$)
Kidney RNA	6	1200	1255	1.01 ± 0.01 –
Adrenal RNA	5	1973	1695	1.16 ± 0.18 –
Muscle RNA	4	1192	1294	0.92 ± 0.12 –
Lung RNA	5	1018	1165	0.86 ± 0.14 –
Thymus RNA	5	1790	2300	0.78 ± 0.18 –

[a] Control horn injected with RNA from untreated immature rats.

TABLE 7

Protein synthesis in isolated uterine horns: effects of uterine and hepatic RNA from estrogen stimulated immature rats instilled into uterine horns in vivo

Source of RNA	No. of rats injected	Experimental horn (cpm/mg prot.)	Control horn (cpm/mg prot.)
Uterus	13	31,610 ± 690	25,500 ± 1,540
Liver	7	23,250 ± 1,030	15,900 ± 1,430

Values are mean ± standard error. Differences are significant, $p < 0.01$, analysis of variance. Uterine horns were cut into small segments and incubated 2 hr at 37°C in 2 ml medium 1066 containing 1 μCi of uniformly labeled L-amino acid mixture.

suggests that the amount of RNA taken up without the addition of polyornithine is enough to saturate the protein synthesizing system. Alternatively, it is possible that polyornithine increases the uptake of proteins but not of RNA, or it may be ineffective when instilled into the uterine lumen rather than when added to a tissue culture preparation.

We have explored the range of tissues from which RNA effective in stimulating uterine growth could be extracted (table 6). Although RNA from the liver of an estradiol treated rat is effective in stimulating protein synthesis in the uterus, RNAs from the kidney, adrenals, skeletal muscle, lung or thymus were completely ineffective. Thus these experiments show that it is possible to prepare an RNA from the hormonally treated uterus, and that after suitable purification this RNA will stimulate protein synthesis when instilled into the uterine lumen in a control animal. It is of interest that RNA from the uterus of estrogen-treated rats does not increase protein synthesis when instilled into the seminal vesicle. Conversely, RNA from the seminal vesicle of testosterone-treated rats does not increase protein synthesis when instilled into the uterus.

In subsequent experiments RNA was instilled into the uteri of 21-day-old rats. The rats were killed 24 hr later, the uteri were removed and sliced uterine horns were incubated two hours in a medium containing a mixture of uniformly labeled amino acids (table 7). In these experiments RNA from the uteri of estrogen-treated rats was instilled into one uterine horn and RNA from the uteri of control rats was instilled into the contralateral horn. The horn treated with RNA from the uterus of an estrogen-treated rat showed a significant increase in protein synthesis when compared to the control contralateral horn. When actinomycin D in high concentration (7.5 μg per horn) was added to the RNA instilled into one uterine horn, the subsequent synthesis of protein by that horn was decreased substantially. Actinomycin at this concentration appears to act by some non-specific cytotoxic effect. However, a lower concentration of actinomycin D (0.75 μg per horn) instilled along with the RNA had a slight stimulatory effect on protein synthesis compared to horns that received RNA alone.This appears to be similar to the "superinduction" of hormone-induced increases in enzyme activity observed when actinomycin

D is given in vivo [44, 45]. One might speculate that the actinomycin D inhibited the synthesis of endogenous messengers, thus facilitating the action of the exogenous RNA. Alternatively, it may block the synthesis of a cytoplasmic repressor of mRNA translation [45] or it may block enzyme degradation [46].

To test the hypothesis the RNA instilled may be acting as a template, about 100 μg of an artificial messenger RNA, polyuridylic, polycytidylic, polyadenylic or polyguanylic acid was instilled into one uterine horn and saline was instilled into the contralateral horn. Twenty-four hours after the polynucleotide was instilled the animals were killed and the uteri were removed and incubated two hours in Medium 1066 containing 1 μCi of amino acid mixture [47]. A horn into which polyuridylic acid was instilled, when subsequently incubated in the presence of phenylalanine, showed a statistically significant increase in the incorporation of phenylalanine into protein compared to the contralateral horn (table 8). Other sections of the uteri were incubated in the presence of another amino acid such as glycine, one not coded for by the polynucleotide; these showed no difference in the rate of incorporation of glycine into protein in the two horns. Uterine horns into which polycytidylic acid had been instilled showed an increased incorporation of proline, but not of lysine, compared to control uterine horns when the two were incubated in vitro. Similarly, uterine horns instilled with polyadenylic acid showed an increased incorporation of lysine but not of glycine and uterine horns into which polyguanylic acid was instilled showed an increased incorporation of glycine, but not of proline, when subsequently incubated with the amino acid in a Medium 1066. The increased incorporation of the coded amino acid (20 to 25%) produced by the instillation of polynucleotides is similar to the increased overall incorporation of a

TABLE 8
Amino acid incorporation into uterine horns following the instillation of synthetic polynucleotides

Polynucleotide instilled	^{14}C-labeled amino acid	Control horn (saline)	Polynucleotide horn	Percent increase
Polyuridylic (8)	Phenylalanine	67.2 ± 1.5	81.9 ± 2.2	22
Polyuridylic	Glycine	14.7 ± 0.9	15.6 ± 0.7	–
Polycytidylic (6)	Proline	23.5 ± 0.8	28.4 ± 1.4	21
Polycytidylic	Lysine	68.2 ± 4.7	59.7 ± 3.8	–
Polyadenylic (14)	Lysine	42.0 ± 0.8	50.6 ± 0.8	20
Polyadenylic	Glycine	16.3 ± 0.6	15.0 ± 0.9	–
Polyguanylic (9)	Glycine	15.9 ± 0.6	18.6 ± 0.8	18
Polyguanylic	Proline	24.9 ± 1.4	23.6 ± 1.8	–

Values expressed as dpm/μg protein, mean ± standard error. No. of experiments in parentheses. Immature (21-day-old) rats; saline instilled in one uterine horn and 100 μg polynucleotide instilled in the contralateral horn. Twenty-four hours later the rats were killed, the uteri were excised and cut into segments. The segments were incubated in 2.0 ml medium 1066 containing 1 μCi of ^{14}C-labeled amino acid for 2 hr with 95% O_2; 5% CO_2.

mixture of labeled amino acids into protein produced by the instillation of RNA isolated from the uteri of estradiol treated rats. The addition of actinomycin D to the polynucleotide instilled into the uterine horn did not inhibit the increased incorporation of coded amino acid induced by the polynucleotide. These results are consistent with the inference from earlier experiments that the RNA instilled may be acting as some sort of informational RNA and that it brings about increased protein synthesis in this way. It is also consistent with the hypothesis that RNA is an intermediate which plays a role in the effects of estrogens and androgens on the metabolism of their respective target tissues.

A series of experiments to determine whether cyclic AMP might be involved in the response of the uterus to estrogen yielded negative results. Uteri excised from 22-day-old rats were cut into slices and incubated for an hour with cytidine-5, ^{3}H and a ^{14}C labeled amino acid mixture. The uterine horns had previously been instilled either with theophyllin, with dibutyryl cyclic AMP, with a combination of the two, or with neither. A comparison of the incorporation of cytidine into RNA and of amino acids into protein showed that none of these treatments had any effect on the synthesis of either RNA or protein (table 9).

In another series of experiments 200 g female rats were injected with estradiol or with propylene glycol and with appropriately labeled nucleotide precursors. The uterine RNA was isolated subsequently and separated by density gradient centrifugation into 4 S, 16 S and 28 S peaks. Each peak was hydrolyzed and the constituent nucleotides were separated on a Dowex Column. Analyses revealed no differences in the base ratios of 28 S RNA of control uteri versus 28 S RNA from estrogen treated uteri (table 10). Similarly there was no difference in the base ratios between 18 S RNA of the control versus 18 S RNA of the estrogen treated or between 4 S RNA of the control and 4 S RNA of the estrogen treated uterine RNA. Thus, the injection of estradiol produced no gross difference in the kind of RNA synthesized

TABLE 9

The synthesis of RNA and protein in isolated uterine horns

	RNA (dpm/mg)	Protein (dpm/mg)
Control	3780 ± 147	21.6 ± 2.7
Theophyllin, 10^{-3} M	3250 ± 140	17.2 ± 0.49
Dibutyryl cAMP, 10^{-3} M	4190 ± 133	17.5 ± 0.45
Dibutyryl cAMP plus theophyllin	3440 ± 368	17.0 ± 0.29
	Mean ± S.E. of 6 experiments	

The uteri of 22-day-old rats were excised, cut into slices and incubated 50 min with cytidine-5-^{3}H and uniformly ^{14}C-labeled amino acid mixture with or without theophyllin and dibutyryl cyclic AMP as indicated. The tissues were removed at the end of the incubation period, rinsed copiously and homogenized in perchloric acid. Protein and RNA were isolated and counted by the method of Fleck and Munro.

TABLE 10
Base ratios of uterine RNA from control and estrogen-treated castrate adult rats

		UMP	GMP	AMP	CMP
28 S RNA	Control	19.5	28.3	22.1	30.1
	Estrogen	20.8	28.7	20.6	29.9
18 S RNA	Control	22.1	25.4	23.7	28.8
	Estrogen	23.1	26.9	21.6	28.3
4 S RNA	Control	25.3	24.6	22.8	27.3
	Estrogen	25.4	24.4	23.1	27.1

RNA was extracted from the uteri of 200 g castrate adult rats 4 hr after injection with estradiol-17β (1 μg) or with saline. The RNA was purified and separated by density gradient centrifugation into its constituent peaks. Each peak was recovered, hydrolyzed, and the nucleotides were separated on a Dowex 50 H^+, 200–400 mesh, 4× cross-linked column and measured by the optical density at their respective maxima.

that was reflected in a change in the base ratio. Earlier experiments in my laboratory by Dr. N. Bashirelahi used the technique of competitive DNA–RNA hybridization to determine whether the injection of dihydrotestosterone led to the production in the prostate of new RNA that differed from the RNA previously present. Although the prostatic RNA synthesized by dihydrotestosterone-treated males did not differ from the RNA produced by control prostates a different type of RNA was produced in the prostates of rats treated with estradiol, a difference that could be detected by competitive DNA–RNA hybridization. The ineffective competition of unlabeled control prostatic RNA against labeled estradiol treated prostatic RNA suggested that some species of RNA are present after estradiol treatment which are absent from the control animal.

Experiments in several laboratories have described a number of specific enzymes in target tissues whose activity increases markedly in response to androgen [48] or

TABLE 11
Glucose-6-phosphate dehydrogenase activity of rat uterine horns

	Control rats		Estrogen-treated rats
Protein content (mg protein/g wet weight)	52	13%	59
G6PDH activity (O.D. units/min × mg protein^{-1})	0.083	45%	0.120

The figures represent the mean value of six experiments. Female rats, aged 21 days, were injected s.c. with either 0.1 ml propylene glycol alone or 0.1 ml propylene glycol containing 5 μg estradiol-17β. 40 hr later they were killed, the uteri were dissected free, homogenized, and centrifuged 15 min at 15,000 g. The supernatant fraction was analyzed for protein content and glucose-6-phosphate dehydrogenase activity.

TABLE 12
Glucose-6-phosphate dehydrogenase activity of rat uterine horns

	RNA from control uteri		RNA from uteri of estradiol-treated rats
Protein content (mg protein/g wet weight)	54		52
G6PDH activity (O.D. units/min × mg protein^{-1})	0.043	37%	0.059

The figures represent the mean values of 24 experiments. Ovariectomized rats weighing 200 g were injected s.c. with either 0.1 ml propylene glycol or 10 μg estradiol-17β in propylene glycol. 16 hr later the rats were killed and RNA was prepared from the uteri using phenol and sodium dodecyl sulfate. Polyvinyl sulfate and dextran sulfate were added to all solutions to inhibit ribonuclease. Female rats aged 21 days were anesthesized and RNA (ca. 70 μg) from control uteri was instilled into one horn and RNA from the uteri of estradiol-treated rats was instilled in the other horn. 24 hr later the rats were killed, the uteri were removed and their G6PDH activity and protein content were measured.

estrogen [49]. The activity of hexokinase in the rat prostate is greatly increased following the administration of testosterone [50]. The activity of glucose-6-phosphate dehydrogenase in the uterus is markedly increased following the administration of estradiol [51, 52]. We carried out a comparable series of experiments and confirmed this finding, obtaining a 45% increase in uterine glucose-6-phosphate dehydrogenase activity 40 hr after the injection of estradiol (table 11). In continuing these investigations we instilled RNA from the uteri of estradiol-treated 200 g female rats in one horn and RNA from the uteri of control 200 g castrate females in the other horn. Twenty-four hours later the rats were killed, the uteri were removed, and their glucose-6-phosphate dehydrogenase activity and protein content were measured. The instillation of uterine RNA from estradiol treated rats caused no increase in the protein content of the uteri, that is, the amount of protein per gram wet tissue, but did lead to an average 37% increase in the glucose-6-phosphate dehydrogenase activity (table 12). This difference is significant at the 0.01 level.

In our most recent experiments the RNA from the uteri of estrogen-treated 200 g castrate rats and RNA from the uteri of control 200 g castrate rats were extracted and separated by sucrose density gradients into three peaks of roughly 28 S, 18 S and 5 S RNA. The corresponding peaks from each of six gradients were pooled, precipitated, repurified, and then instilled into the uteri of 21-day-old female rats. One uterine horn received RNA from a given peak from the uteri of control rats and the contralateral horn received the comparable RNA peak obtained from the uteri of estrogen-treated rats. Twenty-four hours later the rats were killed, the uteri were removed, and their glucose-6-phosphate dehydrogenase activity and protein content were measured. In four replications of this rather complex experimental design the glucose-6-phosphate dehydrogenase activity of the uterus was not altered by the instillation of 5 S RNA from estrogen-treated compared to control rats (table 13). In contrast, the instillation of either 18 S or 28 S RNA from the

TABLE 13
Glucose-6-phosphate dehydrogenase activity of rat uterine horns

Type of RNA	Amount instilled (μg)	Uterine horn instilled with	
		RNA from control uteri	RNA from uteri of estradiol-treated rat
28 S	40	0.033	0.043
18 S	20	0.040	0.050
5–6 S	20	0.042	0.044

RNA was isolated from the uteri of 200 g castrate female injected 16 hr previously with 10 μg estradiol-17β in 0.1 ml propylene glycol or with propylene glycol alone. The RNAs were separated into three fractions by centrifugation on sucrose density gradients and comparable fractions from the several gradients were pooled and reprecipitated with ethanol and NaCl. The RNAs were redissolved in saline. The RNA peak from the uteri of estrogen-treated rats was instilled into one horn of the uterus of a 21-day-old female and the RNA peak from the control uteri was instilled into the contralateral horn. 24 hr later the rats were killed, the uteri were removed and homogenized and glucose-6-phosphate dehydrogenase activity and protein content were measured.

uteri of estrogen treated rats led to a greater activity of glucose-6-phosphate dehydrogenase in the uteri of the recipient rats than in the uterine horn receiving the comparable type of RNA from control uteri. The average increase with the 18 S RNA was 45%, the average increase with 28 S RNA was 36%. In another series of experiments (table 14) the control horn was instilled with saline and again the instillation of both 28 S and 18 S RNA led to increased glucose-6-phosphate dehydrogenase activity.

The increased activity of glucose-6-phosphate dehydrogenase in the absence of any overall increase in total protein would suggest that the RNA had stimulated the

TABLE 14
Glucose-6-phosphate dehydrogenase activity of rat uterine horns

Type of RNA	Amount instilled (μg)	Uterine horn instilled with	
		Saline	RNA
28 S	72	0.036	0.047
18 S	35	0.039	0.048
5–6 S	53	0.041	0.042

RNA was isolated from the uteri of 200 g castrate females injected s.c. with 10 μg estradiol-17β in 0.1 ml propylene glycol 16 hr previously. The RNA was separated into 3 fractions by centrifugation on sucrose density gradients and comparable fractions from the several gradients were pooled and reprecipitated with ethanol and NaCl. The RNA was redissolved in saline and instilled into one horn of a 21-day-old female; an equivalent volume of saline was instilled in the contralateral uterine horn. 24 hr later the rats were killed, the uteri removed and homogenized and glucose-6-phosphate dehydrogenase activity and protein content were measured.

synthesis of certain proteins, but not all. Thus, these experiments are also consistent with the hypothesis that some of the RNA produced by the uterus in response to injected estradiol is informational RNA carrying the code for the synthesis of specific enzymes such as glucose-6-phosphate dehydrogenase. The separation of RNA by sucrose density gradient centrifugation is admittedly a crude process at best. When the 18 S peak or the 28 S peak is subsequently analyzed by electrophoresis on agarose acrylamide gels each can be shown to contain sizable amounts of other types of RNA. Thus, the biological activity found in the 28 S peak could be due to 16 S RNA.

In summary, there is evidence from a number of laboratories using a number of biological systems [28–31, 37–42, 47] that exogenous RNA can be taken up by cells and lead to increased protein synthesis, to increased activity of certain enzymes or to altered cellular structure. The experiments render most unlikely the hypothesis that these effects are due to minute amounts of contaminating steroids remaining in the RNA preparations, but suggest strongly that they result from the RNA itself. They are consistent with the hypothesis that a hormone is taken up in its target tissue by a specific receptor, transferred to the nucleus and there initiates the production of new RNA or increases the production of certain types of RNA. At least part of the RNA is template RNA that provides biologic information for the synthesis of specific enzymes.

References

[1] P. Karlson, Perspectives Biol. Med. 6 (1963) 203.
[2] E.V. Jensen and H.I. Jacobson, Recent Progr. Hormone Res. 18 (1962) 387.
[3] D. Toft and J. Gorski, Proc. Natl. Acad. Sci. U.S. 55 (1966) 1574.
[4] A.J. Eisenfeld and J. Axelrod, J. Pharmacol. Exptl. Ther. 150 (1965) 469.
[5] J. Kato and C.A. Villee, Endocrinol. 80 (1967) 567.
[6] K.M. Anderson and S. Liao, Nature (London) 219 (1968) 277.
[7] S. Fang, K.M. Anderson and S. Liao, J. Biol. Chem. 244 (1969) 6584.
[8] B.W. O'Malley, D.O. Toft and M.R. Sherman, J. Biol. Chem. 246 (1971) 1117.
[9] A. Munck, C. Wira, D.A. Young, K.M. Mosher, C. Hallahan and P.A. Bell, J. Steroid Biochem. 3 (1972) 567.
[10] G.J. Chader, personal communication.
[11] A.R. Means and J. Vaitukaitis, Endocrinol. 90 (1972) 39.
[12] H. Rajaniemi and T. Vanha-Perttula, Endocrinol. 90 (1972) 1.
[13] P. Freychet, J. Roth and D.M. Neville, Jr., Proc. Natl. Acad. Sci. U.S. 68 (1971) 1833.
[14] B.W. O'Malley, Biochemistry (Washington) 6 (1967) 2546.
[15] B.W. O'Malley, W.L. McGuire, P.O. Kohler and S.G. Korenman, Rec. Progr. Hormone Res. 25 (1969) 105.
[16] M. Sluyser, J. Mol. Biol. 22 (1966) 411.
[17] M.A. Goldberg and W.A. Atchley, Proc. Natl. Acad. Sci. U.S. 55 (1966) 989.
[18] L. Jurkowitz, Arch. Biochem. Biophys. 111 (1965) 88.
[19] M.E. Mitchell, The interaction of rat liver chromatin with steroid hormones as measured by ^{3}H actinomycin D binding, Thesis, M.D. with honors in biochemistry, Harvard University, Boston, Mass. (1969).

[20] K. Marushiga and J. Bonner, J. Mol. Biol. 15 (1966) 160.
[21] R. Haselkorn, Science 143 (1964) 682.
[22] N. R. Ringertz and L. Bolund, Biochem. Biophys. Acta 174 (1969) 160.
[23] J. Gorski and N.J. Nelson, Arch. Biochem. Biophys. 110 (1965) 284.
[24] C.E. Sekeris and N. Lang, Life Sci. 3 (1964) 169.
[25] W.D. Wicks and C.A. Villee, Arch. Biochem. Biophys. 110 (1965) 284.
[26] P.M. McRoberts and C.A. Villee, Abstract, Session 4B (Hormone action), No.121, Endocrine Society Meetings in St. Louis, Mo. (1970).
[27] J.H. Hamilton, Proc. Natl. Acad. Sci. U.S. 49 (1963) 373.
[28] A.M. Mansour and M.C. Niu, Proc. Natl. Acad. Sci. U.S. 53 (1965) 764.
[29] S.J. Segal, B.W. Davidson and K. Wada, Proc. Natl. Acad. Sci. U.S. 54 (1965) 782.
[30] A. Unhjem, A. Attramadal and J. Solna, Acta Endocrinol. 58 (1968) 227.
[31] P. Galand, N. Dupont-Mairasse, F. LeRoy and J. Chretien, In: Basic actions of sex steroids on target organs, Eds. P.O. Hubinout, F. LeRoy and P. Galand (S. Karger, Basel, 1971) 227–239.
[32] W.E. Sherry and C.S. Nicoll, Proc. Soc. Exptl. Biol. Med. 126 (1967) 824.
[33] P. Bernhard and W. Krampitz, Z. Geburtshilfe Gynakol. 156 (1960) 1.
[34] G. Levander and P. Normann, Acta Obstet. Gynecol. Scand. 34 (1955) 366.
[35] G. Levander, Induction phenomenon in tissue regeneration (Williams and Wilkins, Baltimore, 1964).
[36] M.C. Niu and L. Mulherkar, J. Embryol. Exptl. Morph. 24 (1970) 33.
[37] T. Fujii and C.A. Villee, Proc. Natl. Acad. Sci. U.S. 57 (1967) 1468.
[38] C.A. Villee and T. Fujii, In: Hormones in development, Eds. Max Hamburgh and E.J.W. Barrington (Appleton-Century-Crofts, New York; Education Division, Meredith Corporation, 1971) Ch. 48, 610–630.
[39] T. Fujii and C.A. Villee, Proc. Natl. Acad. Sci. U.S. 62 (1969) 836.
[40] M. Fencl and C.A. Villee, Endocrinol. 88 (1971) 279.
[41] O.H. Lowry, N.J. Rosenbrough, A.L. Farr and R.J. Randall, J. Biol. Chem. 193 (1951) 265.
[42] P.J. Tuohimaa, S.J. Segal and S.S. Koide, J. Steroid Biochem. 3 (1972) 503.
[43] H.J.P. Ryser, Science 159 (1968) 390.
[44] L.D. Garren, R.R. Howell, G.M. Tomkins and R.M. Crocco, Proc. Natl. Acad. Sci. U.S. 52 (1964) 1121.
[45] G.M. Tomkins, T.D. Gelehrter, D. Granner, D. Martin, Jr., H.H. Samuels and E.B. Thompson, Science 166 (1969) 1474.
[46] F.T. Kenney, In: Mammalian protein metabolism, Vol. IV, Ed. H.M. Munro (Academic Press, New York and London, 1970) 131–176.
[47] C.A. Villee and J.M. Loring, Endocrinology 88 (1971) 212.
[48] R.L. Singhal and G.M. Ling, J. Physiol. Pharmacol. 47 (1969) 233.
[49] V.I. Noack and H. Schmidt, Endokrinologie 53 (1968) 291.
[50] R.S. Santti and C.A. Villee, Endocrinology 89 (1971) 32.
[51] B.C. Moulton and K.L. Barker, Endocrinology 89 (1971) 1131.
[52] B.C. Moulton and K.L. Barker, Endocrinology 91 (1972) 491.

Niu and Segal (eds.), The role of RNA in reproduction and development
North-Holland Publ. Co., 1973

The role of RNA in the differentiation of presumptive ectoderm from urodele embryos

Naoi SASAKI and M.C. NIU
Embryological Laboratory of Biology Department, Kyushu University, Fukuoka, Japan, and Department of Biology, Temple University, Philadelphia, Pa. 19122, USA

RNA from calf kidney was used as an inducer for the study of embryonic differentiation. The activity of pure kidney RNA was very low, like that of infectious RNA from tobacco mosaic virus. In order to manifest functional potentiality of kidney RNA experiments were carried out in association with another substance, the saline extract of bone marrow from guinea pig (BME). When BME was "sandwiched" between presumptive ectoderm (PE) from California newt, *Taricha torosa,* for 2 hr and BME removed mechanically, PE would not undergo differentiation. On the other hand the 12-hr-treated PE developed structures typical of spinocaudal induction. Should the 12-hr-BME-treated PE be further exposed to kidney (K)-RNA, some of the spinocaudal structures, e.g. neural tissue, notochord, somite, did not develop or were inhibited, while others, e.g. pronephros and connective tissue formation increased significantly. If PE was exposed simultaneously to BME and K-RNA for 2 hr, it differentiated. The kinds of tissue formed in PE resembled more those formed after the 12-hr-BME plus K-RNA than those formed after the 12-hr-BME alone. Therefore, the inductive capacity of K-RNA in both series appears to be greater than that of BME.

PE from *Ambystoma tigrinum* was similarly treated with BME. Three and seven hour treatments rendered PE responsive specifically to the action of brain (B)- and K-RNAs respectively. The 3-hr-BME-treated PE differentiated into archencephalic types of structure and the 7-hr series into spinocaudal tissues. In the 3-hr series, B-RNA stimulated and K-RNA inhibited brain formation. On the other hand, B-RNA was less effective in the 7-hr than the 3-hr series, but still promoted neural and inhibited notochord and connective tissue formation. In contrast, K-RNA inhibited the formation of neural tissue, notochord and somites but greatly stimulated the formation of pronephros and connective tissue. These data show clearly that kidney-RNA induces the formation of organ specific tissue while it inhibits that of other structures. The ability of K-RNA to stimulate tubule formation was sensitive to pancreatic RNase. The responsive state of PE changed with time apparently in response to the treatment with BME.

The acquisition of inductive capacity was analyzed using PE of the Japanese newt, *Triturus pyrrhogaster.* Cultivation of the 3-hr-BME-treated BE for 12 hr made it a potent agent for spinocaudal induction. The active substance of the BME-treated PE was sensitive to RNase and actinomycin-D thus suggesting that it is RNA.

1. Introduction

Early experiments have shown that the ribonucleoprotein (RNP) released by the in vitro developing posterior chordamesoderm (presumptive tail mesoderm) is capable of initiating differentiation of the presumptive ectoderm (PE) into an outgrowth of neural crest or an outgrowth of muscle cells or both. The type of outgrowth produced is determined by the age of the culture at which the PE is introduced [1]. Thus, in the early days of cultivation, the chordamesoderm cells undergo active divisions (growth phase) and the RNP released would induce the formation of neural crest derivatives. This type of induction is heterogenetic and is typified by the classical example, primary embryonic induction. When the chordamesoderm cells at later days of cultivation enter the phase of differentiation, i.e., reach the stage of myoblasts and/or muscle cells, the RNP released would induce the formation of muscle cells. This type of induction is homeogenetic and also highly specific. It is apparently mediated through informational RNP. The active component of the RNP is RNA [1, 2]. However, Yamada [3] and Tiedemann [4] have not been able to confirm the specific role of RNA in the differentiation of PE. In order to test whether or not this discrepancy can be resolved, we decided to adopt the method employed by Yamada and his co-workers, namely the sandwich technique.

It is well known that the saline extract of bone marrow from guinea pig (BME) induced the formation of spinocaudal derivatives. According to the analytical data of Yamada [3], BME contains approximately 5–10% of RNA. If, therefore, the RNA component of the BME were responsible for the inductive activity, it would seem that additional RNA treatment of the BME-treated PE should result in an enhancement or an alteration of the BME effect. The enhancement of the BME effect depends entirely upon the time of exposure to BME. If the BME treatment is so brief that the BME-induced differentiation may occur poorly or not at all, then addition of RNA could initiate the differentiation. On the other hand, should the BME treatment be optimum, i.e., capable of initiating spinocaudal induction, subsequent exposure to RNA could affect the BME-induced potentiality of differentiation. Therefore, our experimental strategy is first to find the threshold time that will initiate the minimal effect of BME on PE and the optimum time that will initiate the maximal effect. Second after removal of BME from the minimally and the optimum BME-treated PE the effect of RNA on both PEs will be tested. Third, after removal of BME at predetermined times, a differential response of the PE to RNAs from two organs will be tested. Fourth and last, with the removal of BME, PEs will be cultured in Niu-Twitty solution and their inductive property will then be analyzed. The results of these experiments showed that (i) kidney-RNA is responsible for the differentiation of the 2-hr-BME-treated PE into spinocaudal tissue (ii) kidney-RNA is capable of diverting the BME-directed differentiation to a new avenue leading to the formation of tubules with and without other tissues and (iii) the inductive capacity of the BME-treated PE is sensitive to the treatment with pancreatic ribonuclease and actinomycin D.

2. Materials and methods

Egg clusters of the California newt, *Taricha torosa,* were supplied at the cleavage stage by Mr. Robert Little of Stanford University, those of the tiger salamander, *Ambystoma tigrinum,* by Mr. Glenn Gentry of Nashville, Tennessee; those of axolotl, *Ambystoma mexicanum,* by Prof. R.R. Humphrey of Indiana University and those of the Japanese newt, *Triturus pyrrhogaster,* from Fukuoka, Japan. When the eggs reached the early gastrula (stage 10–11), the clusters were washed with boiled spring water. Individual eggs were removed mechanically from the jelly capsule and sterilized by placing them in 0.05% of chloramine-T in Niu-Twitty (N–T) solution for 3 min. After rinsing with N–T solution they were used as donors of presumptive ectoderm (PE). Strips of excised PE were folded into double layers with or without a tiny piece of BME. The sandwich thus formed was cultured at different lengths of time in N–T solution.

2.1. Preparation of saline extract from bone marrow (BME) and from kidney (KE) of guinea-pig

Bone marrow was obtained from the femur of a guinea-pig. Both bone marrow and liver were homogenized in saline. The homogenate was centrifuged at 20,000 rpm for 30 min. To the supernatant, 2 vols of ice-cold 95% alcohol was added. The precipitate was packed into pellet by centrifugation (15,000 rpm for 30 min). The pellet was cut into pieces and used within a few days.

2.2. Isolation of brain (B-) and kidney (K-) RNAs

Kidney and brain were obtained from freshly slaughtered calves at a local meat packing company. They were sliced immediately into thin pieces and put into ice-cold sucrose solution (0.25 M and 0.0003 M $CaCl_2$). The procedure and routine laboratory test of RNA purity have already been described elsewhere [6, 7].

2.3. Testing system

As stated previously [6, 7], the BME or KE was used for the purpose of sensitizing the PE. The sandwich explants were opened after 2, 3, 7 or 12 hr and the BME or KE were removed mechanically. One half of the explants served as controls and designated as BME(2, 3, 7 or 12). The other half was treated with B- or K-RNA for 2 hr before culturing for 10 days. These explants were designated as BME or KE(2, 3, 7 or 12) : B- or K-RNA(2).

Microsurgical operations were performed in room temperature 21–23°C. Cultures were maintained at 18°C. They were fixed in Bouin's fluid, sectioned at 8 nm in thickness and stanied with eosin and haematoxylin.

2.4. Preparation of ectodermal inductor

BME was removed from the sandwich explants (PE of *T. pyrrhogaster*) 3 hr later.

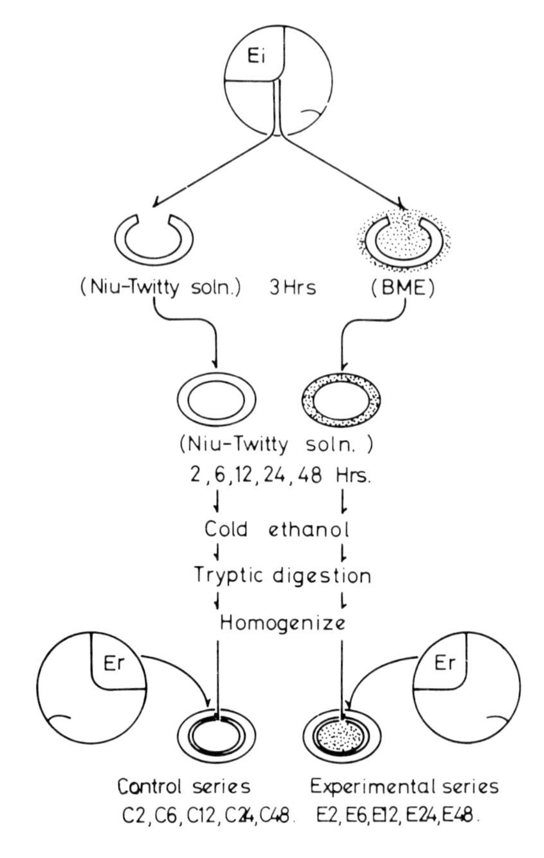

Fig. 1. Diagram of experimental procedure in the preparation of homogenate of the cultured PE and the test of its inductive capacity.

The PEs were transferred into N–T solution for 2, 6, 12, 24 and 48 hr. Then they fixed with 2 vols of ice-cold 95% alcohol and washed with N–T solution. 0.1% trypsin (Difco) in N–T solution was used to digest the contaminant from the effective factor of BME (3). After digestion (30 min), they were washed 3 times and crushed with a glass rod. The pellet was used as an inductor substance. The control received no treatment with BME (fig. 1).

Actinomycin D (Merck Co. Rahway, N.J.) was added to the N–T solution (2/ml) in which the 3-hr-BME-treated PEs were cultured for 2, 6, 12, 24 or 48 hr. After fixing in alcohol, the explants were processed as above.

Both K-RNA and homogenate of 3-hr-BME-treated PE from *T. pyrrhogaster* were incubated with Worthington's pancreatic RNase (10 μg/ml, pre-boiled for 30 min) at 37°C for 30 min. To the digest, rabbit antiserum (10 μg r-globulin) against RNase was added and mixed well. The mixture was kept at 37°C for 1 hr and then 0°C for 1 hr. After centrifugation, the supernatant of K-RNA and sediment of the homogenate were saved for experimental use.

TABLE 1
Time effect of BME on the differentiation of presumptive ectoderm from *Taricha torosa*

Experiments	Cont.	BME (2 hr)	BME (12 hr)	BME (24 hr)	BME (10 days)
No. of explants	32	26	14	11	11
No. of induction	0	8 (31%)	12 (86%)	8 (73%)	11 (100%)
Brain	0	1 (13%) [a]	1 (8%)	0	0
Spinal cord	0	0	2 (17)	2 (25)	2 (18)
Neural tissue	0	7 (88)	4 (33)	3 (38)	2 (18)
Notochord	0	0	9 (75)	3 (38)	5 (45)
Myoblasts	0	0	9 (75)	5 (63)	6 (55)
Pronephros	0	0	4 (33)	2 (25)	2 (18)
Mesothelium	0	0	2 (17)	2 (25)	3 (27)
Mesenchyme	0	2 (25)	10 (75)	5 (63)	8 (73)

[a] Percentage in parentheses is calculated from No. of induction.

3. Results

3.1. Time dependent effect of BME and KE (kidney extract) on PE differentiation

3.1.1. BME

BPEs from *Taricha torosa* and *Ambystoma tigrinum* were used to prepare sandwich explants with BME. At intervals of cultivation, BME was removed from the explants. The PE thus freed was cultured in Niu–Twitty solution for 10 days. The

TABLE 2
Time effect of BME on the differentiation of presumptive ectoderm from *Ambystoma tigrinum*

Experiments	Control	1 hr	3 hr	7 hr	10 days
No. of explants	43	46	58	36	38
No. of induction	12 (28%)	39 (85%)	39 (67%)	30 (83%)	37 (97%)
Fore brain	10 (83)	21 (54)	8 (21)	0	1 (3)
Hind brain	0	3 (8)	6 (15)	3 (0)	3 (8)
Unident. brain	1 (8)	12 (32)	10 (26)	2 (7)	6 (16)
Neural tissue	0	5 (13)	17 (44)	13 (43)	25 (68)
Spinal cord	0	0	0	5 (17)	7 (19)
Notochord	0	0	1 (3)	14 (47)	22 (59)
Myoblasts	0	0	3 (8)	12 (40)	21 (57)
Pronephrotic tubule	0	0	0	7 (23)	20 (54)
Mesothelium	0	1 (3)	3 (8)	21 (70)	18 (48)
Mesenchyme	2 (17)	24 (62)	15 (46)	25 (83)	37 (100)

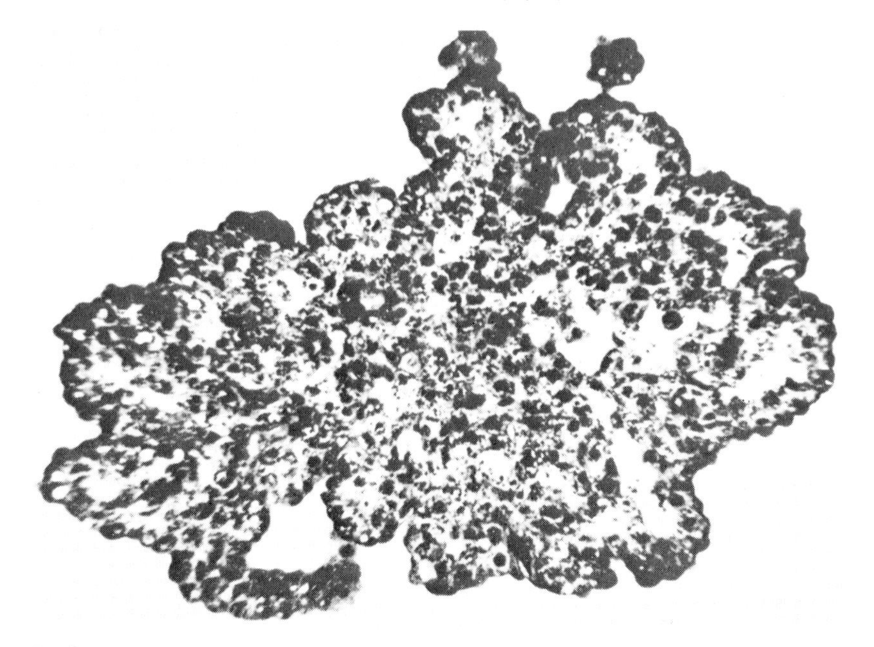

Fig. 2. Photomicrograph of a cross section of a control PE explant (*Taricha torosa*) showing atypical epidermis.

results from histological examination are shown in tables 1 and 2. As can be seen, autoneuralization occurs in *A. tigrinum* but not in *T. torosa* (fig. 2). In both series, the effect of BME varies with the length of contact time between BME and PE. When the contact was 1–2 hr, the PE developed into some neural tissue (brain and neuroid) and some mesenchyme. PEs of *T. torosa* and *A. tigrinum* after contact with BME 12 and 7 hr acquired respectively the same developmental potentiality as the series without removal of BME. (fig. 3).

3.1.2. KE

Sandwich explants were also prepared with PEs from *Ambystoma mexicanum* and KE. The KE was removed after 0.5, 3 and 6 hr and then cultured in Niu–Twitty solution for 10 days. Histological data of the explants are shown in table 3. Apparently KE induced the formation of brain, neuroid tissue and some mesenchyme only.

3.2. Effect of K-RNA on the BME-treated PE

3.2.1. T. torosa

It can be seen from table 1 that after 2-hr contact with BME, PEs of *T. torosa* has not acquired the capacity to undergo differentiation and the effect of BME is manifested fully in PE differentiation only after 12 hr of contact. Whatever the

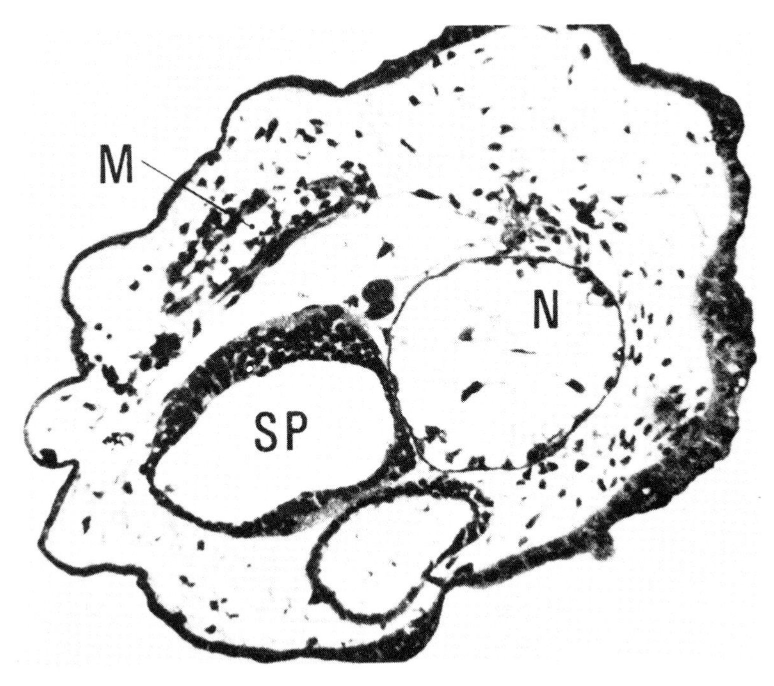

Fig. 3. Photomicrograph of a cross section of an explant (*Ambystoma tigrinum*) from the BME(10) series showing the structures of a spinocaudal induction. M, myotome (muscle); N, notochord; and SP, spinal cord.

TABLE 3

Time effect of guinea pig kidney extract on the frequency and specificity of differentiation of *Axolotl* ectoderm

Duration of contact with kidney extract (hr)	0	0.5	3	6	10 (days)
No. of explants	27	24	17	27	23
Positive cases	9	11	13	21	23
No. of explants containing:					
Fore brain	1	3	3	5	4
Hind brain	0	0	0	7	9
Unspecifiable brain	3	4	6	5	7
Neural tissue	5	5	6	4	5
Sensory placode	0	1	1	10	12
Mesenchyme	0	0	0	2	6

TABLE 4

Effect of BME and K-RNA on differentiation of presumptive ectoderm from *Taricha torosa*

Experiments	K-RNA	BME(2)	BME and K-RNA(2)	BME(12)	BME(12) and K-RNA(2)
No. of explants	26	24	26	14	27
No. of induction	8 (31%)	3 (13%)	13 (50%)	12 (86%)	20 (74%)
Brain	1 (13)	0	0	1 (8)	0
Spinal cord	0	0	0	2 (17)	1 (5)
Neural tissue	7 (88)	3 (100)	7 (54)	4 (33)	0
Notochord	0	0	9 (69)	9 (75)	2 (10)
Myoblasts	0	0	9 (69)	9 (75)	3 (14)
Pronephros	0	0	6 (46)	4 (33)	14 (50)
Mesothelium	0	0	5 (38)	2 (17)	14 (70)
Mesenchyme	2 (25)	1 (33)	8 (62)	10 (75)	20 (100)

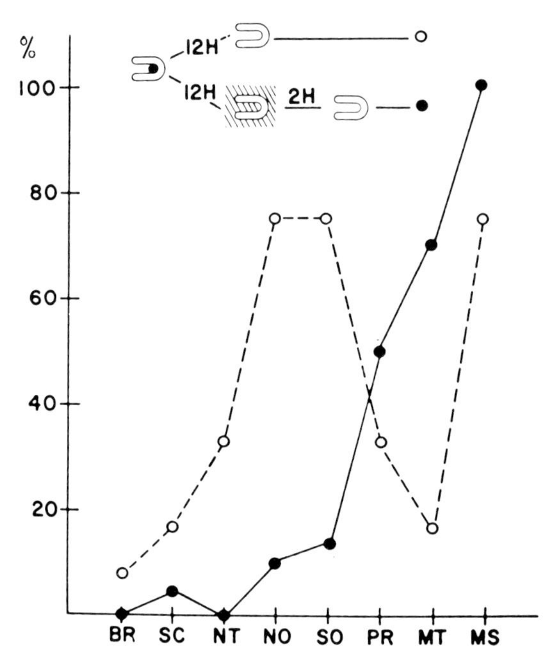

Fig. 4. Two curves showing the percentages of the induced tissues in the BME(12) series (0) and the BME(12) : K-RNA(2) series (0). Experimental procedure is outlined on the top. BR, SC and NT are brain, spinal cord and neural tissue respectively. NO, SO, PR, MT and MS are notochord, somite, pronephros, mesothelium and mesenchyme respectively.

Fig. 5. Photomicrograph of a cross section of an explant (*T. torosa*) from the BME(12): K-RNA(2) series showing pronephrotic tubules (T), mesenchyme (M) and mesothelium (ME).

active component of the BME may be, its chemical nature remains the same regardless of 2, 12 hr in the sandwich explants. When K-RNA (column 1, table 4) and BME (column 2) are applied simultaneously to PEs (column 3), they undergo differentiation in which the formation of various tissue types resembles the effect of treatment with BME(12) (column 4) and the BME(12): K-RNA(2) series (column 5). An apparent difference is the frequency of differentiation which we believe results from the delicate microsurgery. This finding shows clearly that the function of K-RNA and BME(2) is cumulative because either one alone is a poor inducer. Furthermore, brain and spinal cord formation disappeared in this series while the formation of pronephros and mesothelium is higher than the BME(12) series. These properties show that the cumulative effects of K-RNA and BME(2) tend more toward the BME(12) : K-RNA(2) type of effect than the BME(12) type. It seems, therefore, that K-RNA plays a key role in the differentiation of ectoderm into pronephros and connective tissue.

TABLE 5
The effect of B- and K-RNA on the 3-hr-BME-treated PE from *A. tigrinum*

Experiments	BME(3)	BME(3) + B-RNA(2) [a]	BME(3) + K-RNA(2) [a]
No. of explants	58	30	29
Positive cases	39 (67)	25 (83) –2.6, 0.2>*P*>0.1	20 (69) –0.03, 0.9>*P*<0.8
Fore brain	8 (21)	15 (60) –10.26, *0.01 < P < 0.001*	1 (5) –10.25, *0.01 > P < 0.00*
Hind brain	6 (15)	2 (8) –0.77, 0.5>*P* <0.3	2 (10) –0.33, 0.7>*P*<0.5
Unident. brain	10 (26)	7 (28) –0.04, 0.8>*P*<0.7	9 (45) –2.27, 0.2>*P*<0.1
Neural tissue	17 (44)	5 (20) –3.77, *0.1>P<0.05*	9 (45) –0.01, 0.9>*P*<0.8
Spinal cord	0	0	0
Notochord	1 (3)	0	0
Myotome (muscle)	3 (8)	0	0
Pronephros	0	0	0
Mesothelium	3 (8)	3 (12) –0.34, 0.5>*P*<0.3	4 (20) –1.72, 0.2>*P*<0.1
Mesenchyme	18 (46)	16 (64) –1.95, 0.2>*P*<0.1	17 (85) –7.98, *0.01>P<0.001*

[a] Comparison with BME(3) series [8].

Of most importance is the fact that when the 12-hr-BME-treated PEs were incubated with K-RNA for 2 hr, the BME-induced developmental pathway was diverted to a new avenue (fig. 4). As a result, the formation of brain, spinal cord, neural tissue, notochord and myoblasts was inhibited severely or completely while the formation of pronephros (fig. 5) and connective tissue increased significantly.

3.2.2. A. tigrinum

3.2.2.1. Three-hour-BME-treated PE. Forebrain (B) and K-RNAs were separately used to treat PEs immediately after the removal of BME from the sandwich explants. The effect of the 2 RNAs on the ectodermal differentiation is shown in table 5. As can be seen, the frequency of differentiation in the control, B-RNA and K-RNA-treated series is practically the same. From a statistical point of view, B-RNA has specifically promoted the formation of forebrain at the expense of neural tissue. On the contrary, K-RNA promoted the development of mesenchyme at the expense of forebrain.

3.2.2.2. Seven-hour-BME-treated PE. The results of this series are summarized in table 6. Again the frequency of differentiation in the control, K-RNA- and B-RNA-treated series is practically the same. However, K-RNA abolishes brain, spinal cord and notochord formation, inhibits neural tissue but significantly stimulates pronephros formation (fig. 6). On the other hand, B-RNA abolishes spinal cord, and significantly inhibits notochord, mesothelium and mesenchyme formation.

TABLE 6
Effect of K- and B-RNA on the 7-hr-BME-treated PE from *Ambystoma tigrinum*

Experiments	BME(7)	BME(7) + K-RNA(2)	BME(7) + B-RNA(2)
No. of explants	39	28	20
No. of induction	38 (63%)	21 (75%)	16 (80%)
Fore brain	0	0	0
Hind brain	3 (10)	0	2 (13)
Unident. brain	2 (7)	0	2 (13)
Neural tissue	13 (43)	4 (19), –3.28; *0.1*>*P*<*0.05*	10 (63), –1.53, 0.3>*P*<0.2
Spinal cord	5 (17)	0	0
Notochord	14 (47)	0	3 (19), –3.49, *0.1*>*P*<*0.05*
Myoblasts	12 (40)	8 (38), –0.02, 0.9>*P*<0.8	5 (32), –0.34, 0.7>*P*<0.5
Pronephrotic tubule	7 (23)	10 (48), –3.28, *0.1*>*P*<*0.05*	3 (19), –0.13, 0.8>*P*<0.7
Mesothelium	21 (70)	12 (57), –0.09, 0.5>*P*<0.3	4 (25), –8.53, *0.01*>*P*<*0.001*
Mesenchyme	25 (83)	17 (81), –0.67, 0.5>*P*<0.3	9 (56), –3.98, *0.05*>*P*<*0.02*

3.3. Effects of K-RNA on KE-treated PE

KE-treated PEs from *A. mexicanum* were subsequently incubated with either B-RNA or K-RNA. The results are summarized in table 7. As can be seen, B-RNA

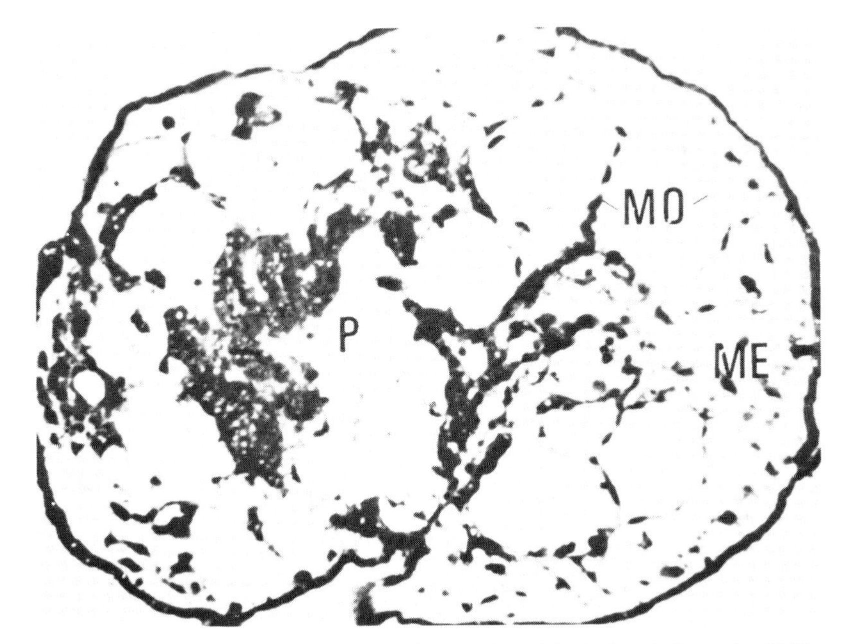

Fig. 6. Photomicrograph of a cross section of an explant (*A. tigrinum*) from the BME and K-RNA(2) series showing pronephrotic ducts (P), mesenchyme (ME) and mesothelium (MO).

TABLE 7
The effects of kidney- and brain-RNA on the conditioned ectoderm of *Axolotl*

Series	KE(1/2)–K-RNA(2)	KE(3)–K-RNA(2)	KE(6)–K-RNA(2)	KE(1/2)–B-RNA(2)	KE(3)–B-RNA(2)	KE(6)–B-RNA(2)
No. of explants	13	30	19	21	19	28
Positive cases	8	30	16	3	3	15
No. of explants containing:						
Fore brain	2	12	9	0	0	5
Hind brain	0	8	1	0	0	3
Unspecif. brain	2	8	2	0	0	2
Neural tissue	4	6	4	3	3	5
Sensory placode	1	13	4	0	0	7
Mesenchyme	0	7	2	0	1	2

promotes the frequency of brain and sense organ formation while K-RNA inhibits them. However, the antagonistic action between them diminishes in the PE treated with KE for 6 hr (fig. 7).

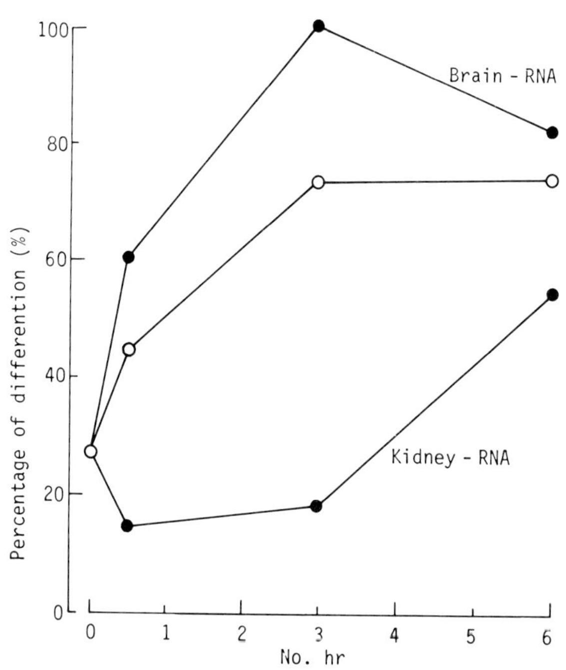

Fig. 7. The relationship between the frequency of the B- and K-RNA-induced differentiation and the time course of the KE-treated PE from *Ambystoma mexicanum*.

TABLE 8
Differentiation of PEs (*T. pyrrhogaster*) induced by the control and the BME-treated PEs at intervals of hours

Series	C2	C6	C12	C24	C48	E2	E6	E12	E24	E48
No. of explants	21	19	21	25	29	22	42	48	31	23
Positive cases	6	4	8	5	10	5	14	25	16	7
No. of explants containing neural structures	6	4	8	5	10	5	12	17	9	7
No. of explants containing mesodermal structures	2	1	2	1	2	2	9	19	8	1
No. of explants containing:										
Fore brain	0	0	2	1	0	0	1	2	2	0
Unspecifiable brain	1	2	1	1	1	2	5	6	2	1
Neural tissue	5	2	5	3	9	3	6	10	5	6
Notochord	0	0	0	0	0	0	0	1	0	0
Muscle, myotome, myoblast	0	0	0	0	0	0	3	3	4	0
Pronephros	0	0	0	0	0	0	0	1	0	0
Mesothelium	0	0	0	0	0	0	2	7	2	0
Mesenchyme	2	1	2	1	2	2	7	13	5	1
Blood cell	0	0	0	0	0	0	1	2	1	0

TABLE 9
The effects of actinomycin D and ribonuclease on the inducing activity of 12 hr-conditioned ectoderm

	Control (E12)	Act.	RNase
No. of explants	48	41	28
Positive cases	25	7	3
No. of explants containing neural structures	17	7	3
No. of explants containing mesodermal structures	19	2	1
No. of explants containing:			
Fore brain	2	0	0
Unspecificable brain	6	1	1
Neural tissue	10	6	2
Notochord	1	0	0
Muscle, myotome, myoblast	3	0	0
Pronephros	1	0	0
Mesothelium	7	0	0
Mesenchyme	13	2	1
Blood cell	2	0	0

3.4. The inductive activity of the BME-treated PE

As shown in the diagram (fig. 1), the homogenate of the PEs from both control and experimental series were tested for inductive activity. The results are given in table 8. From the table it is clear that the most potent inductor (homogenate) is the experimental PE cultivated for 12 hr. The active component of the homogenate was analyzed using actinomycin D and pancreatic RNase. Both were capable of drastically inhibiting the inducing activity of the PE homogenate (table 9).

4. Discussion

The active component of BME is responsible for the ectodermal differentiation into spinocaudal structures. The types of structures developed by the BME-treated PE depend upon the length of the exposure to BME(5). Thus short exposure, e.g., 1–3 hr, results in a type of differentiation (archencephalic induction) which differs from that produced by longer exposure, e.g., 7–12 hr (spinocaudal induction) (tables 1 and 2). Since the alcohol-fixed BME was used in the making of sandwich explants, both archencephalic and spinocaudal inductions must be caused by the same agent. The question is how can a single agent act on PE and lead to different kinds of induction? One possibility is that the bulk of BME, presumably protein, may be involved. Among the factors, two may be of particular relevance: (i) the release of the active agent from BME is time-dependent and (ii) the responsiveness of PE to the active agent varies with time [6, 7].

According to the release hypothesis, the concentration of the active agent released from BME during short exposures is low and during longer exposures is higher, thus reaching the threshold needed for specific differentiation. Subthreshold concentrations would induce neural tissue and mesenchyme. In development, the competence of PE prevails during the gastrula stage. For secondary and/or tertiary inductions, the ectoderm must be conditioned or activated before it becomes responsive to an inducer [9]. The bulk of the inducer, BME, is protein and a small fraction is RNA (5–10%). Whatever the chemical nature of the active agent may be, the rest of the BME may well be functioning as an activator or conditioner. Depending upon the length of exposure, it places PE in such a physiological state that it will respond readily to the active agent. As a result, 1–3-hr-BME-treated PEs of *A. tigrinum* responds to the active agent by developing into archencephalic structures while the 7-hr-treated series develops into spinocaudal structures (table 2). In order to show experimentally the existence of the two responsive states, both B- and K-RNAs were separately applied to PEs immediately after the removal of BME. It can be seen in table 5 that B-RNA promotes brain formation at the expense of neural tissue while K-RNA inhibits brain formation but promotes mesenchyme formation. Table 6 shows that K-RNA inhibits the formation of brain, spinal cord, neural tissue and notochord but promotes the formation of pronephros and connective tissue. Furthermore, the 7-hr-treated PEs respond poorly to B-RNA, presumably due to loss of competence in neuralization [10], but B-RNA still inhibits connective tissue formation. Therefore, the competence of PE changes with time. When BME brings PE to the state sensitive to the brain-forming agent, e.g., B-RNA, B-RNA supersedes the action of the active agent. On the other hand, at the state sensitive to the kidney-forming agent, e.g., K-RNA, the action of the active agent interacts with that of K-RNA, thus diverting the BME-induced pathway of differentiation (spinocaudal induction) to a new avenue (renal tubule and connective tissue formation).

To further show the interaction between the active agent and exogenous-RNA, BME treated PEs from *T. torosa* were used. PE gives a minimal reaction to pure K-RNA. Upon short contact with BME, this PE can hardly undergo differentiation. However, when the sandwich explant is prepared in Niu–Twitty solution containing K-RNA and kept 2 hr in the same solution, the PE alone would develop into spinocaudal tissues (table 4, column 3). This result has led the authors to believe that K-RNA is responsible for the induced differentiation and BME functions in some unknown way as a sensitizer of PE. Were the PE properly sensitized, for example the 12-hr-BME-treated PE of *T. torosa,* K-RNA would interact with the active agent of BME. As a result, the BME-induced developmental potentiality of PE will be altered in the same way as PE of *T. tigrinum* described above. The K-RNA-induced increase of pronephros formation is abolished by the treatment of pancreatic RNase. Of particular interest is the observation that the kinds of tissues formed in the BME : K-RNA(2) series by and large resemble those in the BME(12) : K-RNA(2) series more than the BME(12) series. The authors has taken this point to support the argument that K-RNA causes embryonic induction [12].

The function of pure RNA is poor in living tissue. In combination with protein, its activity could increase up to 1000 times, for instance, the composition of tabacco mosaic virus is 94% protein and 6% RNA. Upon dissociation into RNA and protein, the active fraction is RNA [11]. The infectivity is one thousandth of the intact viral particles. However, the infectivity can be restored by the restitution of viral particles using RNA of one strain with protein from another, or vice versa. The reconstituted virus will produce new particles containing RNA and protein of the RNA strain [13]. Similarly, PE reacts to pure thymus-RNA poorly. When albumin is used together with RNA, the inducing activity increased significantly [2]. All these proteins are important in the manifestation of RNA function. The manner by which they work is not known at the present time.

PE from *T. pyrrhogaster* does not undergo autoneuralization. After contact with BME for 3 hr, it acquired the capacity to differentiate into trunk-mesodermal derivatives, but the frequency was low. Attempts to transfer this acquired property were made. Success came from the BME-treated PEs which were further cultured in N–T solution for 12 hr (table 8). As can be seen, the homogenate of C_{12} caused the formation of brain and neural tissue [14] and E_{12} initiated the formation of brain and trunk-mesodermal tissues as well. Remarkably the quality and quantity of the trunk-mesodermal tissues do not differ from those of the 3-hr-BME-treated PE. It appears therefore that the homogenate mediates the transfer of the BME action to the reacting PE. The active component of the homogenate has not been analyzed chemically because of the small size of the PE explants. However, RNase abolishes its activity. Furthermore, there is no inductive activity in those explants cultured in N–T solution containing actinomycin D, thus resulting in the loss of differentiation.

References

[1] M.C. Niu, Cellular mechanism in differentiation and growth, Ed. Rudnick (Princeton Univ. Press, (1956) 155–171.
[2] M.C. Niu, Proc. Natl. Acad. Sci. U.S. 44 (1958) 1264.
[3] T. Yamada, In: A symposium on chemical basis of development, Eds. McElroy and Glass (Johns Hopkins University Press, 1958) 217–238.
[4] H. Tiedemann, In: The Biochemistry of animal development, Vol. 2, Ed. Weber (Academic Press, New York, 1967) 3–55.
[5] L. Saxen and S. Toivonen, Primary embryonic induction (Academic Press, New York, 1962).
[6] M.C. Niu and N. Sasaki, Exptl. Cell Res. 64 (1971) 57.
[7] M.C. Niu and N. Sasaki, Exptl. Cell Res. 64 (1971) 65.
[8] R.A. Mather, Statistical analysis in Biology (Methuen, London, 1966).
[9] V.C. Twitty, In: Analysis of development, Eds. Willier, Weiss and Hamburger (Saunders, Philadelphia, 1955) 402–414.
[10] P.D. Nieuwkoop, Acta Embryol. Morphol. Exptl. 2 (1958) 13.
[11] A. Gierer and G. Schramm, Z. Naturforsch. 11B (1956) 138.
[12] M.C. Niu, In: Axenic mammalian cell reactions, Ed. Tritsch (Marcel Dekker Press, New York, 1969) 155–180.
[13] H. Fraenkel-Conrat, J. Am. Chem. Soc. 78 (1956) 882.
[14] N. Sasaki and S. Iyeiri, Inducing activity of the mesodermalized ectoderm of *Triturus pyrrhogaster,* manuscript prepared for publication (1973).

Session Five

Nucleic Acid-Induced Changes in Living Systems

Chairman

MARSHALL NIRENBERG

National Heart and Lung Institute,
National Institutes of Health
Bethesda, Maryland

Niu and Segal (eds.). The role of RNA in reproduction and development
North-Holland Publ. Co., 1973

Transforming RNA as a template directing RNA and DNA synthesis in bacteria*

Mirko BELJANSKI* and Michel PLAWECKI†
Institut Pasteur, Paris, France

Attempts were made to approach the mechanism by which the transfer of information is carried by transforming RNA from *Escherichia coli*. Transforming RNA, capable of inducing inherited changes in recipient cells, is used in vitro as a template by two distinct enzymes which mediate the transfer of information from RNA to RNA and DNA respectively. Using transforming RNA as a template, polynucleotide phosphorylase insensitive to rifampicin, synthesizes a product which was characterized as being a "copy" of template RNA. Reverse transcriptase, which can be physically separated from DNA polymerase, transcribes the transforming RNA into a complementary DNA product. According to hybridization experiments, DNA from *Agrobacterium tumefaciens* transformed by *E. coli* transforming RNA seems to contain one or less than one copy of complementary DNA.

1. Introduction

Discoveries of particular RNA species in chromosomes [1], plasmids [2], plants [3–5] and bacteria [6, 10] have opened a new avenue to study the transfer of information in biological systems. Thus "viroid RNA" [3, 5], free from proteins (causing diseases in plants) and transforming RNA inducing transformation in bacteria [6, 7] are both excellent candidates for Temin's provirus [8] and protovirus hypotheses [9]. Our investigations have established that genetic information can be transferred to different bacterial species by a specific transforming RNA found as a product excreted into the culture medium of showdomycin resistant mutants of *Escherichia coli* [6, 7, 11]. Transformants, which appear at a high rate, exhibit physiological and biochemical changes of unexpected magnitude [7] especially illustrated by transformed *Agrobacterium tumefaciens* B_6 [7, 13]. These trans-

* His work was supported by the Centre National de la Recherche Scientifique, the Ligue Nationale contre le Cancer, Energie Atomique and the Pasteur Institute.
** Mirko Beljanski is Maître de Recherches of the CNRS.
† Michel Plawecki is a research fellow of the Ligue Nationale française contre le Cancer.

Abbreviations: RNase, pancreatic ribonuclease; DNase, pancreatic deoxyribonuclease; RNA, ribonucleic acid; DNA, deoxyribonucleic acid; XDP, ribonucleoside-5′-diphosphate; XTP, ribonucleoside-5′-triphosphate; d-XTP, deoxyribonucleodide-5′-triphosphate; Sho-R, showdomycin resistant; PNPase, polynucleotide phosphorylase.

formants have acquired new characteristics expressed in profoundly altered ribosomal RNA (rRNAs), ribosomal proteins and certain enzymes [7, 11, 12]. Unexpectedly, the transformants of the oncogenic strain *A. tumefaciens* B_6 have partially or completely lost their capacity for tumor induction in plants [7].

In order to explain the origin of transforming RNA and the process by which it induces *inherited changes* in recipient bacteria we raised the two following questions:

(1) Does the transforming RNA direct RNA or DNA synthesis in the presence of specific enzymes?

(2) Does the transformation by RNA operate through an RNA to DNA pathway?

The present work deals with studies conducted in order to determine whether transforming RNA can be used in vitro as a template by two distinct enzymes which mediate the transfer of information from RNA to RNA and to DNA respectively. Pathways through which transforming RNA could transform the recipient bacteria are discussed.

The most striking property of transforming RNA (excreted RNA from *E. coli* Sho-R) is that at a low concentration (0.1 μg/ml) it rapidly induces in *E. coli* the transformation of the whole recipient population, leading to stable transformants which express new biological properties [6, 7, 12]. This same RNA fraction also transforms *A. tumefaciens* with high efficiency into partial and complete transformants with characteristics which have never been observed before in wild types of the same species [7, 13]. It is also remarkable that partial transformants of *A. tumefaciens* become completely transformed during further growth in the absence of transforming RNA [15].

The techniques for isolation, purification and characterisation of transforming RNA as well as the procedure for transformation of bacteria have been described elsewhere [6, 10]. It should be added that recipient bacteria are usually harvested before the stationary phase of growth and that the synthetic medium used for transformation should contain a rather low amount of carbon source in order to achieve rapid and efficient transformation of recipient cells.

2. *In vitro replication of transforming RNA*

Excreted RNA and episomal RNA* [6, 10], both carrying the genetic information for transformation differ from all other RNA species found in *E. coli.* In particular, their base ratio G+A/C+U of 1.70 to 2.0 differs considerably from that of RNA's from the wild type (G+A/C+U = 1.0). This characteristic indicates that purine-rich RNA is not complementary to DNA of the same bacterial species. This was already shown by the absence of hybridization between ribosomal RNA's rich in purines from *E. coli* Sho-R and DNA of the same strain. This characteristic implies that the mechanism by which the transforming RNA is synthesized and replicated has to be

* Episomal RNA is a transforming RNA bound to DNA.

different from that proposed for other RNA species. Thus DNA-dependent RNA polymerase should not be the enzyme involved in replication of the transforming RNA. In fact, in vitro experiments show that this is the case.

As an enzyme which would replicate the transforming RNA, polynucleotide phosphorylase (PNPase) was a likely prospect for two reasons. First, PNPase from wild type bacteria grown in the presence of showdomycin has modified properties, i.e. in the presence of equivalent amounts of all four ribonucleoside-5′-diphosphates (XDP) it synthesizes a polymer whose content of purine nucleotides exceeds that of pyrimidine nucleotides [14]. Second, PNPase from *E. coli* M 500 Sho-R, synthesizes in vitro an AGUC polymer in which the amount of purines is twice that of pyrimidines. It should be recalled that "episomal RNA" and excreted RNA from M 500 Sho-R contain purine nucleotides in excess and these two RNA fractions possess an equivalent genetic potential toward recipient cells [6, 10].

Preliminary experiments showed that transforming RNA was not used as a template by DNA-dependent RNA polymerase. We considered the possibility that it was used as a template by the PNPase. If it was, the product synthesized by wild type PNPase in the absence of template RNA might have a different base composition from that synthesized by the same enzyme in the presence of transforming RNA. We were thus led to search for conditions in which PNPase from wild type bacteria (wild enzyme) would replicate the transforming RNA in vitro. Although there is an endogenous activity of PNPase in the absence of any kind of RNA, addition of transforming RNA to the reaction mixture results in a several fold stimulation of the enzyme activity (fig. 1). Remarkably, the stimulating effect with transforming RNA is observed only if all four XDP's are present in the reaction mixture. No effect is observed in the presence of all four XTP's. High enzyme purity is not required, since crude or 250-fold purified preparations respond in the same fashion to the presence of transforming RNA [14].

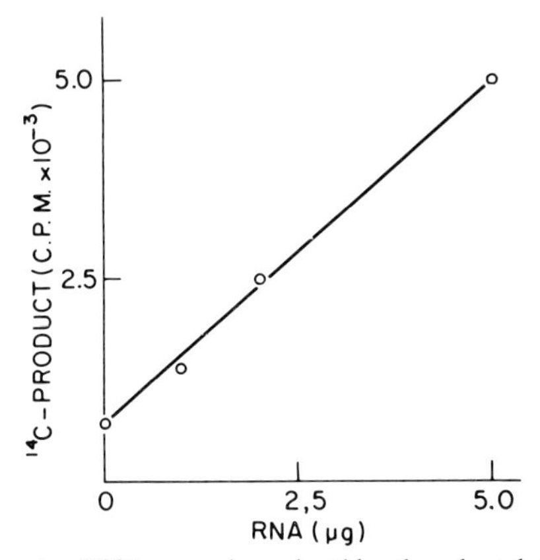

Fig. 1. Effect of transforming RNA on polynucleotide phosphorylase activity. Incubation conditions, see legend to table 1. Transforming RNA excreted by showdomycin resistant *E. coli* was used. Time, 30 min at 36°C.

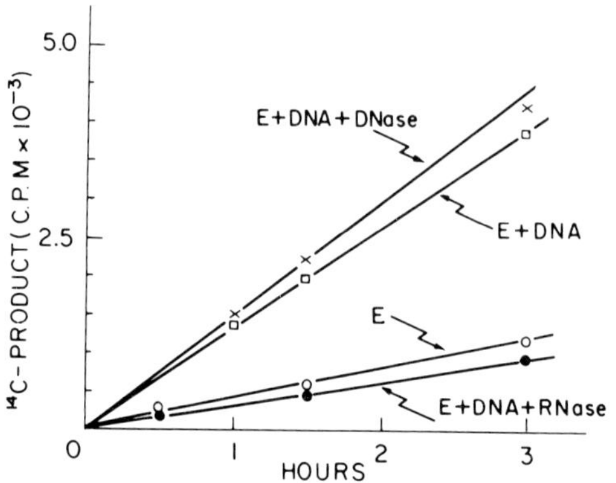

Fig. 2. Effect of episomal RNA on polynucleotide phosphorylase activity. PNPase was isolated and purified from *E. coli* Hfr wild type as described [14]. Incubation conditions, see legend to table 1. All four ^{14}C-XDP were used. Aliquots taken at indicated intervals were mixed with TCA (5%) and the precipitate washed on GF/C glass filters. After drying, the radioactivity was determined in a Packard liquid spectrometer.

The effect of various bacterial RNA species on the endogenous activity of PNPase has been tested. None of these RNA's was stimulatory [14]. However DNA isolated from wild type *E. coli* exhibits a strong stimulatory effect on PNPase activity (fig. 2). But such a DNA preparation has to be purified under appropriate conditions [10] in order to contain no more than 7–8% of RNA (orcinol reaction).* It still carries the episomal RNA from which the excreted RNA originates. The stimulatory effect remains even if the DNA is extensively treated with pancreatic DNase.

TABLE 1
Replication of episomal RNA by polynucleotide phosphorylase in the presence of rifampicin

	Nanomoles of ^{14}C-nucleotides incorporated in 10 min	
	4-XDP	4-XTP
Complete	5.80	0.76
+ rifampicin 10 μg	5.85	0.80
+ rifampicin 20 μg	5.70	–
+ rifampicin 40 μg	5.80	0.70

Incubation mixture contains per 0.2 ml: Tris (pH 8.0) 100 μM; Mg^{2+} 2 μM; XDP 0.25 μM of each, 2.5×10^4 cpm. DAN from *E. coli* (wild type) 20 μg. DNase 10 μg. Enzyme (G-200 Sephadex) [14] 100 μg; rifampicin added at the beginning. Time of incubation: 30 mn at 37°C (except when indicated). ^{14}C-product was precipitated with trichloroacetic acid, washed, filtered on GC/F glass filters, dried and radioactivity determined in a Packard liquid spectrometer.

* If DNA preparation contains more than 10% of RNA it does not stimulate the activity of PNPase.

TABLE 2
^{14}C product synthesized by polynucleotide phosphorylase in the presence of RNase

	Moles per 100 moles of nucleotides	
	^{14}C-product with enzyme alone + RNase	^{14}C-product with enzyme + DNA + RNase
A	23.5	31.0
G	50.0	40.5
C	14.3	14.3
U	12.1	14.2
G+A/C+U	2.79	2.51

^{14}C-polymer was synthesized in the presence of RNase (20 μg/0.2 ml) without and with DNA carrying the episomal RNA (see legend to table 1). ^{14}C-polymer was precipitated with TCA, washed several times with TCA (5%), hydrolyzed with N HCl at 100°C for 1 hr. Degraded material was separated and analysed [43].

The effect of increasing concentrations of template RNA is given in fig. 1. With purified PNPase (50–100 μg) the rate of polymer synthesis is linear during several hours. Fragmented RNA (after treatment by pancreatic RNase) has completely lost its stimulating activity. The ^{14}C labelled product, synthesized in the presence of RNase (independently if template RNA is present or not) is mainly made from AMP and GMP nucleotides (table 2).

The absence of effect of *rifampicin* on PNPase activity (table 1) clearly shows

TABLE 3
Base analysis of transforming RNA (associated with DNA) and ^{14}C RNA product

Nucleotides	Moles per 100 moles of nucleotides		
	Polymer with enzyme alone	Polymer with enzyme + DNA	Episomal RNA (*E. coli* wild type)
A	23.4	29.7	29.6
G	26.7	35.2	34.4
C	21.5	17.6	18.0
U	28.4	17.5	18.0
G+A/C+U	1.01	1.80	1.73

The base ratio of transforming RNA (episomal RNA) was determined using a Dowex column as previously described [10]. ^{14}C-polymer synthesized in vitro (see legend to table 1) was precipitated and washed several times with 5% TCA, hydrolysed with 0.3 N NaOH and separated on a Dowex column (20.000 cpm per sample) in the presence of 1 mg of unlabelled ribosomal RNA hydrolysed with 0.3 N NaOH [21]. Ratio of radioactive nucleotides is presented.

that the synthesis of polymer is not catalyzed by DNA dependent RNA polymerase. It is well established that rifampicin inhibits the activity of this latter enzyme [15].

3. *Analysis of the ^{14}C product synthesized in the absence and presence of transforming RNA bound to DNA*

3.1. Base ratio analysis

The ^{14}C labelled product was synthesized with wild type PNPase in the presence of equivalent amount of ^{14}C-labelled XDP and in the presence and absence of transforming RNA. The ^{14}C-product was separated from the enzyme as described in the legend to table 3, and washed. The dialysed ^{14}C product was degraded by alkali and the nucleotides separated on a Dowex-column.

Table 3 shows that ^{14}C-product synthesized in the presence of transforming RNA (episomal or excreted RNA) has base ratios close or identical to those found for transforming RNA itself. In the absence of template RNA, base ratios are completely different. These results show that the replication of the transforming RNA is accomplished by PNPase through a mechanism in which complementarity of bases is not observed.

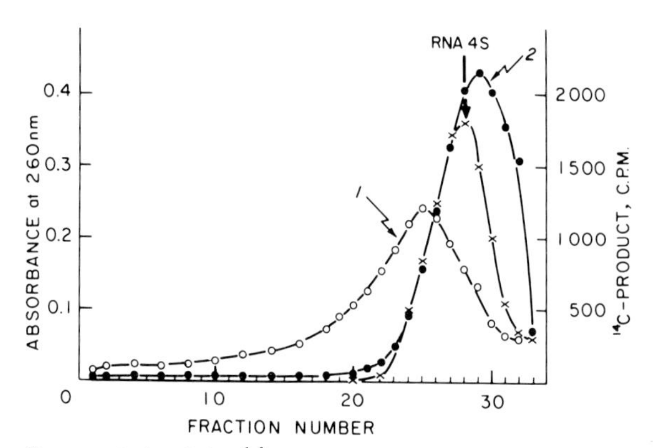

Fig. 3. Sucrose gradient analysis of the ^{14}C-product made with polynucleotide phosphorylase. The ^{14}C product was synthesized as described in the legend to table 1 using four labelled ^{14}C-ribonucleoside-5′-diphosphates. The ^{14}C product was extensively dialysed against 0.2 M KCl in distilled water and sedimented in a 5–20% linear sucrose gradient at 4°C in the Spinco L SW_{39} rotor at 25,000 rpm for 16 hr. Fractions were collected and the radioactivity determined. Transfer RNA^{met} was included as marker. (1) = ^{14}C polymer synthesized in the absence of template. (2) = ^{14}C polymer synthesized in the presence of DNA + DNase.

3.2. Sucrose gradient analysis

Fig. 3 illustrates the profile of the ^{14}C-polymer synthesized in the absence and presence of template (DNA + DNase). 4 S RNA was included as internal marker in a linear sucrose gradient. One sees that ^{14}C polymer synthesized upon template sediments at a position different from that of the ^{14}C polymer made in the absence of template. Further analysis showed that the ^{14}C product is not bound to DNA, that it is not double stranded [16]. However it is rather resistant to pancreatic RNase due to the fact that it contains an excess of purine nucleotides.

4. *Transcription of transforming RNA into DNA by reverse transcriptase*

Transformation of bacteria by transforming RNA raised the following question: could the transforming RNA be transcribed into DNA? To answer this question it was necessary to determine (a) to what extent transforming RNA was contaminated by DNA and (b) to search for an enzyme fraction which would transcribe the RNA into DNA.

4.1. Does the transforming RNA contain DNA?

Transforming RNA purified as previously described [5] was analysed for DNA content. Even when large amounts of RNA were used the diphenylamine reaction was negative (table 4). However when the excreted RNA was isolated from culture medium in which *E. coli* Sho-R had been grown in the presence of ^{14}C-thymidine, radioactivity was detectable in the RNA preparation and corresponded to a DNA content of around 0.05% on the basis of the specific radioactivity of newly synthesized bacterial DNA. No further attempts were made to characterise the radioactive material. Transforming RNA, centrifuged in Cs_2SO_4 gradient, sedimented in the region of RNA and no detectable amount of U.V. material or radioactivity was found in DNA density region (fig. 4). A small contaminating DNA product, if present in the transforming RNA, could be considered as a product of transcription of the transforming RNA.

4.2. Search for reverse transcriptase

In our preliminary experiments we found that an enzyme fraction, present in the 105,000 *g* supernatant from *E. coli*, was capable of polymerizing deoxyribonucleotides in the presence of transforming RNA used as template, and this suggested that the synthesized product was a DNA like material [17]. It was particularly interesting to find out if such an enzyme fraction from *E. coli* and *A. tumefaciens* would distinguish the transforming RNA from all other RNA species from various sources

TABLE 4
DNA content of template RNA(?) and RNA content of DNA

	Ratio 260/280	U.V. absorption at 260 nm	Orcinol reaction	Diphenylamine reaction
Transforming RNA from *E. coli*	2.06	100 μg	162 μg	not detectable [a]
5.5 S RNA from *Alcaligenes faecalis*	2.15	206 μg	204 μg	not detectable [a]
DNA from *E. coli*	2.1	280 μg	26 μg	256 μg

[a] 400 μg of RNA (on the basis of U.V. absorption) were used for diphenylamine reaction.

Several samples and different concentrations of transforming RNA and *Alcaligenes faecalis* 5.5 S RNA [41] were analysed for DNA content. Agreement between U.V. determination and orcinol reaction are presented. DNA isolated under appropriate conditions [10] always contains RNA.

and especially from DNA. The behavior of reverse transcriptase from *A. tumefaciens* toward *E. coli* transforming RNA was of great interest, since this RNA induces inherited changes in *A. tumefaciens.*

5. *Physical separation of reverse transcriptase from DNA dependent DNA polymerase*

The observation that transforming RNA directed DNA synthesis in the presence of a soluble fraction from *E. coli* [17] strongly suggests that this fraction contained a

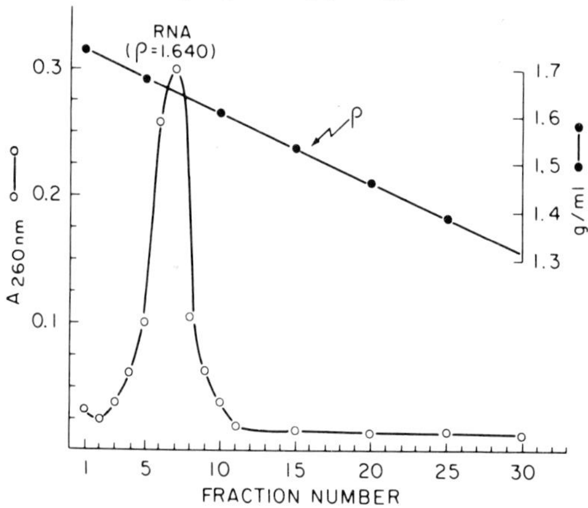

Fig. 4. Centrifugation of transforming RNA in cesium sulfate equilibrium gradient. Transforming RNA (25 μg) in Tris buffer was mixed with Cs_2SO_4 (1.8 g pH 7.3 fin. vol. 3.1 ml) and centrifuged at 20°C (30,000 rpm) for 64 hr in Spinco SW_{39} rotor. Fractions were collected and analysed for refractive index, absorbance at 260 nm and acid-precipitable radioactivity.

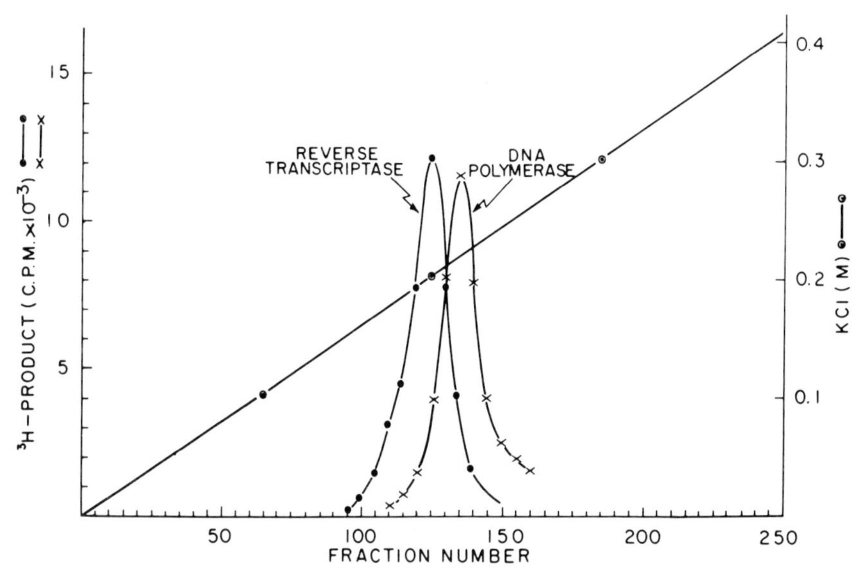

Fig. 5. Chromatography of reverse transcriptase and DNA dependent DNA polymerase on DEAE-cellulose. *E. coli* ML 30 Sho-R cells were grown and disrupted as described [21]. A crude cell extract was centrifuged at 20,000 *g* in the SS1 Servall centrifuge for 30 min. The pellet was discarded. The supernatant was precipitated with $SO_4(NH_4)_2$ (70%) and the precipitate was collected by centrifugation and dialysed against Tris-HCl buffer 10^{-2} M (pH 7.6) and mercaptoathanol (10^{-4} M) containing $MgCl_2$ 10^{-3} M and KCl 0.06 M. This dialysed preparation was chromatographed. 4 g of proteins were applied to the DEAE column (54 × 3 cm) equilibrated with Tris-HCl buffer 10^{-2} M (pH 7.6). After washing the column with the same buffer, it was eluted with a linear gradient: Tris 10^{-2} M (pH 7.6) – Tris 10^{-2} M (pH 7.6) containing 0.5 M KCl. 5 ml fractions were collected and the enzyme activity was assayed with 0.05 ml of each fraction under conditions described in the legend to table 6. Blank values around 100–200 cpm were subtracted. Transforming RNA (2 μg) and thymus DNA (2 μg) were tested as templates.

TABLE 5

Effect of potassium ions on reverse transcriptase activity from *E. coli* and *A. tumefaciens*

Concentration in potassium chloride (M)	Per cent of activity (cpm)	
	E. coli	*A. tumefaciens*
0.01	100	100
0.02	100	74
0.03	97	35
0.04	96	12
0.06	80	9

Incubation conditions, see legend to table 6. ^{3}H d-CTP was used. Excreted RNA (1 μg) from *E. coli* ML 30 Sho-R); enzyme fraction (DEAE, see fig. 6), 50 μg.

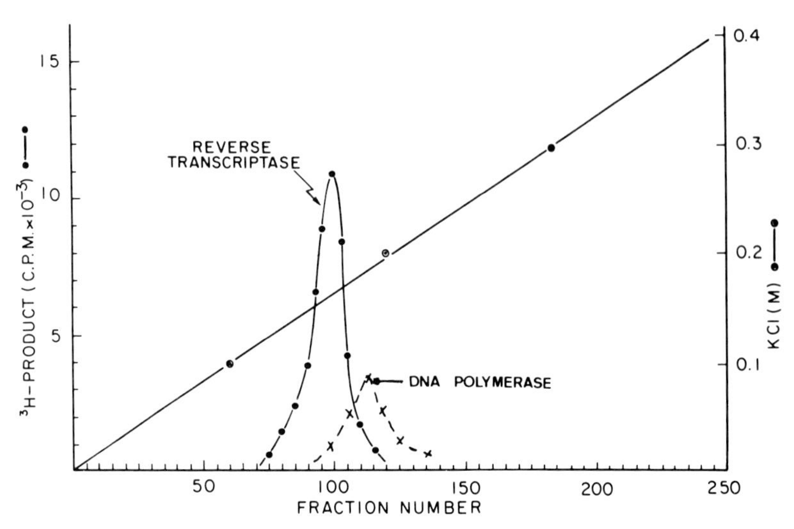

Fig. 6. Chromatography of RNA dependent reverse transcriptase from *A. Tumefaciens* B_6 on DEAE cellulose column. Bacteria were grown in rich medium [10] and the cells broken in the French Press. An ammonium sulfate fraction was prepared as described for *E. coli* in the legend to fig. 5. Chromatography conditions were identical to those described in the legend to fig. 5.

TABLE 6

^{3}H product formation from deoxyribonucleotide-5'-triphosphates in the presence of reverse transcriptase from *E. coli* and *A. tumefaciens*

Reaction mixture	*E. coli* enzyme (pmoles)	*A. tumefaciens* enzyme (pmoles)
Complete	276	30
– transforming RNA	< 1	< 1
– $MgCl_2$	< 1	< 1
– dGTP	72	7
– dCTP	65	8
– dGTP, dCTP, dTP	< 2	< 1
+ RNase (50 μg) preinc.	53	6
+ DNase (10 μg)	< 1	< 1

Incubation mixture (0.2 ml) contains: $MgCl_2$, 2 μM; Tris-HCl buffer (pH 7.65), 25 μM; each deoxyribonucleoside-5'-triphosphate, 5 nmoles + ^{3}H d-TTP (100,000 cpm); RNA 2 μg; enzyme fraction (DEAE) 40 μg (see fig. 5). Incubation 20 min at 36°C. TCA is added to precipitate the ^{3}H product. Precipitates were washed with TCA 5% filtered on Whatman GC/F glass filters, washed, dried, and the radioactivity measured in a Packard liquid spectrometer.

TABLE 7
Activity of reverse transcriptase in the presence of different polymers

	E. coli enzyme (pmoles)	*A. tumefaciens* enzyme (pmoles)
Complete transforming RNA	206	32
Alcaligenes faecalis RNA (5.5 S)	71	–
E. coli rRNA (23 S + 16 S)	< 1	< 1
$tRNA^{met}$	< 1	< 1
rA-dT (pH 7.65)	36	–
rA-dT (pH 9.0, Mn^{2+})	67	–
polyovirus RNA	< 1	< 1
myeloblastosis virus RNA	< 2	–
poly AGUC	< 1	< 1

Incubation conditions, see legend to table 6. ^{3}H d-GTP was used.

reverse transcriptase activity. Evidence for the presence of such an enzyme is presented here. Fractionation of an *E. coli* soluble fraction (see legend to fig. 5) on a DEAE cellulose column shows the presence of an enzyme which actively uses the transforming RNA (and not DNA) as template for DNA synthesis. Separated from DNA dependent DNA polymerase this enzyme behaves like reverse transcriptase found in the RNA viruses. Fig. 6 shows that *A. tumefaciens* also contains a reverse transcriptase and that the activity of DNA dependent DNA polymerase in eluates from DEAE column is very low under our fractionation conditions.

A. tumefaciens reverse transcriptase uses transforming RNA from *E. coli* as a template for DNA synthesis, although with much less efficiency than the *E. coli* enzyme. In contrast to the *E. coli* enzyme, reverse transcriptase from *A. tumefa-*

TABLE 8
Activity of reverse transcriptase and DNA polymerase in the presence of DNA

	E. coli reverse transcriptase (pmoles)	*E. coli* DNA dependent DNA polymerase (pmoles)
Thymus DNA	16	198
Thymus DNA + RNase	8	186
A. tumefaciens B_6 DNA	8	168
A. tumefaciens B_6 DNA + RNase	6	174
E. coli DNA	21	18
E. coli DNA + RNase	7	20
Transforming RNA	206	7

Incubation conditions, see legend to table 6. DNA used 2 μg; ^{3}H d-GTP (100,000 cpm) was used. Transforming RNA 1 μg.

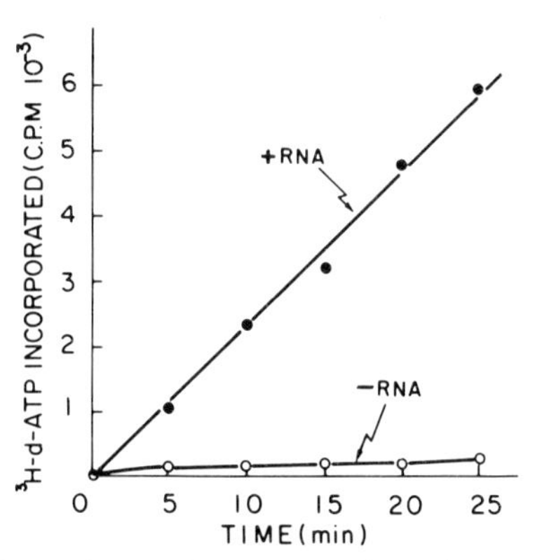

Fig. 7. Synthesis of ^{3}H-DNA product in the presence of transforming RNA as function of time. Incubation conditions, see legend to table 6.

ciens is strongly inhibited by potassium chloride even at rather low concentrations (table 5). However actinomycin D and rifampicin have similar inhibitory effects on the activity of reverse transcriptase from both *A. tumefaciens* and *E. coli.*

6. *Specificity of RNA dependent DNA polymerase for RNA templates*

Reverse transcriptase transcribes the transforming RNA into a DNA like product most actively in the presence of all 4 d-XTP and Mg^{2+} ions (table 6). The template activity of transforming RNA is destroyed by treatment with RNase. DNA polymer does not accumulate in the presence of DNase. Among various RNA or DNA templates tested (tables 7 and 8), transforming RNA is the best template for reverse transcriptase, although 5.5 S RNA (rich in messenger of transforming RNA) serves as active template for polymerisation of the d-XTP into TCA precipitable material. Time course of enzyme activity is illustrated in the fig. 7.

The amount of ^{3}H-product synthesized in 20 min corresponds roughly to 20% of that of template RNA, although we do not know if all fractions of transforming RNA are transcribed or exclusively one small fraction.

7. *Inhibitory effect of actinomycin D, rifampicin and N-demethylrifampicin on reverse transcriptase activity*

It has already been shown that actinomycin D [18] and some rifampicin derivatives [28] inhibit the activity of viral RNA dependent DNA polymerase, i.e. of reverse transcriptase. Our results with bacterial reverse transcriptase are similar to those found for the viral enzyme. Actinomycin D at a rather low concentration inhibits the DNA synthesis upon transforming RNA when added to the incubation mixture.

TABLE 9

Activity of reverse transcriptase in the presence of actinomycin D and N-demethyl-rifampicin

	E. coli enzyme		*A. tumefaciens* enzyme	
	pmoles	% inhibition	pmoles	% inhibition
Complete	199	–	26	–
+ actinomycin D 2 μg	120	40	21	18
+ actinomycin D 4 μg	87	56	9.2	61
+ actinomycin D 8 μg	86	55	4.1	84
+ rifampicin 10 μg	106	47	–	–
+ rifampicin 20 μg	62	68	8.1	69
+ N-demethylrifampicin 10 μg	108	45	–	–
+ N-demethylrifampicin 20 μg	89	55	–	–

Incubation conditions, as described in the legend to table 6. ^{3}H d-GTP was used, transforming RNA from *E. coli* ML 30 Sho-R (4 μg). Enzyme fraction (DEAE cellulose) 40 μg. Actinomycin D, rifampicin and N-demethylrifampicin were added at the beginning of the reaction.

Somewhat higher concentrations of rifampicin and N-demethylrifampicin also inhibit RNA dependent DNA synthesis (table 9). In addition to that further evidence (fig. 5) shows that reverse transcriptase is not DNA dependent DNA polymerase.

TABLE 10

Base analysis of DNA synthesized by reverse transcriptase from *E. coli* and *A. tumefaciens*

	Moles per 100 moles of nucleotides			
	Transforming RNA (*E. coli*)	^{3}H-DNA (*E. coli* enzyme)	^{3}H-DNA (*A. tumefaciens* enzyme)	DNA (*E. coli*)
A	31.0	17.3	17.7	24.5
G	33.0	18.8	18.8	24.8
C	18.0	30.8	30.3	24.6
U (T)	17.8	33.1	32.1	26.1
	$\frac{G+A}{C+U} = 1.76$	$\frac{C+T}{G+A} = 1.74$	$\frac{C+T}{G+A} = 1.71$	$\frac{C+T}{G+A} = 1.01$

Incubation conditions (see legend to table 5). Each ^{3}H-dXTP (100,000 cpm) was used in equivalent amount. Acid precipitable material was washed several times. Dialysed ^{3}H-DNA product was hydrolysed in 70% formic acid for 30 min at 173°C and the hydrolysate chromatographed on Whatman 1 paper. Each spot was detected with an ultraviolet lamp, eluted with 0.1 HCl and the radioactivity determined in a Packard liquid spectrometer.

8. Nature of DNA synthesized upon transforming RNA

Data summarized in table 10 and fig. 13 show that transforming RNA is transcribed into the complementary DNA like material in the presence of reverse transcriptase isolated either from *E. coli* or *A. tumefaciens*. This is demonstrated by two sets of results. First, base ratio analysis of the ^{3}H-DNA product (table 10) is close if not identical to that found for transforming RNA. Second, evidence was obtained by annealing the ^{3}H-DNA product with transforming RNA and analysis by equilibrium sedimentation in Cs_2SO_4 density gradient (fig. 13). Synthetic RNA-^{3}H-DNA hybrid sediments in the density region between RNA and DNA as does the enzymatically formed hybrid (see fig. 8).

A synthetic hybrid was not detected when transforming RNA was replaced by ribosomal RNA or bacterial DNA in annealing experiments (fig. 13).

9. Density gradient analysis of the ^{3}H product synthesized in vitro

To provide evidence that the ^{3}H-heterodeoxypolymer synthesized in vitro under optimal conditions is DNA like it was separated from the enzyme by repeated treatments with chloroform, dialysed and analysed [11]. It was also submitted to equilibrium density centrifugation in Cs_2SO_4; unlabelled transforming RNA and DNA from *E. coli* were used.

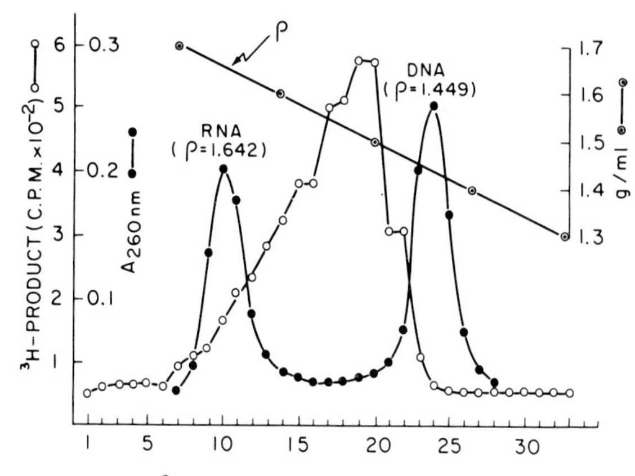

Fig. 8. Centrifugation of the ^{3}H-DNA product synthesized for 10 min (Cs_2 SO_4 equilibrium gradient). The incubation mixture (legend to table 6) in which ^{3}H-DNA was synthesized upon transforming RNA (^{3}H-dTTP and ^{3}H d-GTP were used) was twice treated with chloroform + 0.001 M EDTA, centrifuged, and the aqueous solution dialysed against a solution of 0.1 M KCl in Tris buffer 10^{-2} M (pH 7.6). Around 5500 cpm (TCA precipitable material) was centrifuged (legend to fig. 4). Internal markers, transforming RNA and *E. coli* DNA were included.

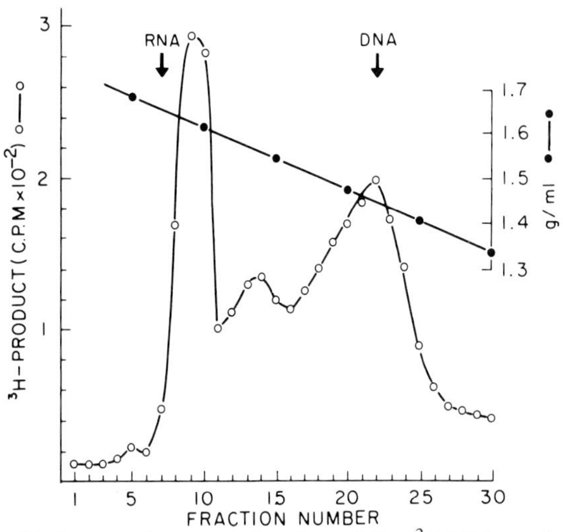

Fig. 9. Cs_2SO_4 equilibrium gradient; centrifugation of the ^{3}H-DNA product synthesized for 30 min. Conditions described in the legend to fig. 8. ^{3}H-DNA product (2300 cpm) was used.

The results of neutral Cs_2SO_4 density gradient sedimentation analysis of the products synthesized during in vitro incubations lasting 10, 30 and 60 min are respectively shown in fig. 8, 9 and 10. In all three cases most of the ^{3}H product is

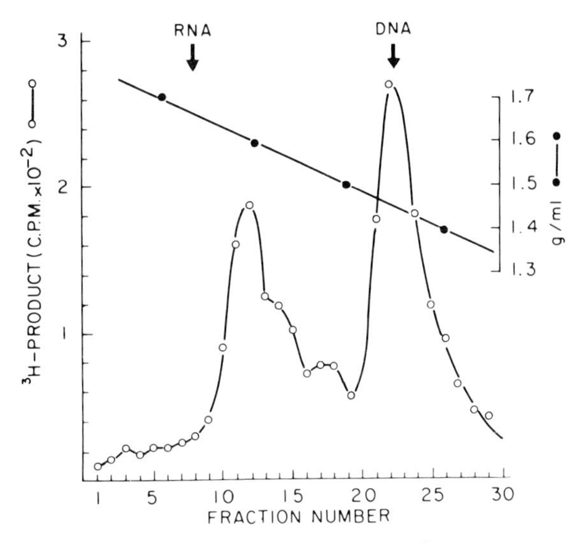

Fig. 10. Cs_2SO_4 equilibrium gradient: centrifugation of the ^{3}H-DNA product synthesized for 60 min. Conditions described in the legend to fig. 8. ^{3}H-DNA product (2400 cpm) was used.

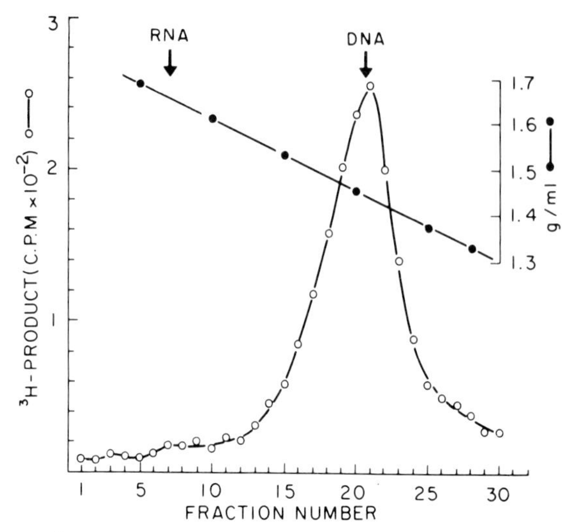

Fig. 11. Profile of the ^{3}H-DNA product in alkaline Cs_2SO_4 equilibrium gradient. The ^{3}H-DNA product was incubated in NaOH 0.3 M at 37°C for 16 hr. For centrifugation see legend to fig. 4.

largely spread over density region between RNA and DNA. Practically no ^{3}H material is found at the density of RNA itself. The heterogenous distribution of the ^{3}H product is due to its existence in the form of RNA–DNA hybrids [19, 20] with different chains lengths.

It is remarkable that the amount of ^{3}H product present in the DNA density

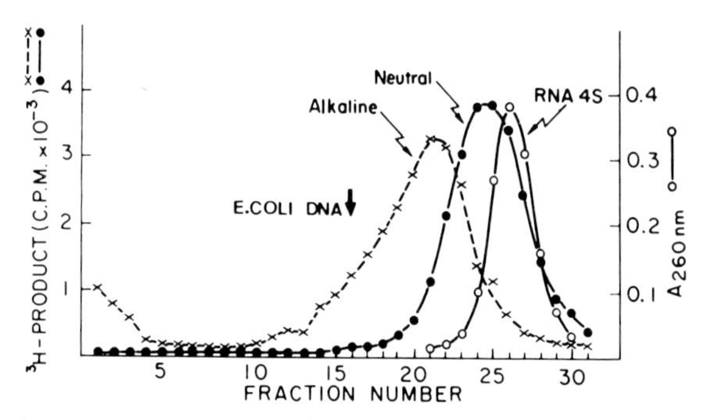

Fig. 12. Sucrose gradient analysis of the ^{3}H-DNA product before and after treatment with alkali. ^{3}H-DNA product was isolated from a 20 min incubation mixture and dialysed as described in the legend to fig. 8. An aliquot was incubated at 37°C with 0.3 N NaOH for 16 hr then layered on a linear 5–20% sucrose gradient containing 0.3 M NaOH + 0.001 M EDTA. A second untreated aliquot was layered on a neutral sucrose gradient. Both gradients were centrifuged at 4°C for 16 hr at 25,000 rpm in the SW 39 Spinco rotor. Fractions were collected, and alkali stable radioactivity determined. The neutral gradient contained tRNA^{met}, as RNA marker.

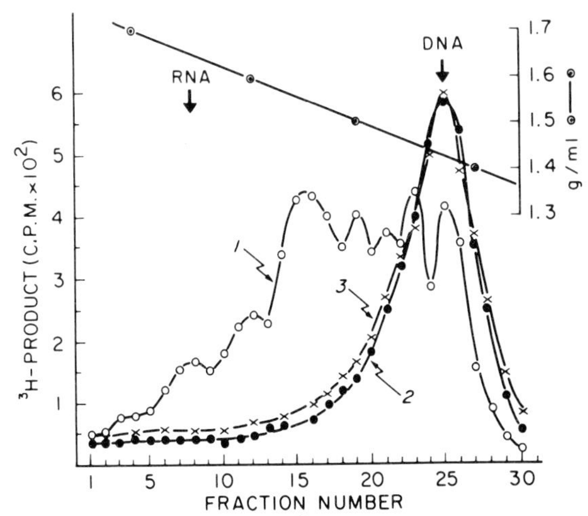

Fig. 13. Synthetic hybrid formation between ^{3}H-DNA and transforming RNA. The ^{3}H-DNA product lasting 10 min was synthesized and isolated from the incubation mixture as described in the legend to fig. 8. Dialysed solution containing the ^{3}H-DNA product was treated with 0.3 N NaOH for 16 hr to eliminate the template RNA, then neutralized before used. The ^{3}H-DNA (5,600 cpm) was incubated at 66°C for 8 hr in 2 SSC: (1) synthetic hybrid, transforming RNA–^{3}H-DNA (50 μg of transforming RNA); (2) ^{3}H-DNA after annealing with *E. coli* DNA (50 μg of *E. coli* DNA); (3) ^{3}H-DNA after annealing with ribosomal RNA(50 μg rRNA).

region increases with the time of incubation. This means that DNA synthesized on the RNA template is progressively released as free DNA.

After treatment of the ^{3}H product–RNA hybrid with alkali, the heterogeneity of the ^{3}H product disappears and the radioactive material is found in the *E. coli* DNA density region (fig. 11). Since the curve (peak) representing the ^{3}H material is rather large this would suggest that the population of DNA is heterogenous.

Sucrose gradient analysis of the RNA–^{3}H-DNA hybrid product before and after treatment with alkali shows (fig. 12) that the alkali-released ^{3}H-DNA from the hybrid sediments between 6 and 7 S. These values correspond to those determined for transforming RNA itself [7]. On this basis one can estimate that the molecular weight of synthesized DNA is close to 1.8×10^5.

10. Search for homology between the ^{3}H-DNA product made in vitro and bacterial DNA

Transcription of transforming RNA into DNA by the reverse transcriptase raised the following question: is there homology between ^{3}H-DNA product synthesized upon transforming RNA and DNA originating from the same strain or from the *A. tumefaciens* transformants?

For hybridization experiments with bacterial DNA the ^{3}H-DNA product was synthesized in the presence of actinomycin D in order to restrict the synthesis to single stranded DNA. Several attempts were made to demonstrate the homology between enzymatically synthesized ^{3}H-DNA product and DNA isolated either from

TABLE 11
Hybridization of ^{3}H-DNA product with bacterial DNA

Input of bacterial DNA (μg)		Input of ^{3}H-DNA (cpm) (0.04 μg)	DNA/ ^{3}H-DNA ratio	Cot [40]	^{3}H-DNA hybridized with bacterial DNA	
					cpm	per cent
E. coli Sho R	100	4200	2.5×10^3	0.8×10^4	280	6.6
	200	4200	5.0×10^3	1.6×10^4	310	7.4
A. tumefaciens transformant	300	4300	7.5×10^3	2.5×10^4	2160	50
	600	4300	1.5×10^4	5.0×10^4	2740	64
A. tumefaciens wild type	300	4200	7.5×10^3	2.5×10^4	145	3.4
	600	4200	1.5×10^4	5.0×10^4	163	3.8

Both ^{3}H-DNA and bacterial DNA dissolved in 0.2 × SSC were denaturated by incubating in 0.3 N NaOH at 37°C for 16 hr. Samples were dialysed against 1 SSC overnight. ^{3}H-DNA (about 0.04 μg, 4300 cpm) was incubated with DNA from various origins in 2 SSC [21] at 66°C for 6 days (in sealed tubes). Separation of single stranded from double stranded DNA was achieved using hydroxyapatite column [39]. ^{3}H-DNA incubated alone gave around 1% of double stranded material. We considered it as negligible. Radioactivity of eluates was determined in a Packard liquid spectrometer. Results are expressed in cpm (per cent) for the ^{3}H-DNA found in hybrid form with bacterial DNA. Recovery from hydroxyapatite column was 85–90%.

E. coli mutant Sho-R or from *A. tumefaciens* transformed by RNA from *E. coli*. The ^{3}H-DNA product was annealed at 66°C ("Cot" values*, table 11) with DNA from *E. coli* Sho-R from which transforming RNA is excreted into culture medium. The most surprising fact is that under described conditions the presence of ^{3}H-DNA–DNA hybrid was hardly detectable when *E. coli* Sho-R or *A. tumefaciens* wild type DNA was used. On the contrary when the ^{3}H-DNA product was annealed with cold DNA from *A. tumefaciens* transformed by *E. coli* RNA, significant binding of labelled DNA was observed (table 11). These results seem to show that in *A. tumefaciens* transformant DNA synthesized upon transforming RNA from *E. coli* is "integrated" into genome DNA. Although it is very difficult to determine the exact amount of the "integrated DNA" into bacterial genome, one can say that there is one or less than one copy per genome. This estimation was calculated by considering the molecular weight of bacterial DNA = 2×10^9 [42] and that of ^{3}H-DNA = 1.8×10^5 (see fig. 13). In addition the above results suggest that transforming RNA is transcribed into DNA when introduced into bacteria belonging to a different species.

11. Discussion and conclusions

Evidence has been presented [7] that an RNA fraction from *E. coli* Sho-R can transfer information to *A. tumefaciens* which is then capable of maintaining the newly acquired characteristics.

* Cot = DNA concentration in moles per liter × time of incubation in seconds [40].

In the present report we attempted to approach the mechanism by which the transfer of information carried by transforming RNA is accomplished. We have suggested elsewhere [10, 11] that transforming RNA, once inside the recipient cells, could be linked by "ligases" to other RNA fractions already associated with bacterial DNA. Thus a large RNA episome could be formed and either replicated, giving rise to RNA's, or transcribed into DNA. This hypothesis was suggested for the reason that in living organisms different molecular mechanisms seem to be involved in genetic inheritance [22]. Transformation by RNA may reveal one of the "hidden" systems in the cells, thus allowing this RNA to be in function.

Results reported here show that the replication of transforming RNA in vitro can be accomplished by a polynucleotide phosphorylase. The activity of this enzyme is strongly stimulated by transforming RNA exclusively in the presence of 4 XDP, while there is no activity with 4 XTP. It is remarkable that under the same conditions transforming RNA's, even that associated with DNA (episomal RNA) is replicated whether or not DNase is present in the incubation mixture. In contrast, RNase abolishes the RNA dependent transcription process. Furthermore, rifampicin and actinomycin D inhibit DNA dependent RNA polymerase but not affect the polynucleotide phosphorylase.

The product synthesized on transforming RNA by PNPase has a base ratio (G+A/C+U about 1.70–2.0), close to that found for transforming RNA. On this basis it should be considered as an identical copy of RNA used as template. Similar results have been recently described for replication of RNA from reovirus. In fact either single stranded or double stranded RNA in vitro is replicated by an enzyme from L-cells infected with reovirus [23]. This system uses the 4 XDP's as substrate, although there is another enzyme which uses the 4 XTP's for replication of reoviral RNA. This observation and ours shows that certain biologically active RNA's can be replicated by appropriate enzymes using the 4 XDP as substrate.

It is remarkable that transforming RNA can also serve as template directing in vitro DNA synthesis by an enzyme which we have identified in bacteria as a reverse transcriptase. Results presented here show that the activity of reverse transcriptase can be distinguished from that exhibited by DNA dependent DNA polymerase. This is true for enzymes partially purified from both *E. coli* and *A. tumefaciens.* Among various RNA's tested, transforming RNA whose directing capacity is destroyed by RNase, is the best template for reverse transcriptase activity. It should be emphasized that an RNA fraction (5.5 S RNA) [17] from *Alcaligenes faecalis* rich in messenger RNA – or perhaps containing episome like RNA – is used as template by reverse transcriptase from *E. coli.* Active DNA preparation used as template for DNA dependent DNA polymerase is inactive with the reverse transcriptase. In some respects reverse transcriptase from bacteria has properties similar to those observed with the enzyme whose existence was predicted and demonstrated by Temin [24] in oncornavirus and in normal uninfected chicken embryos [44]: (1) its activity is RNA dependent; (2) its activity is inhibited by actinomycin D and by rifampicin or N-demethylrifampicin [30]; (3) the ^{3}H-DNA product synthesized in vitro is complementary to the RNA used as template.

From the point of view of transformation of *A. tumefaciens* by *E. coli* transforming RNA, it is important that reverse transcriptase from *A. tumefaciens* transcribes in vitro (and possibly in vivo) *E. coli* transforming RNA into DNA, although with much less efficiency than that exhibited by *E. coli* reverse transcriptase. Something of this sort could happen in vivo during transformation of *A. tumefaciens* by *E. coli* transforming RNA.

The ^{3}H product made in vitro by reverse transcriptase was characterized by several means as being DNA like. The most significant observation is that ^{3}H-product has a base ratio complementary to that of transforming RNA (table 10) whether it is synthesized by *A. tumefaciens* or by *E. coli* reverse transcriptase. The ^{3}H-product is a single stranded DNA (roughly 10% double stranded) which hybridizes efficiently with transforming RNA. However this same ^{3}H-product hybridizes very poorly with bacterial DNA from *E. coli* Sho-R. It was suggested that RNA and reverse transcriptase perhaps constitute a system which may not be directly dependent on the bacterial genome [11].

It was of particular interest to determine whether transforming RNA of *E. coli* was transcribed (in vivo) into DNA during the process of transformation of *A. tumefaciens* and if it was integrated into the DNA of the corresponding transformant. If this was the case one would expect to find a hybrid formed between ^{3}H-DNA synthesized in vitro on transforming RNA and DNA isolated from *A. tumefaciens* transformant. By annealing the ^{3}H-DNA product with DNA from transformant we have detected the existence of one or less than one copy per genome of DNA transcribed from the transforming RNA.

If reverse transcriptase operated in vivo as it does in vitro it would offer tremendous possibilities for a new type of mutation mechanism proposed and demonstrated by Temin [8, 24]. His findings were confirmed and further developed by others [25–30]. Along these lines it is of great importance to report that several workers have shown [31–38] that in vitro initiation for replication of DNA by DNA dependent DNA polymerase isolated from bacteria or higher organisms requires an RNA which is "covalently" associated with DNA itself [37, 38]. This RNA seems not to be synthesized by classical DNA dependent RNA polymerase since the system is not sensitive to rifampicin [37]. RNA required for replication of ϕ-X 174 DNA with *E. coli* enzyme contains an excess of purine nucleotides over pyrimidines (G+A/C+U = 2.4). The existence of such an RNA may be related to our finding that both *E. coli* transforming RNA excreted into culture medium and episomal RNA associated with DNA [10] contain excess purine nucleotides (G+A/C+U about 1.70–2.1). This RNA is transcribed into DNA by the reverse transcriptase from *E. coli* and *A. tumefaciens* but is not used by DNA dependent DNA polymerase. It will be interesting to determine to what extent and in what manner the primer RNA required for DNA replication differs from one system to another. Also, it now seems quite necessary to establish the essential relationship between RNA dependent reverse transcriptase and DNA dependent DNA polymerase, in bacteria namely, if the former functions for initiation of DNA replication.

The dual potentialities of transforming RNA, i.e. replication into RNA and transcription into DNA constitute a possible molecular mechanism in the process of evolution, independently of whether transcribed DNA is integrated in some way into cell genome or remains in the cytoplasm as a cytoplasmic genetic element [11].

References

[1] R.C. Huang and J. Bonner, Proc. Natl. Acad. Sci. U.S. 54 (1965) 960.
[2] D.R. Helinski, D.G. Blair, D.J. Sherrat, M.A. Lovett, Y. Kupersztoch and D.T. Kingsbury, In: The 6th Miles Symposium on Molecular Biology, June 8–9, 1972, Baltimore (in press).
[3] T.O. Diener and W.B. Raymer, Science 158 (1967) 378.
[4] W.B. Raymer and T.O. Diener, Virology 37 (1969) 343 .
[5] J.S. Semancik and L.G. Weathers, Nature New Biology 237 (1972) 242.
[6] M. Beljanski, M. Beljanski and P. Bourgarel, Compt. Rend., ser. D 272 (1971) 2167.
[7] M. Beljanski, M. Beljanski, P. Manigault and P. Bourgarel, Proc. Natl. Acad. Sci. U.S. 69 (1972) 191.
[8] H.M. Temin, Proc. Natl. Acad. Sci. U.S. 52 (1964) 323.
[9] H.M. Temin, J. Natl. Inst. 46 (1971) III.
[10] M. Beljanski, M. Beljanski and P. Bourgarel, Compt. Rend., ser. D 272 (1971) 2736.
[11] M. Beljanski and P. Manigault, In: The 6th Miles Symposium on Molecular Biology, June 8–9, 1972, Baltimore (in press).
[12] M. Beljanski, A. Kurkdjian and P. Manigault, Compt. Rend., ser. D 274 (1972) 356.
[13] M, Beljanski, P. Bourgarel and M. Beljanski, Proc. Natl. Acad. Sci. U.S. 68 (1971) 491.
[14] M. Plawecki and M. Beljanski, Compt. Rend. 273 (1971) 827.
[15] W. Wehrli, F. Knusel, K. Schmid and M. Staehelin, Proc. Natl. Acad. Sci. U.S. 61 (1968) 667.
[16] M. Plawecki, unpublished data, Ph. Thesis in preparation.
[17] M. Beljanski, Compt. Rend. 274 (1972) 2801.
[18] I.M. Verma, G.F. Temple, H. Fan and D. Baltimore, Nature 235 (1972) 163.
[19] J. Hurwitz and J.P. Leis, J. Virol. 9 (1972) 116.
[20] S. Spiegelman, A. Burny, M.R. Das, J. Keydar, J. Schlom, M. Travnicek and K. Watson, Nature 227 (1970) 563.
[21] M. Beljanski, P. Bourgarel and M. Beljanski, Ann. Inst. Pasteur 118 (1970) 253.
[22] P. Grasse, L'évolution du vivant. Matériaux pour une nouvelle théorie transformiste (Albin Michel, Paris) in press.
[23] G. Schochetman and S. Millward, Nature New Biology 239 (1972) 77.
[24] H.M. Temin and S. Mizutani, Nature 226 (1970) 1211.
[25] D. Baltimore, Nature 226 (1970) 1209.
[26] M. Rokutanda, H. Rokutanda, M. Green, K. Fujinaga, R.K. Roy and C. Gurgo, Nature 227 (1970) 1.
[27] J. Ross, H. Aviv, E. Scolnick and P. Leder, Proc. Natl. Acad. Sci. U.S. 69 (1972) 264.
[28] R. Gallo, Nature 234 (1971) 194.
[29] M. Hill and J. Hillowa, Nature New Biology 237 (1972) 35.
[30] R.C. Ting, S.S. Yang and R.C. Gallo, Nature New Biology 236 (1972) 163.
[31] K.G. Lark, J. Mol. Biol. 64 (1072) 47.
[32] P. Bazzicalupo and G.P. Tocchini-Valentini, Proc. Natl. Acad. Sci. U.S. 69 (1972) 298.
[33] D. Brutlag, R. Schekman and A. Kornberg, Proc. Natl. Acad. Sci. U.S. 68 (1971) 2826.
[34] J.G. Stavrianopoulos, J.D. Karkan and El. Chargaff, Proc. Natl. Acad. Sci. U.S. 68 (1971) 2207.

[35] W. Keller, Proc. Natl. Acad. Sci. U.S. 69 (1972) 1560.
[36] J.G. Stavrianopoulos, J.D. Karkas and E. Chargaff, Proc. Natl. Acad. Sci. U.S. 69 (1972) 2609.
[37] R. Schekman, W. Wickner, O. Westegaerd, D. Brutlag, K. Geider, L.L. Bertsch and A. Kornberg, Proc. Natl. Acad. Sci. U.S. 69 (1972) 2691.
[38] J.D. Karkas, J.G. Stavrianopoulos and E. Chargaff, Proc. Natl. Acad. Sci. U.S. 69 (1972) 398.
[39] D.B. Cowie, R.J. Avery and S.P. Champe, Virology 45 (1971) 30.
[40] R.J. Britten and D.E. Kohne, Science 161 (1968) 529.
[41] M. Beljanski and P. Bourgarel, Compt. Rend. 266 (1968) 845.
[42] J. Watson, Molecular biology of the gene (W.A. Benjamin, New York) 277.
[43] J. Marmur, C. Brandon, S. Neubort, M. Erlich, M. Mandel and J. Konvicka, Nature New Biology 239 (1972) 68.
[44] H.M. Temin, C.Y. Kang and S. Mizutani, In: The 6th Miles Symposium on Molecular Biology, June 8–9, 1972, Baltimore (in press).

Niu and Segal (eds.). The role of RNA in reproduction and development
North-Holland Publ. Co., 1973

RNA-mediated transformation in *Pneumococcus* *

Audrey EVANS
Department of Radiation Biology, Case Western Reserve University, Cleveland, Ohio 44106, USA

In a provocative and stimulating report, Dr. Beljanski has described experiments which indicate that some of the modified RNA molecules produced by showdomycin resistant mutants of *E. coli* have a template function with certain enzymes. He has shown that under appropriate conditions, these RNA's serve as templates with the enzyme polynucleotide phosphorylase and direct the synthesis of additional (complementary?) RNA molecules. Just as intriguing was his demonstration that purified extracts derived from mutant and wild strains of *E. coli* catalyzed the synthesis of DNA complementary (in part) to the active RNA fractions. The latter enzyme activity strongly resembles those effected by reverse transcriptases of viral origin. However, the degree of relatedness between these enzymes from different sources remains to be established. It would also be of interest to know whether the bacterial "reverse transcriptase" is a unique constitutive protein whose concentration and biosynthetic capacities are elevated by RNA or if the normal activities of some enzyme (like DNA polymerase I) are repressed by RNA with concomitant expression of a latent biosynthetic potential.

In addition to these remarkable template properties, the reactive molecules were shown to have transforming activity [1]. Purified RNA extracted from showdomycin resistant strains can confer showdomycin resistance upon sensitive recipients. The transformants produced the purine rich RNA species characteristic of the donor and were stable for the introudced trait. This demonstration of the tranforming ability of bacterial RNA was particularly gratifying to me because it supported certain findings that were made in an earlier but similar study [2]. I shall present a brief, schematic account of that work. A more detailed description containing more recent findings will be published shortly. In the study I'm referring to, the organism employed was *D. pneumoniae* (*pneumococcus*) and the selective marker was sulfonamide resistance. The sulfonamide drugs are structural analogs of para aminobenzoic acid (PAB), the normal substrate for the enzyme folic acid synthetase and thereby inhibit the synthesis of folic acid. Mutants resistant to the various sulfona-

* Discussant's paper to the contribution by M. Beljanski and M. Plawecki.

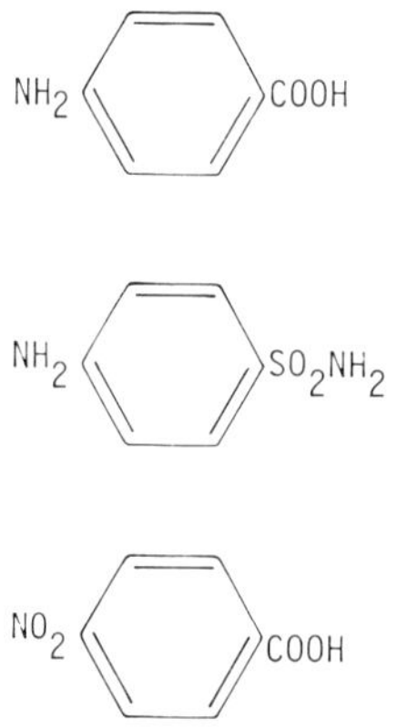

p-aminobenzoic acid	PAB
sulfanilamide	SA
p-nitrobenzoic acid	PNOB

PAB ⟶ FOLIC ACID
(folic acid synthetase)

RECIPIENT		DONOR	
PHENOTYPE	SA RESISTANT PNOB SENSITIVE	PHENOTYPE	SA SENSITIVE PNOB RESISTANT
GENOTYPE	d	GENOTYPE	d^+

INPUT NUCLEIC ACID = $RNA_{d^+}DNA_d$ HYBRID

mides from modified folic acid synthetases. The particular drugs used in these experiments and the pertinent biochemical reaction are shown in fig. 1. The RNA donor molecules were purified extracts of exponentially growing cells of the wild strain of pneumococcus which is resistant to para nitrobenzoic acid (PNOB) and sensitive to sulfanilamide (SA). The recipient cells had the reciprocal drug resistance properties (PNOB sensitive; SA resistant). It was found that under the conditions employed, the recipients would not interact with RNA unless it was bound to DNA. Therefore, RNA was annealed to DNA and the input donor material was an RNA–DNA hybrid. Moreover, as indicated in fig. 2, the genetic marker was present solely on the RNA component of the hybrid structure. The biological activity of these donors was greatly affected by RNase and DNase. No transformants were obtained if the hybrid donors were treated with either nuclease. However, when untreated hybrid molecules were used as donor, rather novel transformant clones were produced. These are depicted in fig. 3. As indicated in the figure, the transformants which developed originally in a selective medium containing PNOB, were

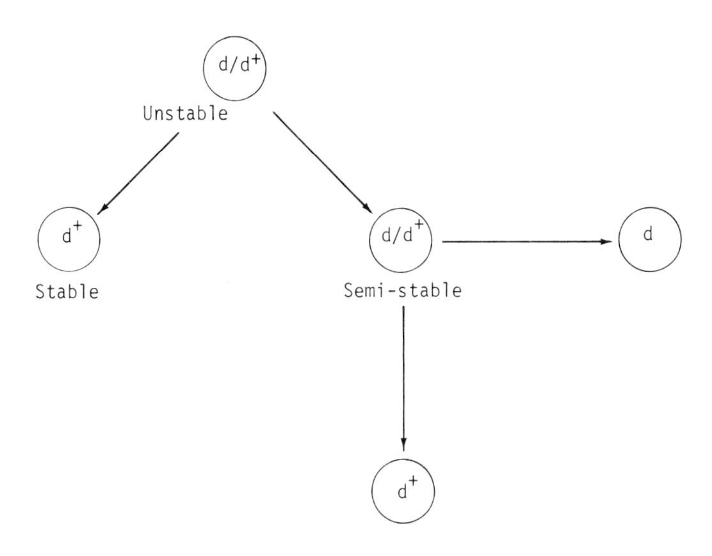

also resistant to sulfanilamide. All cells of the transformant clones were heterogenotes exhibiting both donor and recipient resistance properties. The early transformants were unstable and after a few passages in drug free media, they segregated other cell types. One of the segregant classes was stable and manifested only the donor resistance trait. The other class of segregants had the same phenotype as the early transformants but were able to maintain the dual resistance property for many more generations. After awhile, they segregated stable donor cell types. Less frequently, the semi-stable heterogenotes lost the donor trait and cells identical to the recipient were obtained. So it appears that information imparted to cells by RNA molecules can have diverse fates. It can be lost entirely or it can be propagated in three ways: unstable, semi-stable and stable. The persistence of the information in the semi-stable mode implies that it is replicated for awhile in some as yet unknown manner. Furthermore, the formation of stable clones strongly indicates that the input information (which originally resided on RNA) has been incorporated into the host genome possibly by conversion of RNA into DNA. Perhaps it is worth considering whether the variable cellular responses are reflecting certain biosynthetic reactions generated by the entry of functional RNA into cells. Dr. Beljanski's experiments suggest how this might be achieved and furthermore lead to the production of different kinds of information bearing molecules (RNA, DNA, and RNA–DNA hybrids). These in turn can engage in a variety of reactions with interesting consequences for the cell. Therefore, it might be important to determine the effect that certain structural and compositional features of RNA have upon its fate. Of equal value, is the role of recipient cells in interpreting the various messages (original, repressive or stimulatory) which are imprinted upon and conveyed by RNA – the operational messenger molecule. Investigations roughly along these lines have already begun using various bacteria and the mold *Neurospora.* Hopefully studies with less complex organisms will provide useful insights into the cell–RNA interactioss of higher organisms.

References

[1] M.M. Beljanski, M. Beljanski and P. Bourgarel, Compt. Rend. 272 (1971) 2107.
[2] A.H. Evans, Proc. Natl. Acad. Sci. U.S. 52 (1964) 1442.

Niu and Segal (eds.), The role of RNA in reproduction and development
North-Holland Publ. Co., 1973

Requirement of informational molecules in heart formation

A.K. DESHPANDE, L.C. NIU and M.C. NIU
Department of Biology, Temple University, Philadelphia, Pa. 19122, USA

Nuclear RNA (nRNA), cytoplasmic (cRNA) and poly A attached RNA (mRNA) were isolated from chicken hearts. Post-nodal pieces, obtained by transecting the definitive primitive streak of stage 4 chick blastoderm 0.6 mm posterior to Hensen's node, were exposed to Pannett–Compton solution with and without various heart RNAs. All explants were changed daily to fresh culture medium for 4 days. Pulsating tube or tissue was found in the heart-RNA-treated series but not in the control and brain-RNA-treated series. The formation of beating tissue in nRNA, cRNA and mRNA was 48, 52 and 88% of the differentiated explants respectively. The heart-forming capacity of mRNA was concentration dependent and sensitive to pancreatic ribonuclease.

Histological examination of all explants showed that the control series had practically no tissues other than red blood cells, mesenchyme and epithelial cells and in contrast all heart RNA-treated series had in addition to heart tissue, neuroid, notochord, somite and tubule. The development of the latter tissues was independent of the RNA concentration employed and thus considered to be non-specific. This was demonstrated experimentally in the series using heart-RNA filtrate and RNase-digested mRNA. In these experiments, no beating tissue developed but the types of tissues formed were similar to those in the mRNA treated series.

Electron microscopic study of the beating heart showed cellular structure typical of cardiac muscle cells.

1. Introduction

Early experiments have shown that with RNA treatment the posterior third of the primitive streak of stage 4 chick blastoderm was capable of developing into highly organized tissues. The types of tissues formed in 24 hr of cultivation varied according to the tissue source of RNA [1]. Heart, kidney and brain-RNAs induced respectively the formation of vesicular, tubular and neural tissues. The vesicular heart was not beating. However, when the heart-RNA treated explant was cultured for longer periods of time, pulsating tissue appeared [2]. Similarly, the anterolateral pieces of chick blastoderm could be induced to form beating tissue [3]. In order to determine the active component of heart-RNA, the authors have examined the heart-RNA induced heart formation systematically. In the first place, chicken heart tissue was separated into nuclear and cytoplasmic portions and the RNAs isolated from them were tested for their heart-forming potentiality; secondly, poly A attached RNA (mRNA) was prepared by filtering whole heart-RNA through a millipore filter and the filter-adsorbed mRNA was tested for heart forming potentiality; thirdly, the specificity of heart-RNA in heart-formation was examined by comparing brain-RNA, heart-RNA minus mRNA, and RNase-digested mRNA; fourthly, chicken heart-RNA was compared with calf heart-RNA; fifthly and finally, the beating tissue was examined by electorn microscopy. The aim of this communication is to present experimental results showing that the heart-forming capacity of cytoplasmic RNA is superior to nuclear RNA; that the active component of cytoplasmic RNA is apparently mRNA; that heart-mRNA is specific for heart formation and brain-RNA for brain formation; that calf-heart-mRNA is practically as potent as chicken heart-mRNA; and that the fine structure of the beating tube and/or tissue is typical of cardiac muscle cells.

2. *Materials and methods*

Fertilized eggs of White Leghorn chicken obtained from Shaw's Hatchery, West Chester, Pa., were incubated at 38°C to get the definitive primitive streak stage (stage-4, Hamburger and Hamilton [4]). The blastoderms were explanted and the area opaque was trimmed off. The post-nodel piece (PNP) was obtained by transecting the area pellucida 0.6 mm behind the Hensen's node (fig. 1), so that the presumptive heart-forming areas were excluded [5, 6]. The method of in vitro cultivation is modified after New [7], Chauhan and Rao [8]. This PNP was transferred to the yolk-free vitelline membrane (VM) previously mounted around the glass ring, with hypoblast facing up. The excess Pannett–Compton solution [9] (PC) was carefully removed and the PNP was uniformly flattened on VM. One milliliter of the nutritive medium was added to the watch glass. The nutritive medium used was a whole egg extract prepared by thoroughly mixing whole egg with 50 ml Ringer solution. The homogenate was centrifuged at 2000 rpm, for 30 min at 4°C. The

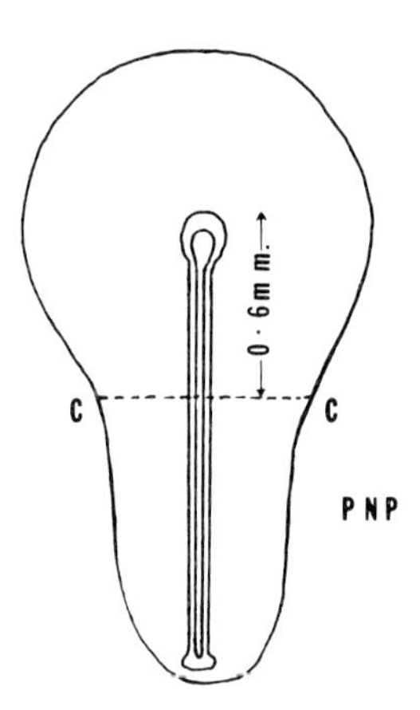

Fig. 1. Diagram of the stage-4 chick blastoderm showing the level of the transection (C).

supernatant was mixed with an equal volume of PC solution. The pH of the medium was 7.6 to 7.7.

2.1. Preparation of heart-RNA

Chicken hearts were excised immediately after slaughtering in Penn Poultry & Egg Co., and pooled in ice cold isotonic sucrose solution (0.25 M, 0.001 M $MgCl_2$). Unless otherwise specified, preparation of RNA was carried out in the cold room at 4°C. Hearts were cleaned free of fat and connective tissue. Batches of 70 g were finely chopped with scissors and blended in a Waring Blender with 4 vols of sucrose solution (0.32 M and 0.003 M $MgCl_2$ and 0.5% diethyl pyrocarbonate). During high speed blending for 30 sec, 2 more vols of sucrose solution were added and the process continued for 90 sec. The homogenate was filtered through 1, 2 and 4 layers of cheese cloth and then through 2 layers of flannel. The filtrate was centrifuged at 2000 rpm, for 30 min at 4°C. The supernatant fluid was suctioned off and saved for isolation of cytoplasmic RNA (cRNA). The sediment was used for the isolation of nuclear RNA (nRNA). The isolation procedures of heart- and brain-nRNA and heart-cRNA have previously been described [10].

2.2. Isolation of poly A-attached RNA (mRNA)

Chicken hearts in groups of 70 g were finely chopped with scissors and blended in a Waring Blender at high speed for 60 sec with 3 vols of a pH 8 buffer (0.025 M EDTA, 0.1 M NaCl and 0.5% sodium dodecyl sulfate (SDS)). Three more vols of the buffer were added and the process continued at low speed for 60 sec. The homogenate was filtered through 1 and 2 layers of cheese cloth. To the filtrate an equal volume of buffer-saturated phenol (pH 8) was added, and immediately mixed in a Lourdes mixer at 60 V for 5 min at room temperature. The mixture was centrifuged at 2400 rpm for 1 hr at 10°C, and the top aqueous layer was saved. To the interphase and phenol layer, 0.5 vol of saline was added, mixed well and centrifuged as before for 30 min. The supernatant fluid was combined with the previous and once again extracted with 0.5 vol of the buffer-saturated phenol. The aqueous layer was recovered after centrifugation to which 2 vols of 95% ethanol were added and kept in deep freeze (–20°C) over night. The nucleic acid preparation from the freezer was centrifuged. The precipitate was dissolved in saline and reprecipitated with 2 vols of ethanol after 1 hr in deep freeze. This process was repeated 2 times in order to get rid of SDS. The precipitate from the last wash was dissolved in a small volume of distilled water. The concentration of nucleic acid was adjusted to 1 mg/10 ml of ice cold solution A (Tris 0.01 M, KCl 0.5 M, $MgGl_2$ 0.001 M, pH 7.6) and filtered through a 47 mm millipore filter (0.45 mm), presoaked in buffer, under negative pressure [11]. The filter was cut into small pieces and soaked in solution B (0.5% SDS, Tris 0.01 M, pH 9; 1 ml of solution B for every 2 mg of nucleic acid) for 30 min at room temperature. Afterwards, solution B was poured into another flask and filters rinsed with 0.5 vol of solution B. The two washes (eluate) were combined. To the combined solution NaCl (up to 0.2 M) and 2 vols of 95% ethanol were added and kept in deep freeze (–20°C) overnight. The bulk of the nucleic acid passed through the millipore filter. To the filtrate 2 vols of 95% ethanol were added and kept in the deep freeze. Both eluate (mRNA) and filtrate (fRNA) fractions were washed 3 times with 95% ethanol for the removal of SDS.

Both filtrate and eluate fractions from deep freeze were centrifuged. The precipitates were dissolved in saline. Glycogen was removed by centrifugation (Sorvall RC-2) at 2°C, 18,000 rpm for 20 min. All preparations (nRNA, cRNA, fRNA and mRNA) were treated with DNase (RNase free). The enzyme was denatured by chloroform containing 1% isoamyl alcohol. All RNAs were reprecipitated by 2 vols of ethanol and washed twice. The sediment was dissolved in saline. U.V. spectrophotometric examination of RNA solution revealed that all preparations were typical of nucleic acids. Tests with the Lowry procedure showed that the nRNA contained about 2%, mRNA, cRNA, and fRNA 1% of protein. The Dische reaction indicated that 1–1.5% DNA was present in nRNA and mRNA but none in cRNA and fRNA.

2.3. *RNA-treatment*

The PNPs were prepared as described above with 1 ml of nutritive medium added outside the VM. RNA was dissolved in PC solution at a concentration of 80 O.D./ml, and applied 0.1 ml inside and 0.1 ml outside the VM. The RNAs used were nRNA, cRNA, fRNA, and mRNA of chicken heart, nRNA of chicken brain and mRNA of calf heart. In experiments with ribonuclease, the heart mRNA (40 O.D./ml) was digested with preboiled pancreatic RNase (1 mg/ml) at 37 °C for 1 hr. RNase activity was neutralized by rabbit antiserum against RNase. The PNPs receiving PC solution were used as controls. All cultures were incubated at 37.5°C. The medium was changed every 24 hr with fresh egg extract medium. Observations were recorded daily. On the 5th day (120 hr) of cultivation some of the beating explants were recorded by movie camera and others were fixed in Bouin's fluid. A few explants were kept as whole mounts and the rest were examined histologically.

2.4. *Electron microscopy*

Six explants with RNA induced pulsating tissue were fixed in buffered glutaraldehyde (0.1 M Na cacodylate, pH 7.4) for 1 hr at 0°C. They were washed in the same buffer containing isotonic sucrose, at 30 min intervals and post-fixed in precooled 1% OsO_4 in cacodylate buffer for 1 hr at 0°C. After dehydration through graded series of ethanol, the tissue was embedded in Araldite. Thin sections were obtained by using a glass knife on a Porter-Blum Ultra Microtome M T-2. The sections were mounted on 200-mesh grids coated with formvar and carbon. They were stained with lead citrate and examined in a Philips EM-300 microscope.

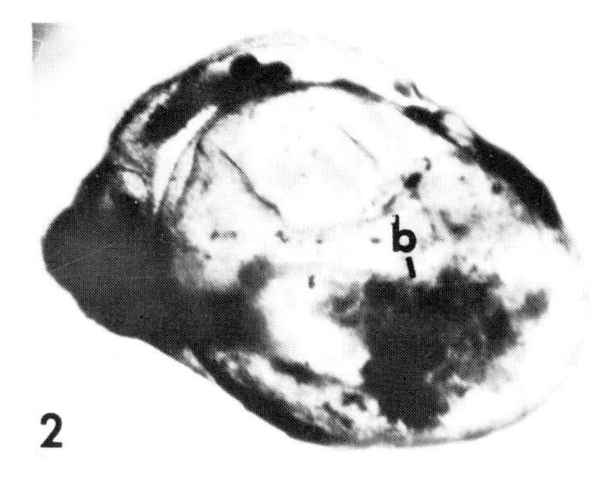

Fig. 2. Whole mount of a PNP cultured in the egg-extract medium for 5 days. Note the absence of the axial structures. b: Accumulation of R.B.C.s. × 20.

3. Results

3.1. Control series

After 24 hr of incubation, the primitive streak of the explants disappeared. Most explants rounded up in 48 hr. Patches of red blood cells developed on the third day. However, twitching tissue had never been seen. Fig. 2 depicts an explant on the fifth day. Histological examination of 48 PNPs revealed that erythrocytes and hemopoietic tissue were invariably present (figs. 5 and 6) and mesenchyme appeared as undifferentiated condensation or loose network. Two of the 48 PNPs developed neural and/or chordal tissue (6%).

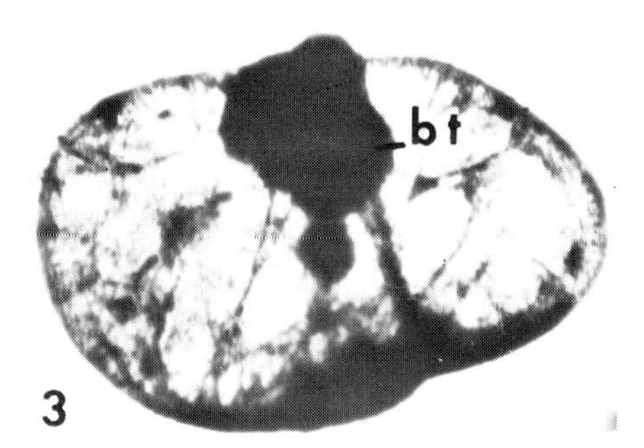

Fig. 3. Whole mount of the PNP treated with chicken heart-mRNA and cultured for 5 days. bt: Pulsating cardiac tissue. Note well developed blood vessels filled with R.B.C.s. ×20.

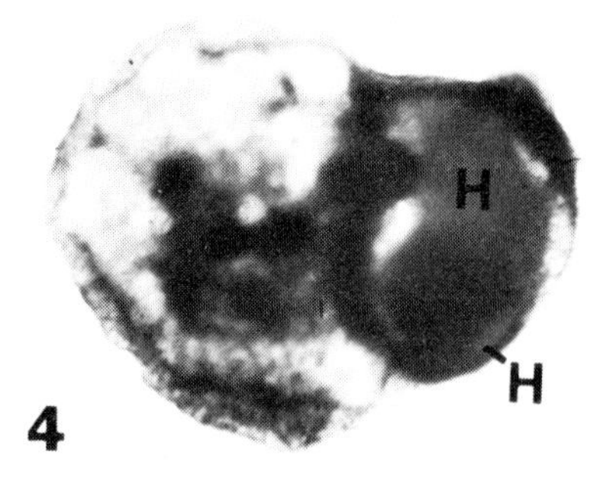

Fig. 4. Whole mount of the PNP treated with chicken heart-mRNA and cultured for 5 days. Note the prominant heart tube (H). The heart tube was beating from 4th to fixation at the end of the 5th day. ×20.

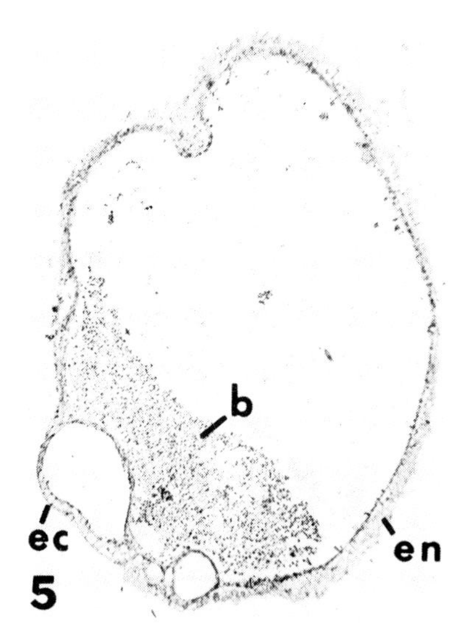

Fig. 5. Photomicrographic section of a control PNP. b: red blood cells; ec: ectoderm; en: endoderm. ×80.

3.2. Experimental series

At the end of 24 hr of incubation, the explants of the experimental series appeared more compact than the control. A small median protuberance extended out at the cut surface and the primitive streak persisted. Soon after the disappearance of the protuberance, the explants would transform into shapes of various kinds. Red patches developed in the vesicular explants on the third day and twitching area(s) on the following day. The twitching evolved to rhythmic beating on the fifth day. The beating rate averaged 20–25 per min. It slowed down gradually during observation

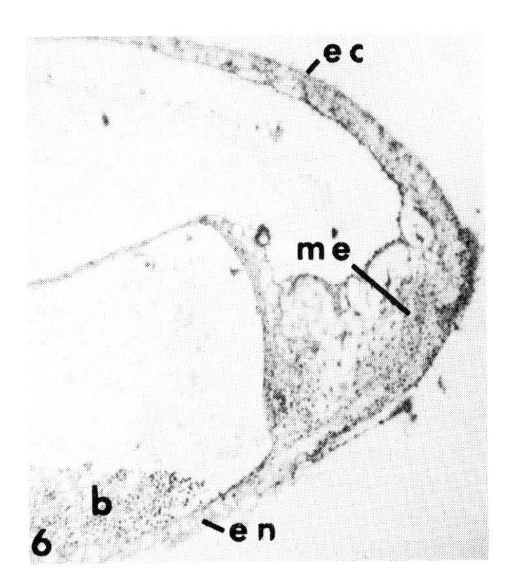

Fig. 6. Photomicrographic section of a control PNP. b: red blood cells; ec: ectoderm; en: endoderm; me: condensed mesenchyme. ×96.

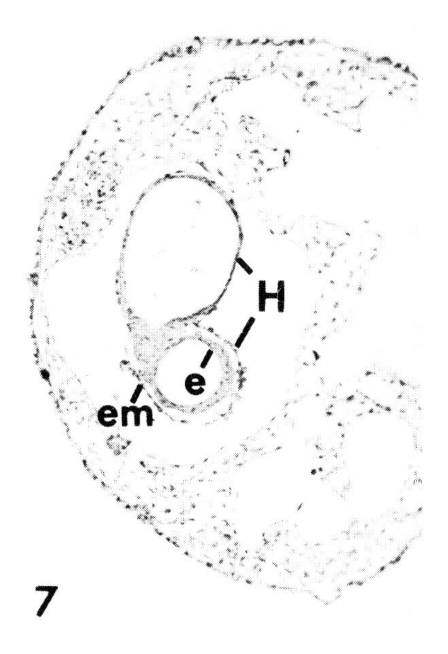

Fig. 7. Section of a chicken heart-mRNA treated PNP passing through bent heart tube. H: tubular heart; em: epimyocardium e: endocardium. ×90.

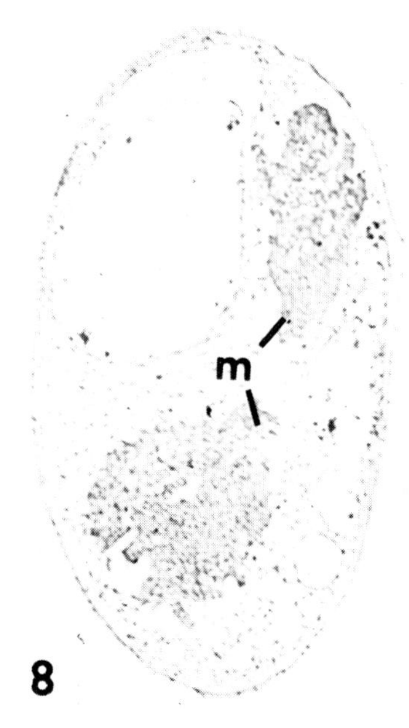

Fig. 8. Section of a chicken heart-mRNA treated PNP showing extensive compact beating tissue. m: cardia myoblasts. ×90.

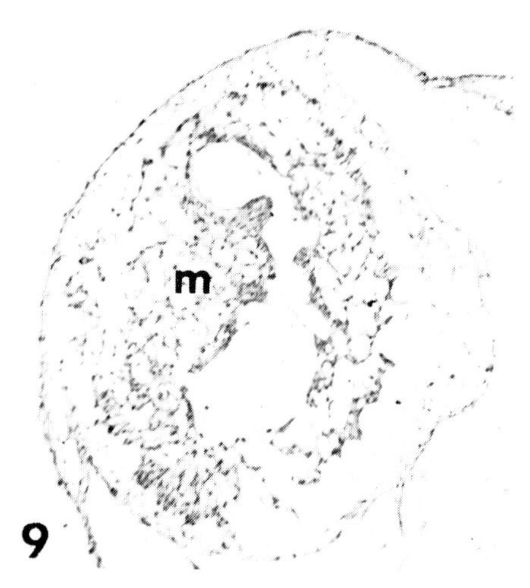

Fig. 9. Section of the PNP treated with mRNA, with loosely arranged beating tissue. m: cardiac myoblasts. ×90.

at room temperature. In a few cases blood vessels established connection with the beating heart tube and thus circulation of the red blood cells was observed. The explants with pulsating tube and patch of tissue are shown respectively in figs. 3 and 4. At the beginning of this study, the authors had used rhythmic beating as the criterion to show the presence of cardiac tissue. After the beating explants were sketched, they were examined histologically. The heart tube was easy to identify (fig. 7). With the sketch of each beating explant and aided by the histology of the beating tissue from chorio-allantic grafts of presumptive heart-forming tissue [12], the authors had positively identified the compact (fig. 8) and loose (fig. 9) type of cardiac tissue. In accordance with the newly acquired information, it was possible to identify the non-beating cardiac tissue without difficulty.

3.2.1. Chicken heart-nRNA

Nineteen of the 51 PNPs were beating and 9 of the 19 were tubular. Histological observations showed 40 of the 51 explants underwent differentiation (78%). The distribution of various organs is tabulated in table 1. As can be seen, the combined frequency of beating and non-beating tissue is 75%, a rate showing that every 3 out of 4 differentiated explants develop cardiac tissue. The tubule and neuroid formation are 23 and 30% respectively.

3.2.2. Chicken heart-cRNA

Twelve of the 37 PNPs were beating and 5 of the 12 were tubular. Data from the 37 serial sections showed that 62% of this series differentiated. The combined

TABLE 1

Developmental potentiality of chick blastoderm (PNPs) in presence and absence of RNA, cultured in vitro for 5 days (120 hr)

Series	No. of PNPs differentiated and no. cultured	No. of differentiated PNPs showing development of various organs					
		Cardiac tissue					
		Beating	Non-beating	Tubule	Somite	Notochord	Neuroid
Control (PC)	3/48 (6%)	0	0	0	0	2	2
Heart-nRNA	40/51 (78%)	19 (48%)	11 (27%)	9 (23%)	3 (8%)	2 (5%)	12 (30%)
Heart-cRNA	23/37 (62%)	12 (52%)	6 (26%)	0	0	5 (22%)	5 (22%)
Brain-nRNA	20/23 (87%)	0	0	7 (35%)	1 (5%)	6 (30%)	17 (85%)

TABLE 2
Distribution of the differentiated organs in PNP explants treated with RNAs

Kinds of organs differentiated	Heart-nRNA (40 of 51 PNP explants differentiated – 78%)	Heart-cRNA (23 of 37 PNP explants differentiated – 62%)	Brain-nRNA (20 of 23 PNP explants differentiated – 87%)
Beating heart only	12 (6 – tube)	11 (5 – tube)	0
Beating heart with neuroid	3 (3 – tube)	0	0
Beating heart with tubule	2	0	0
Beating heart with neuroid and tubule	2	1	0
	Total (19) 48%	Total (12) 52%	
Non-beating cardiac tissue only	8 (8 – tube)	4 (3 – tube)	0
Non-beating cardiac tube with notochord	0	1	0
Non-beating cardiac tissue with tubule and/or somite	3	0	0
Non-beating cardiac tissue with notochord	0	1	0
	Total 11 (28%)	Total 6 (26%)	
Tubule only	1	0	1
Tubule and somite	2	0	0
Notochord and tubule	0	0	1
Notochord and somite	0	0	1
Notochord only	0	1	0
	Total 3 (8%)	Total 1 (4%)	Total 3 (15%)
Neuroid and tubule	0	0	4
Neuroid and notochord	2	2	3
Neuroid, notochord and tubule	0	0	1
Neuroid	5	2	9
	Total 7 (18%)	Total 4 (17%)	Total 17 (85%)

frequency of heart formation was 78% and that of the neuroid and notochord formation was 22% each (table 2). The lack of tubule and somite formation in the cRNA series suggested that the function of cRNA was more restricted than that of nRNA.

3.2.3. Chicken brain-nRNA

Twenty of the 23 underwent differentiation (87%). None of them developed beating tissue. The frequency of neural, notochord and tubule formation was 85%, 30% and 35% respectively (table 1).

The organ formation was re-classified under 3 headings (table 2): (i) explants with heart and associated tissues (top), (ii) explants with tubule, somite and/or notochord (middle) and (iii) explants with neural and associated tissues (bottom). It can be seen that the frequency of the heart-cRNA and heart-nRNA induced heart formation is respectively 78 and 76% and that of neuroid formation is 18 and 17%; in contrast, the brain-nRNA induced neural tube (fig. 10) or neuroid formation has

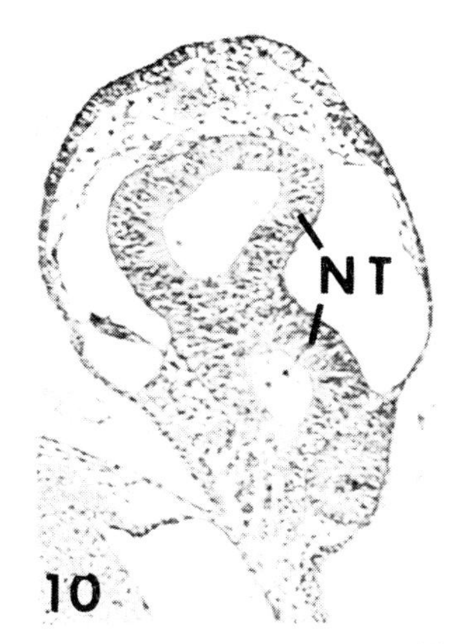

Fig. 10. Section of the PNP treated with chicken brain-nRNA. Note the predominant neural differentiation. NT: neural tubes. × 100.

a frequency of 85% and the mesodermal organ formation 15%. Therefore, the heart-RNA is specific in heart formation and the brain-RNA is specific in brain and other neural tissue formation.

3.2.4. Chicken heart-mRNA (poly A attached RNA)

Twenty-five explants were treated as above with mRNA. Practically all differentiated (96%). Twenty-two of the 25 explants (88%) developed beating tube (8) and tissue (14). The rhythmic pulsation averaged 22 per min. Neural and tubule formation occurred at the frequency of 60 and 40% respectively. Apparently mRNA increased not only heart formation but also neural and tubule formation. The increase in the tubule and neural formation was not specific, and probably due to the presence of poly A. The requirement of mRNA in the heart formation was further shown by the following experiments:

3.2.4.1. The absence of heart-forming capacity of fRNA. This was the fraction of heart-RNA with mRNA selectively removed. Three of the 28 fRNA-treated PNPs developed beating tissue (11%). The frequency of other tissue formation was found similar to that of heart-mRNA series. The similarity in non-specific organ formation between the fRNA and mRNA series manifested a point of great interest, namely, mRNA was able to initiate the formation of specific (heart) and non-specific structures. The capacity of fRNA to initiate heart-formation might be due to mRNA contaminant.

3.2.4.2. The dependence of heart-formation on mRNA concentration. Seven concentrations of mRNA were used to treat 124 PNPs (80, 60, 40, 20, 10, 5, and 0.25

TABLE 3
The concentration effect of heart-mRNA on the development of excised chick blastoderm

Concentration used (O.D./ml)	No. of PNPs	% of Dif'n	No and % (in parentheses) of development of various organs					
			Beating heart	Beats/ min	Neur-iod	Noto-chord	Somite	Tubule
80	25	96	22 (88)	22	15 (60)	– (–)	2 (8)	10 (40)
60	13	100	10 (77)	22	6 (46)	2 (15)	3 (23)	7 (54)
40	24	92	18 (75)	21	7 (29)	2 (8)	1 (4)	8 (33)
20	24	88	15 (63)	15	17 (71)	8 (33)	1 (4)	10 (42)
10	14	91	6 (43)	22	11 (79)	– (–)	2 (14)	1 (7)
5	13	91	4 (31)	11	7 (54)	– (–)	1 (8)	4 (31)
¼	11	81	2 (18)	18	5 (45)	– (–)	1 (9)	4 (36)
Filtrate[a] –80	28	81	3 (11)	20	9 (32)	2 (8)	3 (13)	10 (36)
0 (controls)	48	6	– (–)	–	3 (6)	2 (4)	– (–)	– (–)

[a] This is the fraction of whole heart RNA that passes through millipore filter.

O.D. per ml). First beating occured in all series on the fourth day. The beating rate was counted on the fifth day and averaged 20 per min. Examination of serial sections of all 124 explants revealed the presence in the explants of a variety of tissues: beating tube, beating tissue, neuroid, notochord, somite and tubule. Their distribution is summarized in table 3. As can be seen, the frequency of heart formation is directly proportional to the concentration used (fig. 13) and the concentration of mRNA does not affect the general frequency of differentiation (column 3, table 3) nor the frequency of other organ formation (last 4 columns, table 3). Furthermore, heart tubes were found only in 80 (8 out of the 22 beating), 60 (3 of 10), 40 (9 of 18), and 20 (6 of 15) O.D. series.

3.2.4.3. Destruction of the specific mRNA function by RNase. Twelve PNPs were treated with mRNA-digest. None developed beating tissue. In contrast, the corresponding concentration of intact mRNA (40 O.D. per ml) initiated heart formation in 75% of the explants. The loose type of cardiac tissue was found in one of the 12 explants. The differentiation of the other tissues remained the same: neuroids (83%), notochord (25%), somite (8%), and tubule (66%). This result suggested that mRNA (poly A-attached RNA) possessed both specific and non-specific function. While specific function was sensitive to RNase, the non-specific aspect was retained after treatment with RNase.

3.2.5. Calf heart-mRNA

The calf heart-mRNA was also used to treat PNPs. Preliminary experiments showed that 22 of the 27 explants developed pulsating tissue (82%). These calf

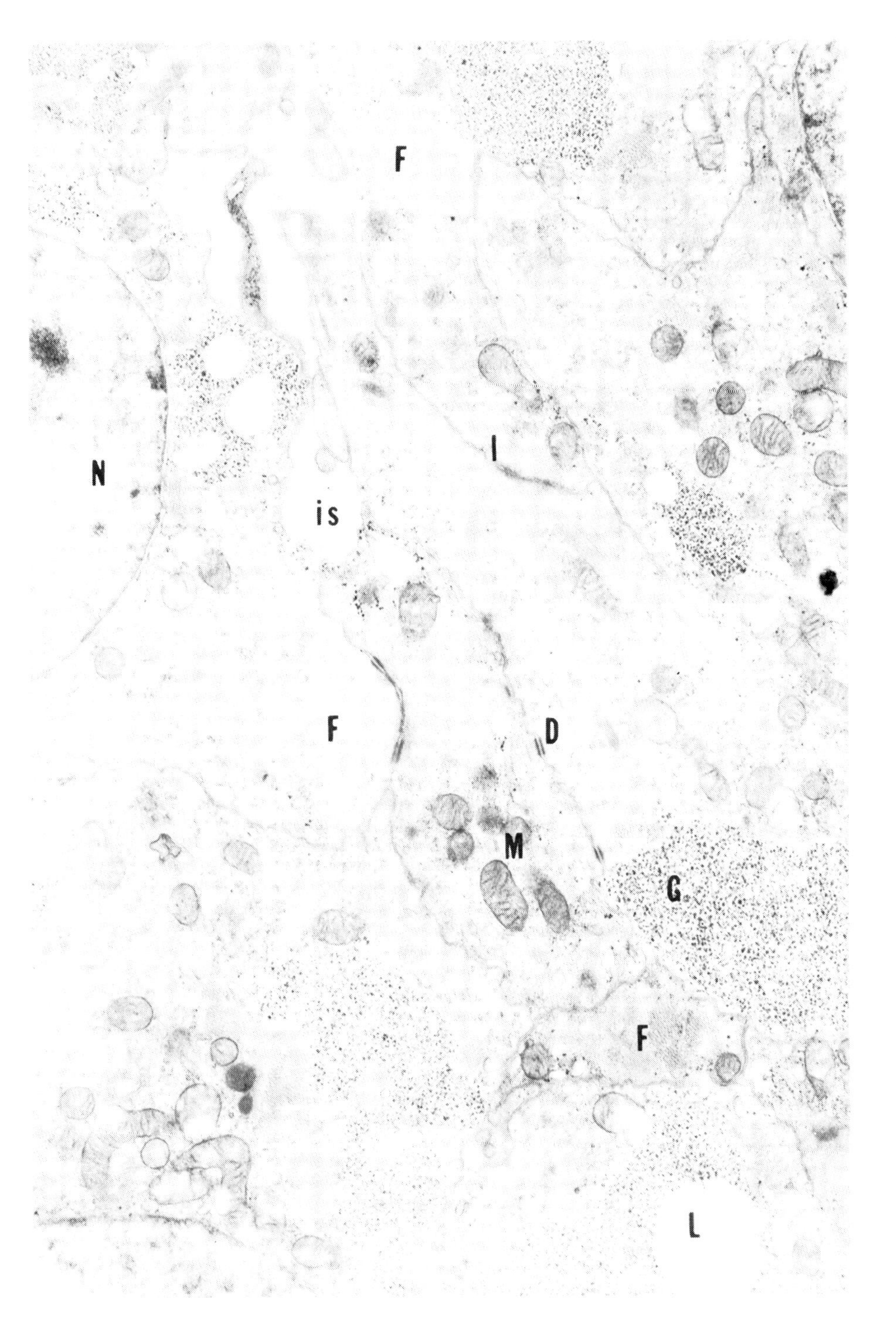

Fig. 11. Electron micrograph of cells from the heart-RNA-induced beating tissue. F: myofibrils; g: Golgi body; G: glycogen; I: intercalated disc; is: intercellular spaces; M: mitochondria; Z: Z-band; N: nucleus; L: lipids; D: desmosome. ×16,830.

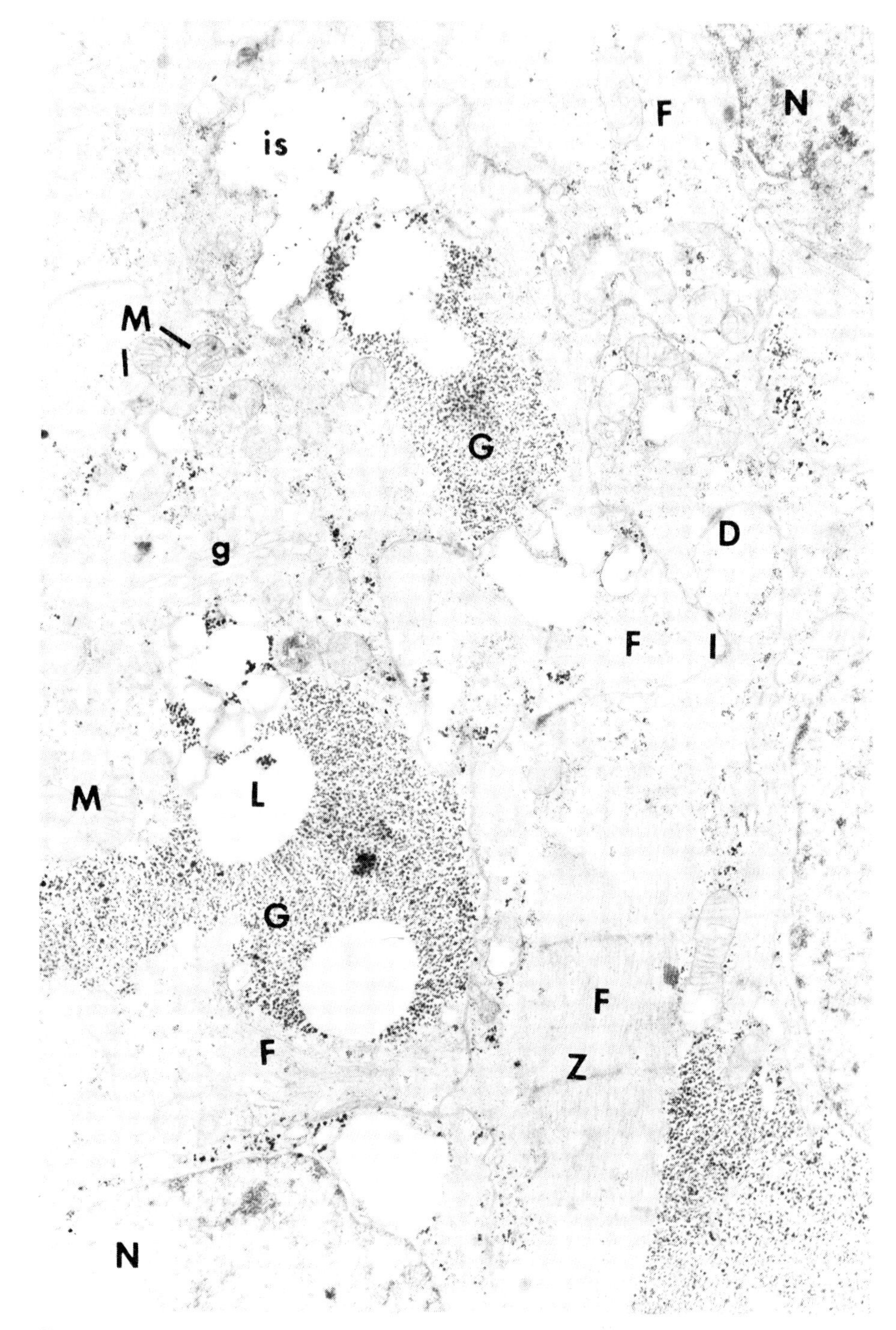

Fig. 12. Electron micrograph of cells from the heart-RNA-induced beating tissue. Legends same as fig. 11. ×11,220.

heart-mRNA induced beating tissues and an equal number of the control explants were employed separately in trial tests for the isozyme, lactic dehydrogenase. While lactic dehydrogenase was not detectable in the control explants, the calf heart-mRNA treated explants had the isozyme IV, a property of the chicken heart.

It has long been known that at early stage of heart development, myofibrils and cross striations could hardly be discernable under light microscope in the beating tissue [13, 14]. Under the electron microscope, however, they should be seen. Fig. 11 depicts an electron microgram of cardiac cells in which myofibrils began to appear at random. Between adjacent cells there were well defined intercalated discs (I) and desmosomes (D), and in the cells there were large numbers of individual glycogen granules (G) accumulated in large masses or pools. These accumulations were usually associated with lipid droplets (L). With the progress of cyto-differentiation, the cardiac cells were impregnated fully with orderly as well as at random arrangements of myofibrils, accumulated and dispersed glycogen granules, and mitochondria (fig. 12).

4. Discussion

The post-nodel piece of the stage 4 blastoderm, transected at a level 0.6 mm posterior to Hensen's node, contained some myosin-producing cells [15] and also some DNA labelled cells [16] that would make their way to the heart in normal development. The fact that these cells were unable to manifest themselves in the explants of control series was probably due to one of the two possibilities: (i) the number of the presumptive heart-forming cells was too small to differentiate in vitro [5] and (ii) the self-differentiating capacity of the explants was so labile that the formation of tissues other than hemopoietic tissue and red blood cells would seldom occur [5, 8, 12]. When the explants were incubated with the culture medium containing heart-nRNA or -cRNA overnight at 37.5°C, both RNA treated explants developed into pulsating heart and other tissues. The frequency of heart formation was dominant. Cardiac tissue was identified by its rhythmic pulsation and fine structure characteristic of chick embryonic cardiac muscle cells. Tubules and somites were not found in the cRNA treated series (table 1). The development of fewer kinds of induced tissues suggested that the function of cRNA was more limited or specific than those of nRNA. If so, cRNA would be expected to induce a higher frequency of heart formation alone in the explants. As can be seen in table 2, 13 out of the 23 explants had heart tissue only (65%) in cRNA series and 20 out of the 40 (50%) in the nRNA series. If only beating hearts were taken into consideration, the frequency would be 46% and 30% respectively.

The greater cRNA specificity as compared to nRNA suggested that the active component of cRNA might be mRNA. Messenger RNA was partially purified by passing through a millipore filter, taking advantage of adsorption in the nitrocellulose membrane of poly A and poly A attached to the 3′ end of the mRNA. After

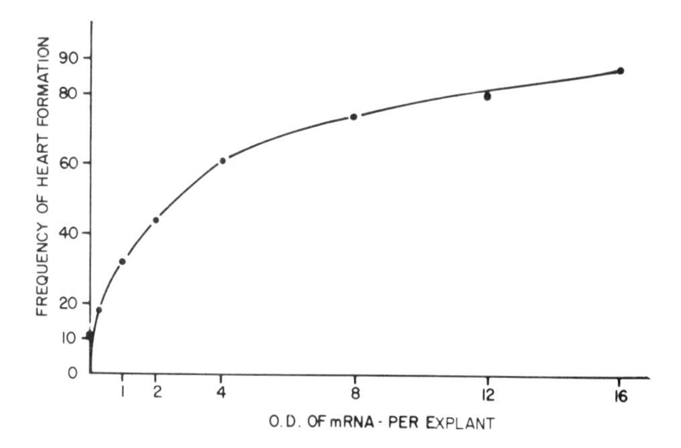

Fig. 13. Graph showing the concentration dependent development of beating hearts in mRNA-treated -PNPs.

incubation with the poly A-attached RNA (mRNA), the PNPs developed into cardiac tissue (88%) and other tissues (lesser percentages). The frequency of heart formation depended upon the concentration of mRNA used (fig. 13) while those of the other tissues were not affected (table 3). This finding supported the requirement of the informational molecule, mRNA, in the development of embryonic heart. Additional support came from the results of three further experiments: (i) filtrate fRNA could induce heart formation at very low frequency (11%). This was lower than the series using 0.25 O.D. per ml of mRNA (table 3). It seemed, therefore, that the heart-forming capacity of fRNA was likely to be due to mRNA contamination: (ii) brain-nRNA was incapable of initiating heart formation and (iii) pancreatic RNase destroyed the heart-forming capacity of mRNA.

Early investigators had used excised portion of stage 4 chick blastoderm in the study of heart formation. Some intended to show the self-differentiating potentiality of the presumptive heart-forming areas [12], while others wanted to induce heart formation by treating undifferentiated explants with heart-RNA [3]. In both cases, beating tissues developed, but none of them organized into beating heart tube as reported in this communication. The reason was two-fold: (i) the tissue was too small to allow organogenesis [5, 17] and (ii) the concentration effect of mRNA on heart formation showed that heart tube developed only in concentration of 20 O.D. per ml and above.

Heart-fRNA was a mixture of ribosomal RNA, 5 S RNA, and 4 S RNA. The functional difference between mRNA and fRNA was the specificity inherent to mRNA. So far as the inductive capacity was concerned, both were equally potent in involking non-specific tissue formation. The chemical basis of mRNA's unspecific function is not known, but probably is related to the presence of poly A. The dual function of mRNA was further shown in an experiment using mRNA digests. In this experiment, the specific function of mRNA was abolished and the non-specific part stimulated the formation of non-specific tissues. Accordingly the chemical(s) for non-specific induction covered a wide range of molecules.

The capacity of nRNA to induce a high frequency of heart formation indicated the possibility that nRNA contained informational molecules. These molecules were functionally similar to mRNA. They differed in the length of nucleotide sequence. In first place, mRNA had poly A attached to the 3′ end. Secondly there is increasing evidence showing that the informational molecules of nRNA are huge and heterogeneous and thus presumably the precursors of mRNA [18, 19].

The manner by which mRNA acted upon the PNP cells was not clear. In view of the finding that calf heart-mRNA induced heart formation at the frequency of 82% versus 88% of the chicken heart-mRNA, however, the authors have approached the problem by asking which kind of heart is developed in the calf heart-mRNA treated PNPs: is it chick or calf type? A definitive answer to this question will undoubtedly help in understanding the mode of mRNA action. Let us suppose that the calf heart-mRNA induces the differentiation of PNP into calf heart. This could mean that mRNA primed the synthesis of DNA as proposed by Kang and Temin [20], or the synthesis of the calf heart specific proteins (enzymes). On the contrary, we might assume that calf heart-mRNA induces chick heart formation. In this case, the mRNA would have to activate the chick genome. As a result, the new mRNA synthesized would be of the chick type and would lead to the production of chick structural proteins. Our strategy to resolve this question involves (i) the use of immunochemical techniques to determine species specific protein in the induced heart and (ii) the analysis of the isozyme form of lactic dehydrogenase. In a preliminary run, the lactic dehydrogenase of the calf heart-mRNA induced beating tissue was found to be of the chicken type.

Acknowledgements

This work was supported by research grants from the Population Council and The National Foundation. The authors wish to thank Dr. R.L. Miller for his valuable assistance in cinematographic recording and Mrs. Sharon L. Howard for her voluntary assistance in histology.

References

[1] S. Sanyal and M.C. Niu, Proc. Natl. Acad. Sci. U.S. 55 (1966) 743.
[2] J. Butros, J. Embryol. Exptl. Morphol. 13 (1965) 199.
[3] M.C. Niu and L. Mulherkar, J. Embryol. Exptl. Morphol. 24 (1970) 33.
[4] V. Hamburger and H.L. Hamilton, J. Morphol. 88 (1951) 49.
[5] D. Rudnick, Anat. Record 70 (1938) 351.
[6] L. Mulherkar, J. Embryol. Exptl. Morphol. 6 (1958) 1.
[7] D.A.T. New, J. Embryol. Exptl. Morphol. 3 (1955) 326.

[8] S.P.S. Chauhan and K.V. Rao, J. Embryol. Exptl. Morphol. 32 (1970) 71.
[9] C.A. Pannett and A. Compton, Lancet 206 (1924) 381.
[10] M.C. Niu and A.K. Deshpande, J. Embryol. Exptl. Morphol. 29 (1973) 485.
[11] G.C. Rosenfeld, J.P. Comstock, A.R. Means and B.W. O'Malley, Biochim. Biophys. Res. Commun. 47 (1972) 387.
[12] M.E. Rawles, Physiol. Zool. 15 (1943) 22.
[13] C.M..Goss, Anat. Record 76 (1940) 19.
[14] F.J. Manasek, J. Morphol. 125 (1968) 329.
[15] J.D. Ebert, In: Thirteenth Growth Symposium (Aspects of synthesis and order in growth), Ed. Rudnick (Princeton University Press, 1953) 69–112.
[16] G.C. Rosenquist, Develop. Biol. 22 (1970) 461.
[17] C. Grobstein and E. Zwilling, Exptl. Zool. 122 (1953) 259.
[18] J.E. Darnell, B.E.A. Maden, R. Soeiro and G. Pagoulatos, In: RNA in development, Ed. Hanley (Utah University Press, 1970) 315–329
[19] R. Williamson, C.E. Drewienkiewicz and J. Paul, Nature 241 (1973) 66.
[20] C.Y. Kang and H.M. Temin, Proc. Natl. Acad. Sci. U.S. 69 (1972) 1550.

Niu and Segal (eds.). The role of RNA in reproduction and development
North-Holland Publ. Co., 1973

Intercellular communication during odontogenic epithelial–mesenchymal interactions: isolation of extracellular matrix vesicles containing RNA*

Harold C. SLAVKIN and Richard CROSSANT
Department of Biochemistry, School of Dentistry, and the Graduate Program in Cellular Molecular Biology University of Southern California Los Angeles, Calif. 90007, USA

One experimental opportunity for obtaining instructive developmental information during epithelial–mesenchymal interactions is in the extracellular matrix interposed between tissue interactants. The embryonic tooth rudiment is a particularly well-suited model system for cytochemical and biochemical analyses of intercellular communication vis-a-vis an extracellular matrix. Tissue-matrix recombination experiments using isolated tooth epithelia or tooth mesenchyme indicates that the acellular matrix is a morphogen and can instruct homotypic tissues to differentiate in vitro. One of the constituents of the isolated matrix (progenitor dentine) is membrane-lined (unit membrane), spherical bodies called matrix vesicles (500–1000 Å in diameter). Data is presented describing the isolation of matrix vesicles (> than 50 million mol. wt.) which contain ribonucleic acid. The identification of extracellular matrix vesicles containing RNA in situ and in vitro may be a highly significant element in the mechanism of intercellular communication during epidermal organogenesis.

1. The tooth as a model system to study secondary embryonic induction

Since the classic experiments of Hans Spemann demonstrating the instructive and organizational capacity of certain tissue regions during gastrulation [1], biologists have repeatedly postulated possible mechanisms of intercellular communication between dissimilar tissues in close proximity during embryonic development [2]. Developmental information of an epigenetic nature appears to be essential to the continued differential expression of gene activity in eukaryotic cells throughout the course of development. More specifically, in the formation of many organ systems (notably those stemming from epithelial–mesenchymal interactions), the conse-

* Discussant's paper to the contribution by A.K. Deshpande.

quences of such intercellular communication has been referred to as secondary embryonic induction where a certain tissue, previously restricted in its developmental potential, guides the differentiation of another tissue type [3, 4].

Several questions are essential in trying to understand induction during organogenesis: (i) Is chemical information actually transferred between heterotypic tissues? (ii) If so, what is the chemical nature of the information transferred? and (iii) In biochemical terms, how do epigenetic chemical factors mediate the alteration of cellular phenotype? The experimental foundations needed to explore these questions were initially provided by the work of Grobstein [5]. He succeeded in isolating precursor epithelium and mesenchyme (by trypsin dissociation) and demonstrated that when recombined in culture across membrane filters, induction was still effective. Numerous other organ systems were later shown not to require direct contact for induction [6, 7]. Implicit in those experiments is the notion that diffusable chemical information is actually transferred between cells during organogenesis [8–11].

Koch demonstrated a trans-filter induction relationship for embryonic mouse tooth mesenchyme and epithelium [12]. His results suggested that the unique tooth morphology of a discrete extracellular matrix, found between differentiating epithelium and mesenchyme, might be an experimentally exploitable source of informational molecules trapped in transit outside the cell. This possibility is particularly attractive since historically the analysis of the chemical nature of information transfer in developing organ systems has produced varied and inconclusive results when whole cell homogenates were used. A myriad of factors, including RNA, have been implicated as specific inductors prerequisite for cellular differentiation during secondary embryonic inductions. Our work is based on the supposition that a specific extracellular morphogen can be isolated from the embryonic tooth matrix which will eventually provide a testable biochemical explanation of secondary embryonic induction.

In this discussion, evidence is given that cell-free tooth matrix does have morphogenetic activity. RNAs, and particularly RNA contained in extracellular matrix vesicles, are present in those active matrix preparations and have been isolated. These findings are especially relevant in light of discussions presented in this symposium which indicate a role for ribonucleic acid in the exchange of information between developing cells. Several points seem clear: (i) RNA viruses transform both prokaryotic and eukaryotic cells, (ii) transformation of cells by RNA viruses established a biological mechanism of stable RNA to DNA information transfer mediated by RNA-dependent DNA polymerases [30], (iii) RNA-dependent DNA polymerase has recently been found in normal uninfected embryonic cells [13] and (iv) various tissue or organ-specific RNAs influence or modulate development when added exogenously in vivo or in vitro [14–18]. In the context of these finding, it remains to be experimentally demonstrated whether RNAs are actually transferred among normal cells during in vivo organogenesis.

After a brief explanation of tooth morphogenesis, a survey of our results on the morphogenetic activity of isolated extracellular tooth matrix and their RNA content is provided from published and preliminary data now in progress.

2. Tooth morphogenesis

In mammals, tooth development begins approximately midway through gestation. Oral ecto-mesenchyme (derived from migratory neural crest cells) condenses and interacts with overlying epithelium, resulting in an inward budding of the epithelium which eventually encapsulates a small mass of mesenchyme. In the lower germinative (or cervical loop) progenitor region, epithelial cells next to the sequestered mesenchyme initiate extracellular matrix formation by establishing a basement membrane. Epithelium and mesenchyme immediately adjacent to this basement membrane organize and undergo a synchronous process of cellular differentiation. A schematic representation of this process if given in fig. 1A.

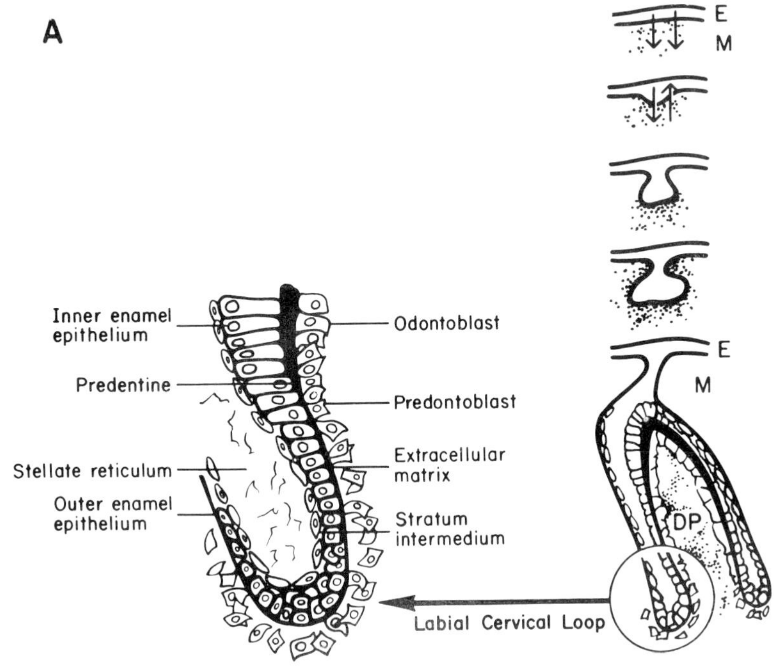

Fig. 1A. Diagrammatic scheme of mammalian odontogenesis. Common to all epidermal organ formations (e.g. thyroid, salivary, and thymus glands, skin, hair and feather formation, scale or nail formation, teeth, etc.) are discrete interactions between epithelia (E) and mesenchyme (M) leading to the formation of a "bud". This development, inclusive of teeth, reflects a critical dependency upon reciprocal, interdependent epithelial–mesenchymal interactions. The cervical loop region contains cells at various stages of differentiation. One unique aspect of tooth formation is the specific differentiation of the interface which subsequently becomes the enamel and dentine organic matrices. (DP) dental papilla.

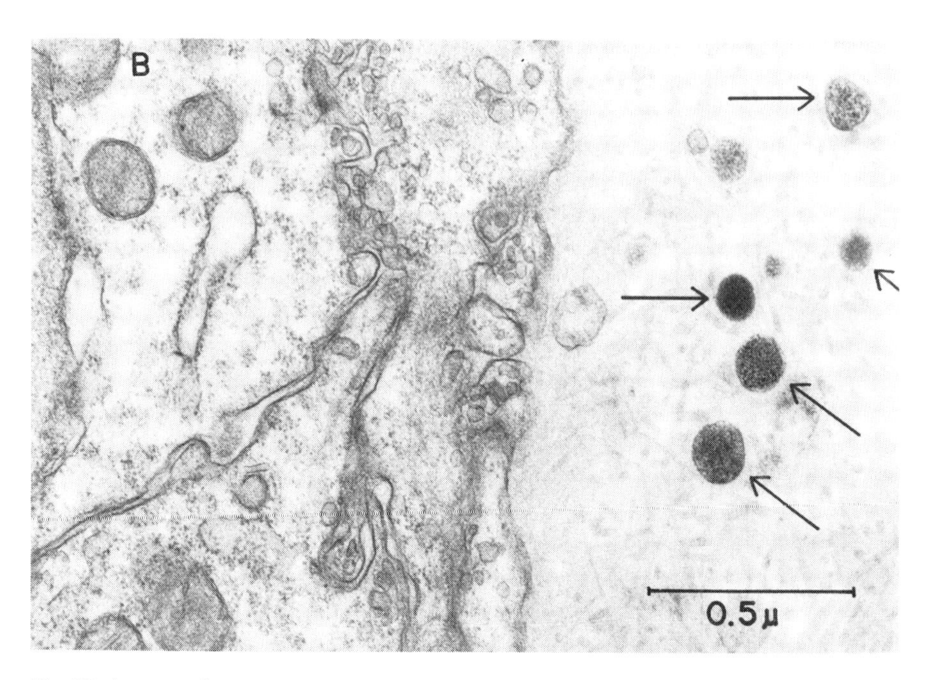

Fig. 1B. A survey electron micrograph of the infranuclear or apical secretory region of the inner enamel epithelia and the forming extracellular organic matrix (progenitor predentine) during embryonic rabbit incisor tooth formation. Clusters of intracytoplasmic secretory vesicles of varying size and electron density appear to migrate towards the basal lamina region in association with the epithelial cells. Numerous matrix vesicles (arrows) were noted as constituents of the predentine matrix. Scale line, 0.5 μ.

In longitudinal section, the embryonic tooth appears as a continuous gradient of cellular differentiation from rapidly dividing to non-dividing, tall columnar merocrine-type secretory cells. The specific protein, carbohydrate and inorganic constituents of these cells contribute to a widening extracellular matrix. Epithelium-derived ameloblasts generate enamel and mesenchyme-derived odontoblasts produce dentine [19].

In the following presentation of experiments, 25-day embryonic New Zealand white rabbit incisors were used. Fig. 1B is a low-power electron micrograph of the germinative region which is isolated by microdissection for these studies.

3. The extracellular matrix possesses morphogenetic activity

The evidence that tooth extracellular matrices contain developmental information is based on tissue recombination experiments using isolated embryonic epithelium, mesenchyme, and cell-free cervical matrix preparations cultured on the chick cho-

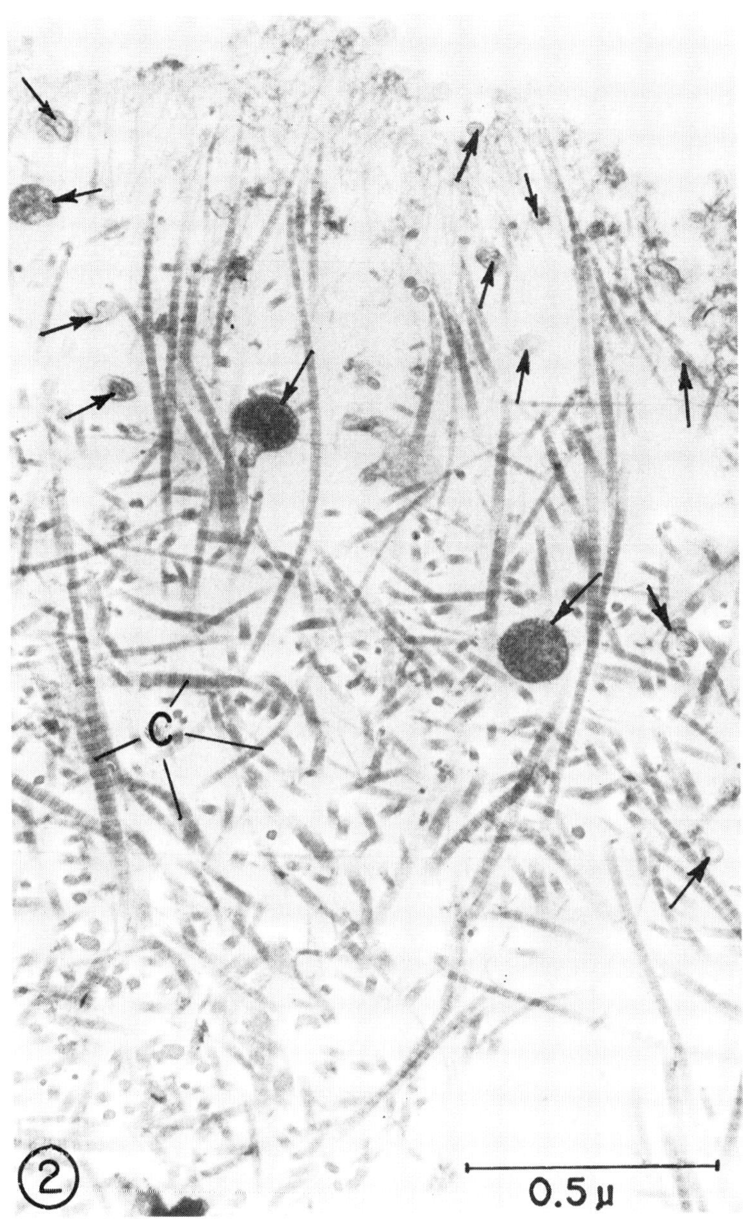

Fig. 2. A representative electron micrograph of the isolated, sonicated matrix. Following dissection of 25-day embryonic New Zealand white rabbit incisor tooth primorida, dental papilla (pulp) was removed, the germinative matrix region (10–20 μm in thickness) was isolated, and this matrix was then sonicated in a calcium- and magnesium-free, phosphate buffered solution at pH 7.4. Note the numerous matrix vesicles (arrows) retained following preparative procedures and the apparent removal of all cell processes. For orientation, the inner enamel epithelium was located in the upper right region prior to microdissection and sonication. A mesenchymal cell process occupied the lower right region. Note the abundant dentine collagen (C). Scale line, 0.5 μ.

rio-allantoic membrane (CAM). Essential to the interpretation of these results is that cervical matrix material can be isolated free from cellular contamination [20]. Low-power sonication in calcium and magnesium free-phosphate buffered saline (CMF-PBS) proved to be an effective technique. Matrices processed in this manner are devoid of visible contamination as examined with light, electron and scanning electron microscopic criteria [19]. In vivo matrix morphology is also preserved after sonication (compare figs. 1 and 2).

Previous experiments demonstrated that mixed suspensions of cervical epithelium and mesenchyme from 25-day embryonic rabbit incisors selectively reassociate and duplicate normal morphogenesis and extracellular matrix formation on the CAM [21, 22]. Cultured suspensions of either cell type alone remain viable, but fail to differentiate.

Homotypic cell suspensions in combination with sonciated matrices, however, do continue to advance developmentally [20]. Epithelium or mesenchyme cultured with sonicated matrices acquires a tall columnar appearance, and a shift in nuclear polarity. In the case of epithelium, production of new basement membrane also occurs. Only those cells in contact with the surface of the matrix undergo these changes. Additional control experiments also demonstrated that homotypic cell populations recombined with freeze-thawed matrices attached to the matrix, but *do not* differentiate. Taken together, these results argue that some morphogenetic component is present in the cervical matrix. This activity is apparently not a substratum recognition phenomenon since freeze-thawed matrices *do not* support morphogenesis.

4. *The extracellular matrix contains RNAs*

Simple organ culture experiments designed to suggest some macromolecular source of the tooth matrix morphogenetic activity (seen in CAM recombination experiments) indicated the presence of extracellular matrix RNAs [23]. Using a variety of radioactive precursors, incubated for different time periods with embryonic tooth rudiments, autoradiographs were prepared. All labeled precursors tested, with the sole exception of tritiated thymidine, moved through both cell types, and a small percentage of the total label incorporated into each cell was transferred into the extracellular matrix [23]. Incorporation of amino acids or sugar amines suggested their utilization in the structural formation of matrix dentine or enamel. Conspicuously, labeled nucleotides (cytosine, guanosine, adenosine and uridine) were each transferred from both cell types and became constituents of the intervening organic matrix (see fig. 2). Most of the grain density over the matrix was RNAse labile. Preincubation of the tooth rudiments with dactinomycin, followed by pulse-chase addition of labeled RNA precursors, also prevented grain density from appearing over the extracellular matrix [24].

A direct test for RNAs in the extracellular matrix utilized phenol-SDS extrac-

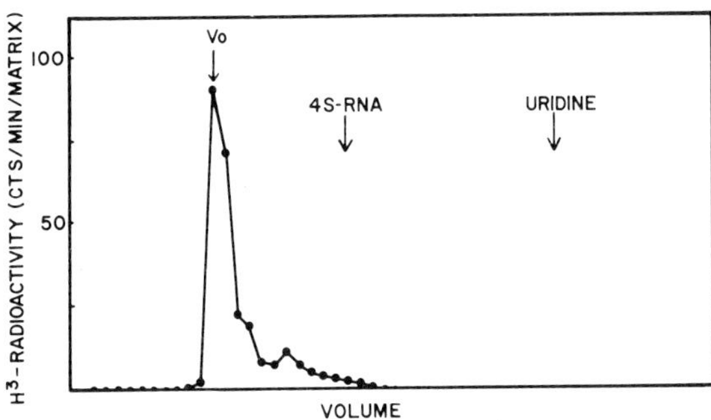

Fig. 3. Sephadex G-100 chromatograph of matrix RNAs labeled in organ culture. RNAs were phenol-SDS extracted from sonciated cervical matrices isolated from embryonic rabbit incisor rudiments incubated in culture with ^{3}H-uridine for 4 hr. After treatment with RNAse A and T_1, all radio activity rechromatographed at the uridine marker position.

tions of sonicated 25-day embryonic matrices previously incubated with radioactive RNA precursors. Control experiments demonstrated that labeled *E. coli* ribosomal RNA, introduced at the time of sonication, was not recovered in the matrix and, therefore, RNA contributions from disrupted cellular contamination would be very minimal. Materials present in the water phase after phenol-SDS extraction at room temperature were ethanol precipitated, dried and chromatographed on Sephadex G-100 (greater than 10^5 mol. wt. exclusion). The RNA isolated from the extracellular matrix consists of two major fractions, both of which are RNAse labile and orcinol positive (fig. 3). Studies with labeled methyl-methionine, cytosine, adenosine, and guanosine produced qualitatively identical results. Approximately 4–7 μg RNA per 24 mg sonicated matrix (dry weight) was recovered [24].

5. Isolation of matrix vesicles containing RNAs

Matrix vesicles (after a term introduced by Anderson [25]) of dimensions 500 to 1000 Å in diameter and varying in electron density and shape were observed within tooth extracellular matrix [26] (see fig. 1B). Serial thin sections indicate that these vesicles are distinct "extracellular organelles" that are discontinuous with cell processes. Electron microscopic cytochemical studies using Indium trichloride staining techniques for locating nucleic acids [27] demonstrated that a small percentage of the total matrix vesicles observed contains RNA (see fig. 5). Our observations of incisor and molar tooth development in situ indicate that at least three types of matrix vesicles (based on morphological criteria) are present within the extracellular matrix. One type of matrix vesicle contains ribosome-like granules. This type of vesicle was observed in the extracellular matrix and within both epithelium and mesenchyme in the differentiating cervical portion of the incisor. Vesicles with ribosome-like granules were not seen in matrix regions where cells had stopped

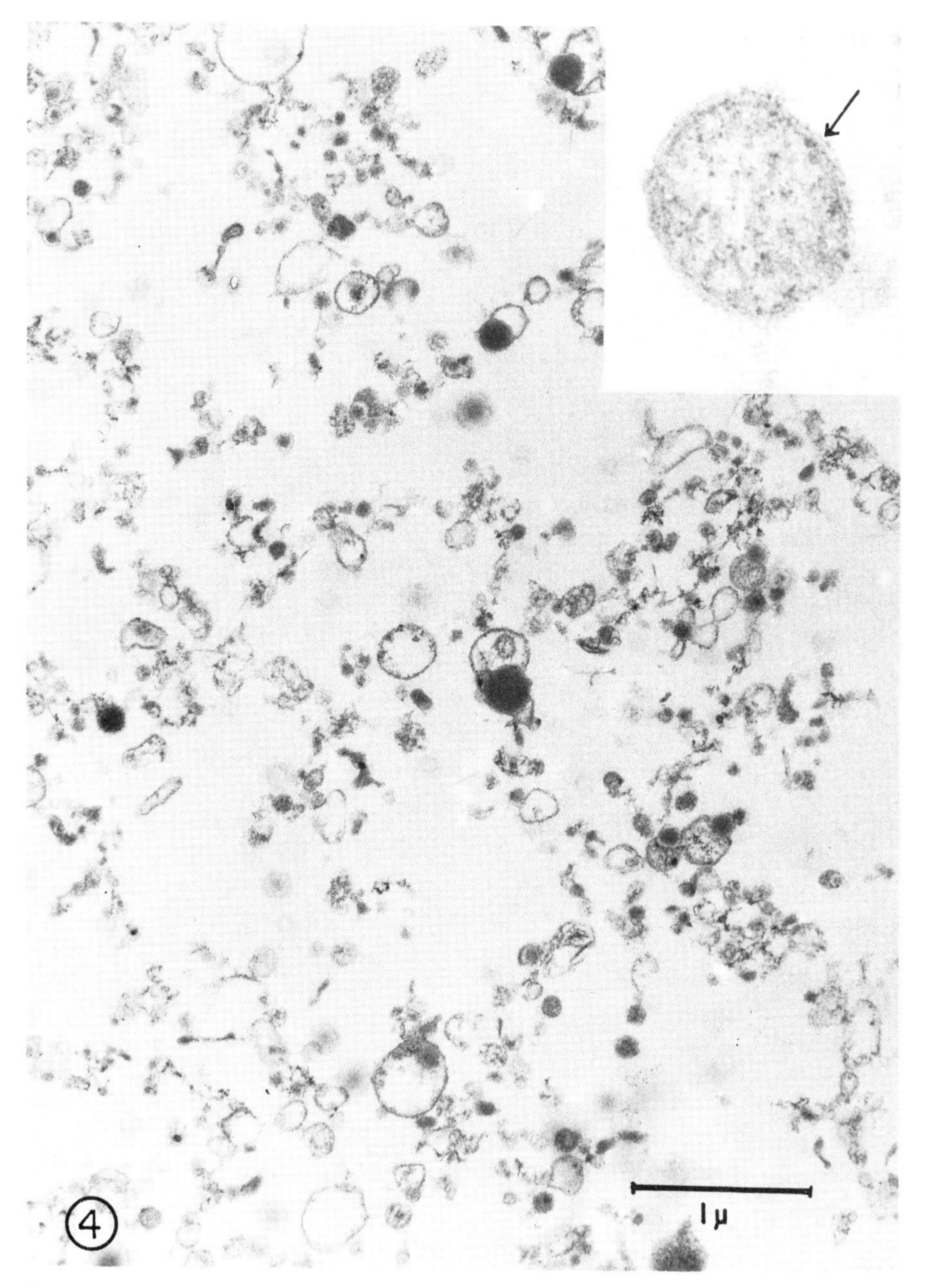

Fig. 4. Electron micrograph of the Bio-Gel A-50 m void volume preparation demonstrating the various types of matrix vesicles. Scale line, 1.0 μ. Insert: Matrix vesicle containing electron dense granules limited by a unit membrane (arrow).

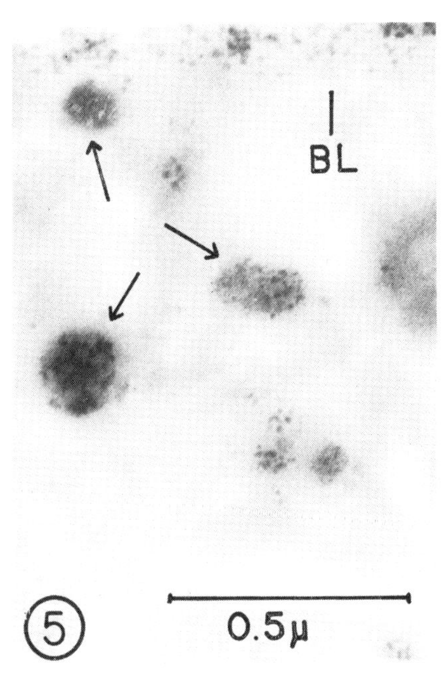

Fig. 5. Matrix vesicles (arrows) in the proximity of the epithelial basal lamina (BL) indicate an affinity for Indium comparable to that of numerous polysomes. Scale line, 0.5 μ.

dividing [26]. In light of these observations which imply that nucleic acid-containing matrix vesicles might be involved in mediating tooth morphogenesis, we developed procedures for the isolation and characterization of matrix vesicles from isolated matrices.

The method used to isolate matrix vesicles is based on the selective solubilization of the tooth matrices (a collagen connective stroma) by bacterial collagenase [28]. Sonicated, extracellular matrices are enzymatically digested for 1 hr at 37°C. The digest supernatant, after a short, low-speed centrifugation, is eluted through Bio-Gel A-50 m (mol. wt. exclusion, 5×10^7) and the void volume collected. Matrix vesicles obtained in this manner included morphological representatives of the various types observed in situ (figs. 2 and 4). RNA content in these matrix vesicles preparations was detected and then analyzed by dimethylsulfate in vitro labeling of phenol-SDS extracted A-50 m void volume fractions [29] followed by Sephadex G-100 column chromatography. Consistent with an association with matrix vesicles, the RNA obtained by this procedure was RNAse-insensitive prior to phenol-SDS extraction. Preincubation of collagenase extracts of sonicated matrices with RNAse A and T_1 for 1 hr at 37°C before A-50 m filtration did not prevent the subsequent G-100 recovery of high molecular weight. A single G-100 excluded peak of DMS-labeled RNA was obtained in these experiments (fig. 6).

The DMS labeled RNA isolated from matrix vesicles is kinetically distinct from

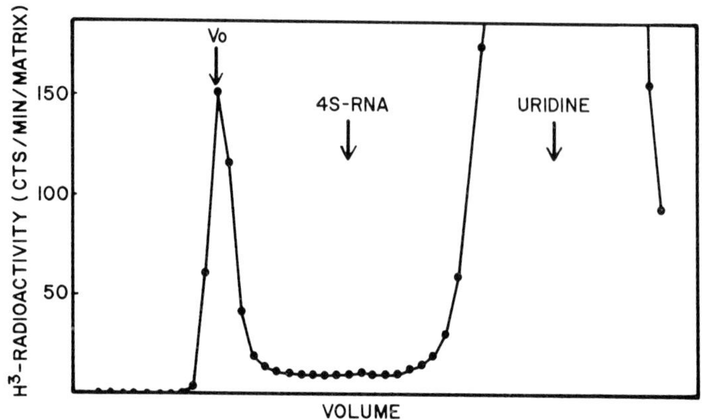

Fig. 6. Sephadex G-100 chromatograph of RNAse-resistant matrix vesicle (Bio-Gel A-50 m void volume) RNAs labeled in vitro with ^{3}H-dimethylsulfate (^{3}H-DMS). Sonicated matrices from 25-day embryonic rabbit incisors were collagenase digested, and the resulting supernatant filtered through Bio-Gel A-50 m. The A-50 m void volume was then incubated with RNAse A and T_1 (1 hr, 37°C), phenol-SDS extracted, reacted for 24 hr at 4°C with 10 μCi ^{3}H-DMS, dialyzed, and then chromatographed on G-100 at 4°C. The appreciable low molecular weight fraction is unreacted ^{3}H-DMS (not removed by dialysis). RNAse A and T_1 digestion (12 hr, 4°C) of an identical aliquot of ^{3}H-DMS-labeled RNA eliminated 80% of the G-100 void volume radioactivity.

the phenol-SDS extracted RNA found within in vitro labeled whole matrices. Attempts to label A-50 m void volume material by incubation in organ culture with radioactive RNA precursors for periods up to 14 hr have proven unsuccessful (unpublished data). In several different organ culture conditions (for periods up to 48 hr) total trichloroacetic acid-precipitable counts incorporated into sonciated matrices were extracted with 70–90% efficiency by collagenase digestion. Matrix vesicle preparations *were not* labeled. Tritiated glycerol (a specific membrane precursor) also *was not* incorporated into the A-50 m excluded volume. These results suggest a stable, slowly turning-over class of RNA contained in membrane-limited vesicles that are found in the extracellular organic matrix.

6. *Concluding remarks*

As yet our data can make no conclusive statement concerning the molecular activities responsible for information transfer during tooth morphogenesis. Our results do, however, restate in a new perspective the question of what is a morphogen. Circumstantial evidence implicating a role for RNA in developmental reactions is very strong. RNA-dependent DNA polymerase is present in normal embryonic tissues and mechanisms for the RNA modulation of gene activity is well documented [30]. Besides the detection of RNA in extracellular matrices showing morphogenetic activity, the observations of RNA-containing matrix vesicles can satisfy other criteria observed and expected in the transfer of genetic information. For example, induction does not occur across Millipore filters as the pore size approaches 0.1 μm; matrix vesicles average 0.05 to 0.1 μm in diameter [4, 31, 32]. Likewise, the RNAse insensitivity of matrix vesicles is compatible with the enzymatic hostility of extracellular environments [33]. Finally, similar appearing extracellular "vesicles" or "bodies" have been reported in other interacting tissue systems [34, 35].

We are presently preparing an in vitro culture system to test the morphogenetic activity of isolated extracellular matrix RNA and vesicles on homotypic cells populations. The tooth model is particularly well suited for histochemical and biochemical analysis of secondary embryonic induction since the epithelial response to developmental activation is the large scale production and export of a specific enamel protein. By the analysis of this protein production in vitro, coupled with the present emerging technology of messenger RNA isolation and characterization, we hope to follow the morpho genetic activity of tooth extracellular matrices well into genetic metabolism within the cell.

Acknowledgements

The authors wish to acknowledge the excellent technical assistance of P. Bringas, V. Mansour, P. Matosian, W. Mino and J. Schwab. The research investigations reviewed in this manuscript were supported by U.S. Public Health Service Research Grants DE-02848, DE-3560 and Training Grant DE-00094 from the National Institute of Dental Research. Dr. Slavkin is a recipient of a Research Career Development Award, No. 5 KO4 DE-41739, U.S.P.H.S. The authors wish to also thank Mrs. Joanne Leynwood and Mrs. Lore Hinton for typing the manuscript.

References

[1] H. Spemann, Embryonic development and induction (Yale University Press, New Haven (Conn.), 1938).
[2] M. Hamburgh, Theories of differentiation (American Elsevier, New York, 1971).
[3] C. Grobstein, Natl. Cancer Inst. Monogr. 26 (1967) 279.
[4] L. Saxen, O. Koskimies, A. Lahti, H. Miettinen, J. Rapola and J. Wartiovaara, In: Advances in morphogenesis Vol. 7, Eds. M. Abercrombie, J. Brachet and T.J. King (Academic Press, New York, 1968) 251–293.
[5] C. Grobstein, Nature (London) 172 (1953) 869.
[6] R. Fleischmajer and R.E. Billingham, Epithelial-mesenchymal interactions (Williams and Wilkins, Baltimore, Md., 1968).
[7] H.C. Slavkin, In: Cellular and molecular renewal in the mammalian body, Eds. I.L. Cameron and J.D. Thrasher (Academic Press, New York, 1971) 221–276.

[8] E.J. Ambrose, In: Cell differentiation, Eds. A.V.S. DeReuck and J. Knight, Little, Brown and Company, Boston, (1967) 110–115.
[9] J. Brachet, In: Comprehensive biochemistry, Vol. 28, Eds. M. Florkin and E.H. Stotz (Elsevier, Amsterdam, 1967) 23–53.
[10] M.C. Niu, Develop. Biol. 7 (1963) 379–393.
[11] H. Tiedemann, In: Morphological and biochemical aspects of cytodifferentiation, Vol. 1, Eds. E. Hagen, W. Wechsler and P. Zilliken (S. Karger, New York, 1967) 8–21.
[12] W.E. Koch, J. Exptl. Zool. 165 (1967) 155.
[13] C.Y. Kang and H.M. Temin, This symposium, page 339.
[14] D.B. Villee and A. Goswami, This symposium, page 73.
[15] M.C. Niu, L.C. Niu and S.F. Yang, This symposium, page 90.
[16] H.Y. Lee and M.C. Niu, This symposium, page 137.
[17] A.K. Deshpande, L.C. Niu and M.C. Niu, This symposium, page 229.
[18] S.J. Segal, This symposium, pages 125 and 270.
[19] H.C. Slavkin and L.A. Bavetta, Developmental aspects of oral biology (Academic Press, New York and London, 1972).
[20] H.C. Slavkin, R. LeBaron, J. Cameron, P. Bringas and L.A. Bavetta, J. Embryol. Exptl. Morphol. 22 (1969) 395.
[21] H.C. Slavkin and L.A. Bavetta, J. Dent. Res. 47 (1968) 779.
[22] H.C. Slavkin, J. Beierle and L.A. Bavetta, Nature (London) 217 (1968) 269.
[23] H.C. Slavkin, P. Bringas and L.A. Bavetta, J. Cell Physiol. 73 (1969) 179.
[24] H.C. Slavkin, P. Flores, P. Bringas and L.A. Bavetta, Develop. Biol. 23 (1970) 276.
[25] H.C. Anderson, J. Cell Biol. 41 (1969) 59.
[26] H.C. Slavkin, P. Bringas, R. Croissant and L.A. Bavetta, Mech. Age Develop. 1 (1972) 139.
[27] M.L. Watson and W.G. Aldridge, J. Biophys. Biochem. Cytol. 11 (1969) 257.
[28] H.C. Slavkin, R. Croissant and P. Bringas, J. Cell Biol. 53 (1972) 841.
[29] H. Akiyoshi and N. Yamamoto, Biochem. Biophys. Res. Comm. 38 (1970) 915.
[30] H. Temin, J. Nat. Cancer Inst. 46 (1971) III.
[31] L. Saxen and E. Saksela, Exptl. Cell. Res. 66 (1971) 369.
[32] S. Nordling, H. Miettinen, J. Martiovaara and L. Saxen, J. Embryol. Exptl. Morphol. 26 (1971) 231.
[33] R. Juliano and E. Mayhew, Exptl. Cell Res. 73 (1972) 3.
[34] A.M. Cohen and E.D. Hay, Develop. Biol. 26 (1971) 578.
[35] M.R. Bernfield and N.K. Wessells, Develop. Biol., Suppl. 4 (1970) 195.

Niu and Segal (eds.). The role of RNA in reproduction and development
North-Holland Publ. Co., 1973

Nucleic acid-induced genetic changes in *Neurospora*

N.C. MISHRA, G. SZABO* and E.L. TATUM
The Rockefeller University, New York, N.Y. 10021, USA

Effects of nucleic acids on the reversion of different genetic markers (*inos*$^-$, *pdx*$^-$ and *rg*$^-$) in *Neurospora* has been described. Allo-DNA was found to increase the frequency of reversion at these 3 loci significantly. Iso-DNA was found to have no such effects. The effect of allo-RNA has been studied only for the *inos*$^-$ locus and the reversion-frequency was found to increase significantly following RNA-treatment. Effects of allo-DNA and allo-RNA were abolished by treatment of the nucleic acid preparations with DNase and RNase respectively. Although these revertants were found to be quite stable, some of the DNA-induced revertants failed to transmit the allo-DNA trait (*inos*$^+$) to the progeny of genetic crosses.

1. Introduction

In comparison with bacteria, the study of transformation in higher organisms is complicated by a host of factors including the diploid nature of the organism, the chromosomal organization of the genetic material and the fact that the genetic analysis of the somatic cells is not a simple process. In this respect, *Neurospora*, a eukaryote which possesse several attributes of the prokaryotic and microbial systems, should provide a good organism for the study of transformation in the higher organisms. Here, we present a preliminary account of our work which provides evidence for the occurrence of nucleic acid mediated genetic changes in *Neurospora*.

In order to approach the problem of transformation we have studied the role of both DNA and RNA in inducing the reversion of different genetic markers in *Neurospora*. The genetic markers included in our study are *inos*$^-$, *pdx*$^-$ and *rg*$^-$. Mutations at *inos*$^+$ and *pdx*$^+$ loci result in a specific biochemical requirement for inositol [1] and pyridoxine [2] respectively, whereas the *rg*$^-$ mutation causes a distinctive colonial morphology in *Neurospora* [3–5]. The choice of these loci in the transformation study is based on their exceedingly low reversion frequencies. The particular inositol requiring mutant (89601), used extensively in our study, has been reported by Giles [6] to have a very low frequency of reversion on exposure

* Present address: The Biology Institute, Medical University, Debrecen, Hungary.

to ultraviolet light or X-ray. An important consideration, in this respect, was the fact that Shockley and Tatum [7] have earlier provided suggestive evidence for transformation at this *inos*$^-$ locus. The ragged (*rg*$^-$) mutant has never been known to revert either spontaneously or in response to ultraviolet light (Mishra, unpublished data). Furthermore, the primary biochemical lesions of *inos*$^-$ and *rg*$^-$ mutations are now known to involve a defective glucose-6-phosphate-cyclase [8] and altered phosphoglucomutases [9, 10] respectively. Therefore, the nucleic acid-induced revertants at the selected loci can be analyzed biochemically for an additional parameter.

2. Materials and methods

2.1. Strains

Neurospora strains used in the present study are listed in table 1. These strains have been previously described [1–6] in detail. Strains with multiple genetic markers (table 1), constructed by appropriate genetic crosses of the original mutants [1–6], were used in order to check against any possible contamination during our experiments. These strains were maintained on appropriately supplemented medium [11]. A large sample of ascospores was analyzed using L-sorbose [12] in the medium as described by Giles [6]. Ascopores from a particular cross were given the necessary heatshock in liquid agar medium (with and without biochemical supplements) at 60°C. The medium containing the ascospores was poured into Petri plates (25 ml/plate). Once the agar solidified these plates were incubated for 40–60 hr and then examined for the colonies growing on medium with or without biochemical supplements.

2.2. Preparation of nucleic acids

DNA was prepared by the method of Marmur [13]; RNA was prepared by phenol extraction and the polyadenylate-rich RNA fraction was obtained by the method of Rosenfeld et al. [14]. *Neurospora* DNA prepared by Marmur's method was fibrous and appeared to be of high molecular weight, but was polydisperse in size (Szabo, unpublished data). The DNA preparations contained some RNA and protein. The RNA preparations were, however, completely free of DNA. DNA and RNA were determined chemically by diphenylamine and orcinol reactions [15] respectively; protein was determined by the method of Lowry et al. [16]. Nucleic acids and proteins were also routinely determined spectrophotometrically. Nucleic acids were dissolved in the appropriate buffer (saline–citrate or saline) [13, 14] before use in experiments.

TABLE 1
Description of the *Neurospora* strains

Strain	Genetic background	Genetic markers	Phenotype	Remarks
1. RL-3-8A	*N. crassa* (Rockefeller)	*rg*$^+$ *inos*$^+$; A (mating-type)	Wild type	*Donor strain* (used for preparation of allo-DNA and RNA)
2. R2506-8-12	*N. crassa*	*rg*$^-$ (R2506); *inos*$^-$ (89601); a (mating-type)	Colonial morphology, requires inositol for growth, defective phosphoglucomutase and glucose-6-phosphate-cyclase	*Recipient strain* (also used for preparation of iso-DNA & RNA)
3. R2473-2-1A	*N. crassa*	*os*$^-$ (R2473); *inos*$^-$ (89601); A (mating-type)	Morphological, requires inositol for growth	*Recipient strain* (and also used in genetic crosses)
4. 89601	*N. crassa*	*rg*$^+$; *inos*$^-$ (89601); a/A (mating-type)	Wild-type morphology, requires inositol for growth	Used in crosses for genetic analysis of the revertants
5. WA	*N. sitophila* (Whitehouse)	*rg–1*$^+$; *pdx*$^+$; A (mating-type)	Wild type (growth not inhibited by deoxypyridoxine)	*Donor strain* (used for preparation of allo-DNA)
6. 18-2A	*N. sitophila*	*rg–1*$^-$ (M-17); *pdx*$^-$ (299); A (mating-type)	Colonial morphology, defective phosphoglucomutase, requires pyridoxine for growth, no growth in presence of deoxypyridoxine	*Recipient strain*

2.3. Designation of nucleic acid preparations

The nucleic acid preparations from the donor strain carrying the wild-type allele(s) of the genetic marker(s) have been designated as allo-DNA and allo-RNA, whereas similar nucleic acid preparations from the mutant recipient strain have been called iso-DNA and iso-RNA.

2.4. Treatment of recipient strains

Conidia or mycelial fragments of the recipient strains were treated with DNA or

RNA (50 μg/ml) in 10 ml of minimal liquid medium [11] appropriately supplemented with inositol or pyridoxine. Saline citrate alone (without nucleic acid) was added to the control cultures. In other control experiments DNase treated DNA or RNase treated RNA was added. The digestion of nucleic acid by DNase or RNase was performed by methods described elsewhere (Enzyme Manual, Worthington Biochemical, Freehold, N.J.). Cultures treated with or without nucleic acids were then grown as shake-cultures for 48 hr at 30°C. Afterwards, the mycelium was harvested, homogenized in 20% sucrose-phosphate buffer (0.2 M) pH 6.5 and then spread on minimal agar and also on appropriately supplemented medium. The plates were incubated at 25°C and after 4–5 days the inositol- or pyridoxine-independent revertants were scored as colonies growing on minimal plates. In the case of the pyridoxine requiring mutant, the cultures were plated on minimal medium containing deoxypyridoxine. The *pdx*$^-$ mutants are killed by this analogue whereas the revertants (*pdx*$^+$), which are independent of this specific biochemical requirement, remain unaffected [17].

3. Results and discussion

3.1. Effect of allo-DNA on genetic reversion

An extensive study has been carried out with the inositol requiring strain. The recipient (*inos*$^-$) cultures were treated with allo-DNA and then examined for the appearance of the wild-type (*inos*$^+$) revertants. Control experiments included the treatment of the recipient culture (a) with iso-DNA or (b) with saline–citrate alone (i.e. without DNA) or (c) with allo-DNA previously digested by DNase. Results of these 4 different kinds of experiments are presented in table 2. The data in the first column of table 2 show that on treatment with allo-DNA, a significant number of experiments resulted in reversion at the *inos*$^-$ locus, whereas in all the control

TABLE 2
Effect of allo-DNA on the reversion of the *inos*$^-$ locus in *N. crassa*

Treatment	Total number of		Revertants	
	Experiments	Colonies ($\times 10^{-6}$)	Total	Frequency [a]
1. Allo-DNA(*inos*$^+$)	55 (37) [b]	405.3	387	0.95
2. Iso-DNA(*inos*$^-$)	4 (1) [b]	335.0	9	0.03
3. Citrate	49 (7) [b]	582.9	23	0.04
4. DNase-digested allo-DNA	5 (0) [b]	10.2	0	0

[a] Frequency expressed as per million.
[b] Number of experiments with revertants.

TABLE 3
Effect of allo-DNA on the reversion of the pdx^- locus in *N. sitophila*

Treatment	Total number of		Revertants	
	Experiments	Colonies ($\times 10^{-6}$)	Total	Frequency [a]
1. Allo-DNA(pdx^+)	5 (4) [b]	9.3	35	3.60
2. Citrate	3 (3) [b]	47.9	13	0.26

[a] Frequency expressed as per million.
[b] Number of experiments with revertants.

experiments the change to inositol-independence was found to occur only in a few cases. Furthermore, when the frequencies of $inos^+$ reversion obtained with the different treatments are compared, the allo-DNA treated cultures showed a marked difference from the controls. Treatment with allo-DNA increased the frequency of reversion at the $inos^-$ locus by about 25–30-fold as compared to the frequency of reversion obtained in control experiments. Results presented in table 2 also show that iso-DNA and the DNase-treated allo-DNA were not effective in inducing $inos^+$ reversion.

The transforming activity of allo-DNA for another biochemical genetic marker (pdx^-) has also been examined; these data are presented in table 3. The reversion frequency of the pdx^- locus was increased by almost 15-fold following treatment of the recipient (pdx^-) with allo-DNA.

The reversion at the rg^- locus was indicated by the wild-type growth of the

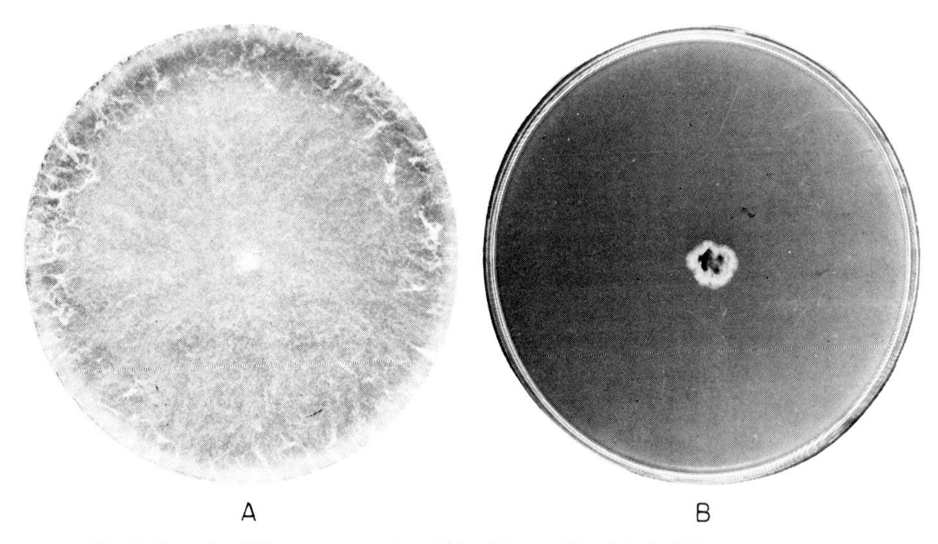

Fig. 1. Growth of *Neurospora* strains: (A) wild-type (RL-3-8A), 40-hr-old culture, and (B) *ragged* mutant (R2506-8-12), 6-day-old culture.

culture and was, therefore, determined by visual inspection. The wild-type and mutant strains showed different growth patterns as seen in fig. 1. The wild-type strain showed rapid growth and covered the whole plate within 40 hr. The mutant strain, however, showed a restricted growth even after 6 days (fig. 1). Seven wild-type revertants (rg^+) were scored among 3000 colonies of the recipient (rg^-) strain treated with allo-DNA. No revertant at the rg^- locus was found among 5000 colonies not exposed to DNA or treated with iso-DNA.

3.2. Effect of allo-RNA on genetic reversion

The rationale for using RNA as a transforming principle derives from more recent work in biology [18–23]. Acquisition of a drug resistance character by sensitive bacteria following treatment with purified RNA preparations has been described in *Bacillus subtilis* [18], pneumococcus [19] and *Escherichia coli* [23]. Recently, the role of exogenous RNA on the pattern of organ development in higher organisms has been described [20–22]. The role of RNA in development is also suggested by the occurrence of the enzyme reverse transcriptase, which can transcribe a RNA template into DNA [25, 26], and also be experiments in which a particular cell can translate the message carried by foreign RNA [27–29]. Results of our experiments in which RNA was used as transforming material instead of DNA are presented in table 4. The recipient cultures (*inos*$^-$) were treated with the polyadenylate-rich RNA fraction and then were examined for the occurrence of stable wild-type revertants (*inos*$^+$). The data presented in table 4 show that in the case of allo-RNA treated cultures all experiments produced *inos*$^+$ revertants, whereas in the controls no revertants were obtained. The allo-RNA-treated recipient cultures showed a 10-fold increase in the reversion frequency when compared to the frequency of reversion obtained in the controls of the DNA experiments (see table 2). The RNA preparations other than the polyadenylate-rich fraction were also found to be active in increasing the reversion frequency at the *inos*$^-$ locus. The activity of the different RNA fractions was completely destroyed by RNase but not by DNase.

TABLE 4
Effect of allo-RNA on the reversion of the *inos*$^-$ locus in *N. crassa*

Treatment	Total number of		Revertants	
	Experiments	Colonies ($\times 10^{-6}$)	Total	Frequency [a]
1. RNA	5 (5) [b]	40.2	11	0.27
2. Saline	3 (0) [b]	81	0	0

[a] Frequency expressed as per million.
[b] Number of experiments with revertants.

3.3. Specific effect of DNA and RNA

The use of strains with multiple markers ensured easy detection of possible contaminants of the experimental cultures. The wild-type revertants, obtained during our experiments, were found to carry all other genetic markers as such. Rarely, isolates appeared having reversion at more than one locus; these might have been co-revertants, but were discarded as possible contaminants. The revertants obtained after nucleic acid treatment can, therefore, be reasonably taken as produced by induced-reversion.

The specific effect of allo-DNA was indicated by the fact that iso-DNA was unable to increase the reversion frequency of the *inos*$^-$ locus. This is also evidenced by the fact that the effect of allo-DNA was abolished by treatment with DNase (see table 2). The specific effect of allo-RNA was suggested by the fact that its activity was lost by treatment with RNase but not by DNase.

3.4. Stability of the revertants

The revertants obtained during our present study have been found to be very stable when propagated vegetatively. Some of these have been maintained on the minimal medium for more than a year to date.

3.5. Genetic analysis

An extensive genetic analysis of the DNA-induced revertants has been performed. The *inos*$^+$ revertants were crossed with a strain carrying the *inos*$^-$ genetic marker; the progeny from such individual crosses were examined for their ability to grow on minimal or inositol-supplemented medium. Results of the random spore analysis of these crosses (revertant (*inos*$^+$) × *inos*$^-$) are presented in tables 5 and 6. In such a cross the wild-type revertants, which originated in the control experiments, showed a normal Mendelian transmission of the revertant (*inos*$^+$) character (see table 5). However, the results of the crosses involving allo-DNA mediated *inos*$^+$ revertants showed aberrant transmission of the allo-DNA trait (table 6). The results of these

TABLE 5
Genetic analysis of the wild-type revertants (*inos*$^+$) obtained without DNA treatment

Cross [a]	No. of progeny	
	inos$^-$	*inos*$^+$
1.	271	302
2.	817	789
3.	70	83
4.	51	28

[a] Revertant × *inos*$^-$.

TABLE 6
Genetic analysis of the wild-type revertants (*inos*$^+$) obtained with allo-DNA-treatment

Cross [a]	No. of progeny		Cross [a]	No. of progeny	
	inos$^-$	*inos*$^+$		*inos*$^-$	*inos*$^+$
1.	90	91	6	126	3
2.	197	151	7	297	3
3.	457	239	8	369	0
4.	374	117	9	700	0
5.	1,608	893	10	20,000	0
			11	100,000	0

[a] Revertant (*inos*$^+$) × *inos*$^-$.

crosses were essentially of three kinds: (1) A normal Mendelian transmission of *inos*$^+$ and *inos*$^-$ characters was seen and thus *inos*$^+$ and *inos*$^-$ phenotypes appeared in 1 : 1 ratio among the progeny. (2) The transmission was not Mendelian – and only a fraction of the progeny inherited the *inos*$^+$ character. (3) The transmission of the *inos*$^+$ character to the progeny was very rare: as for example, only 1 in several thousand progeny analyzed.

The aberrant results of these crosses could be explained by any of the following suppositions: (1) That these revertants were heterokaryotic and carried a large number of *inos*$^-$ nuclei in comparison to *inos*$^+$ nuclei. (2) That the transmission of the allo-DNA-mediated *inos*$^+$ character was somehow prevented at meiosis during the sexual process. In view of the fact that the allo-DNA induced revertants showed no heterokaryosis for *inos*$^+$/*inos*$^-$ loci in conidial or mycelial analysis, these results (genetic crosses showing rare transmission of the allo-DNA trait) could be explained by the second hypothesis (i.e. the transmission of the allo-DNA-mediated *inos*$^+$ character was prevented during meiosis).

It is clear from these data that at least some of the revertants, which were obtained following the treatment with allo-DNA, do not have the characteristics of a typical back mutation. Results presented in this paper suggest a system in which the DNA pieces carrying the genetic information are able to replicate and also manage to find expression during the vegetative cycle, but remain unintegrated and are excluded during meiosis. It is of interest to compare the features of our system with others. An earlier report of transformation [30] at the pyrimidine (*pyr*) locus in *Neurospora* did not exclude the role of spontaneous reversion and selection. Shockley and Tatum [7] in their extensive study have provided suggestive evidence for transformation in *Neurospora*. The reversion frequencies at the selected loci obtained by these authors were much lower than those reported here. The higher reversion frequency in the present study can be ascribed to several factors such as the choice of the particular strains and the better DNA preparation.

Our results do not compare with bacterial transformation particularly in view of the low transformation frequency and the inability of some of the revertants to transmit the allo-DNA-effect to their progeny; also nothing is known about competence in *Neurospora*.

However, the features that we have described are similar to those reported in *Drosophila* by Fox and his collaborators [31], especially in view of the fact that there is no transmission of the allo-DNA-effect to the progeny of the transformed fruit flies. The fact that Fox et al. [31] were unable to find a whole-body transformant in his experiments also suggests that there is no integration of the allo-DNA in germ cells of *Drosophila* [31].

No transformation mediated by RNA has previously been reported in *Neurospora.* It is conceivable that the enzyme reverse transcriptase might be involved in this process. The genetic analysis of the RNA-mediated revertants is not yet available but is being performed and should be of much interest. It is pertinent to our interest that Beljanski, earlier in this symposium, has presented further evidence of RNA-mediated transformation in bacteria.

It is of further interest to describe some of the other effects of DNA seen during transformation in *Neurospora.* A general decrease in fertility and in percentage of ascospore germination was noticed among the DNA-induced revertants. Some of these revertants were also found to have markedly slow growth. In crosses with the wild-type strain, the slow growth characteristics of these revertants segregate as single gene characters, termed modifiers. These modifiers seem to affect the growth pattern of both the ragged and the wild-type strains from crosses. These results are similar to those of Yoshikawa [32] in *Bacillus subtilis* where a mutagenic effect of DNA has been reported to accompany transformation.

In summary, we have presented data to show the occurrence of both allo-DNA and allo-RNA mediated genetic change in *Neurospora.* Further examination of transformation by DNA and RNA in *Neurospora,* particularly in the detailed genetic analysis possible in this eucaryotic micro-organism, should be valuable in elucidating the mechanisms of uptake, integration, and expression of genetic information carried by free nucleic acids. It would seem probable that information so obtained in a relatively simple organism will be pertinent to similar studies with more complex eucaryotic cells and organisms, including mammals.

Acknowledgements

We would like to express our thanks to Dr. M.C. Niu for his valuable help in RNA-preparation. We also thank Miss Anne Hamill for her expert technical help. This work was supported in part by a grant-in-aid from the Research Corporation and by grant RG 16224 from the National Institutes of Health.

References

[1] G.W. Beadle, J. Biol. Chem. 156 (1944) 683.
[2] G.W. Beadle and E.L. Tatum, Proc. Natl. Acad. Sci. U.S. 27 (1941) 499.
[3] D.D. Perkins, Genetics 44 (1959) 1185.
[4] L. Garnjobst and E.L. Tatum, Genetics 57 (1967) 579.
[5] N.C. Mishra and S.F.H. Threlkeld, Genetics 55 (1967) 113.
[6] N.H. Giles, Jr., Cold Spring Harbor Symp. Quant. Biol. 16 (1951) 283.
[7] T.E. Shockley and E.L. Tatum, Biochem. Biophys. Acta 61 (1962) 567.
[8] E. Pina and E.L. Tatum, Biochem. Biophys. Acta 136 (1967) 265.
[9] S. Brody and E.L. Tatum, Proc. Natl. Acad. Sci. U.S. 58 (1967) 923.
[10] N.C. Mishra and E.L. Tatum, Proc. Natl. Acad. Sci. U.S. 66 (1970) 638.
[11] H.J. Vogel, Am. Naturalist 98 (1964) 435.
[12] E.L. Tatum, R.W. Barrat and V.M. Cutter, Sience 109 (1949) 509.
[13] J. Marmur, J. Mol. Biol. 3 (1961) 208.
[14] G.C. Rosenfeld, J.P. Comstock, A.R. Means and B.C. O'Malley, Biochem. Biophys. Res. Comm. 47 (1972) 387.
[15] W.C. Schneider, In: Methods in enzymology, Vol. 3, Eds. S.P. Colowick and N.O. Kaplan (Academic Press, London, New York, 1957) 680–685.
[16] O.H. Lowry, N.J. Rosebrough, A.L. Farr and R.J. Randall, J. Biol. Chem. 193 (1951) 265.
[17] J.C. Rabinowitz and E.E. Snell, Arch. Biochem. Biophys. 43 (1953) 399.
[18] S.C. Shen, M. Hong, R. Cai, W. Chen and W. Chang, Scientia Sinica 11 (2) (1962) 233. Quoted by T. Cheng and R.H. Doi in: Progr. Nucl. Acid. Res. Mol. Biol. 8 (1968) 335.
[19] A. Evans, Proc. Natl. Acad. Sci. U.S. 52 (1964) 1432.
[20] M.C. Niu, Proc. Natl. Acad. Sci. U.S. 44 (1958) 1264.
[21] M.C. Niu, this symposium, pages 90, 137, 183, 229 and 287.
[22] S.J. Segal, O.W. Davidson and K. Wada, Proc. Natl. Acad. Sci. U.S. 54 (1965) 782.
[23] M. Beljanski and P. Bourgarel, C. R. Acad. Sci. Paris D 272 (1971) 2107.
[24] M. Beljanski, M. Beljanski, P. Manigault and P. Bourgarel, Proc. Natl. Acad. Sci. U.S. 69 (1972) 191.
[25] H.M. Temin and S. Mizutani, Nature 226 (1970) 1211.
[26] D. Baltimore, Nature 226 (1970) 1209.
[27] J.B. Gurdon, C.D. Lane, H.R. Woodland and G. Marbaix, Nature 233 (1971) 177.
[28] R.A. Lasky, J.B. Gurdon and L.A. Crawford, Proc. Natl. Acad. Sci. U.S. 69 (1972) 3665.
[29] S.J. Segal, this symposium, pages 125 and 270.
[30] C.A. Shamoian, A. Canzanelli and J. Melrose, Biochem. Biophys. Acta 47 (1961) 208.
[31] A.S. Fox, S.B. Yoon, W.F. Duggleby and W.M. Gelbart, In: Informative molecules in biological system, Ed. L. Ledoux (North-Holland, Amsterdam, 1971) 313–334.
[32] H. Yoshikawa, Genetics 54 (1966) 1201.

Niu and Segal (eds.). The role of RNA in reproduction and development
North-Holland Publ. Co., 1973

Specific and heterospecific transfer of hormone action by mRNA

Sheldon J. SEGAL, Rufus IGE*, Mario BURGOS**,
Pentti TUOHIMAA‡ and S.S. KOIDE

The Population Council, The Rockefeller University, New York City 10021, USA

The action of progesterone in mediating the synthesis of avidin by the chick oviduct can be simulated by the intra-oviductal instillation of nitrocellulose-trapped RNA from hormonally-prepared chick or pigeon oviduct. Similarly, the pigeon oviduct synthesizes avidin in response to chick oviduct RNA. Thus, a heterospecific transfer of hormonal stimulation, through the transfer of progesterone-induced RNA, is demonstrated. This can be achieved also, between vertebrate classes. The oviduct of *Xenopus laevis* synthesizes avidin in response to mRNA extracted from the progesterone-stimulated chick, even though progesterone itself does not cause avidin synthesis in the doses employed. The morphological effects of progesterone, including the stimulation of cilia formation and secretory activity, can also be achieved by the intra-oviductal administration of nitrocellulose-trapped RNA. The biological activity is lost following pancreatic RNase digestion. The 50-fold purification achieved by nitrocellulose chromatography of the total RNA preparation suggests that the activity resides in a messenger RNA fraction.

1. Introduction

There is considerable interest in the mechanism by which steroid hormones elicit characteristic responses in specific target cells. Through the years, various theories have been proposed to explain the basis for the activity of this class of hormone. For example, it has been proposed that steroid hormones may influence membrane permeability of target cells [1], the transport of metabolites [2] or the formation of cyclic AMP [3]. Other theories invoke a stimulatory effect on specific enzyme systems [4], an allosteric effect of steroid-protein macromolecules [5] or the ac-

On leave from:
* University of Ibadan, Nigeria
** University of Cuyo, Mendoza, Argentina
‡ University of Tampere, Finland.

tivation of gene function [6]. Evidence supporting the concept that steroid hormones influence gene function has been based on several experimental systems, including the effects of estradiol on the rat uterus [7] and progesterone on the chick oviduct [8]. In each case, a primary event after hormone stimulation is the rapid increase in the rate of synthesis of messenger RNA. The qualitative nature of the RNA produced differs between the hormone-stimulated and the non-stimulated cell [9]. The effects of these hormones on their respective target cells can be prevented by the inhibition of DNA-dependent RNA synthesis, through the administration of actinomycin-D [10].

In recent years, a new line of experimentation has evolved, providing direct evidence for the role of hormone-stimulated RNA in the conversion of steroid-dependent cells from the quiescent to the active state. Biologically active RNA can be extracted from the uteri of estrogen-stimulated rats and this extract is capable of transforming a non-stimulated rat uterine epithelium to the morphological appearance characteristic of the stimulated state [11]. The possibility that the extract contains a residue of contaminating estradiol molecules cannot be ruled out completely [12]. This explanation, however, would not account for the inactivity of RNA preparations from other organs of estrogen-treated rats or the loss of biological activity when RNA from the uteri of estrogen-stimulated rats is pre-incubated with pancreatic RNase. This observation suggests that after the hormone initiates biosynthesis of RNA, it is not necessarily involved in subsequent steps leading to the morphological changes resultant from hormone stimulation.

Similar observations pertaining to the action of progesterone have been made. Treatment of estrogen-primed chicks with progesterone increases the rate of synthesis by the oviduct of nuclear RNA, induces the appearance of new species of RNA [8] and consequently, evokes the synthesis by the oviduct of the hormone-dependent protein, avidin [13]. The concept that the effect of progesterone is mediated through RNA is supported by the observation that total RNA extracted from progesterone-stimulated chick oviducts can induce avidin synthesis in chicks treated with stilbestrol [14]. Neither RNA from other organs nor oviductal RNA hydrolysed enzymatically in capable of eliciting this biological response. The shell gland, a specialized region of the chick's Mullerian duct system, does not synthesize avidin in response to progesterone, but does produce avidin when treated with RNA extracted from the oviduct of progesterone-stimulated chicks [14]. Tissue minces of chick oviduct synthesize avidin when progesterone is added in vitro to the incubation medium; a total RNA extract of progesterone-stimulated oviducts can duplicate this in vitro effect of thc hormone [14]. Considerable purification of RNA can be achieved by nitrocellulose trapping [15]. This poly A RNA from the oviduct of hormone-treated chicks codes for the synthesis of avidin in a rabbit reticulocyte lysate system [16].

Thus, these are examples of homologous in vivo transfer of the message for avidin synthesis coded in response to progesterone, as well as both homologous and heterologous in vitro systems in which RNA coded for avidin induces the synthesis of this progesterone-dependent protein, even though the hormone is absent. Furthermore, a highly purified, poly A RNA synthesized in the oviduct of one avian species, under progesterone stimulation, can initiate avidin synthesis in another avian species.

2. *Methods*

2.1. Preparation of RNA

Female chicks were received in the laboratory from a commercial hatchery on the first post-hatching day. They were treated with stilbestrol by subcutaneous injection of 0.5 mg per 0.05 ml sesame oil daily for five days (days 2–6 post-hatching). On day 7, the chicks were administered 5 mg progesterone in 0.25 ml propylene glycol by subcutaneous injection. Chicks thus treated, usually in batches of 100, were sacrificed 12 hr after the single injection of progesterone, since this is the approximate time of maximum RNA synthesis in the estrogen-primed, progesterone-treated chick [8]. The oviducts were dissected out, separated from the larger, muscular shell gland and frozen in glass vessels over dry ice. Extraction of poly A RNA was based on the method of Brawerman et al. [15]. Chick or pigeon oviducts were homogenized in 10–15 ml of buffer per gram of tissue. The buffer system contained 100 mM Tris-HCl pH 7.6; 0.5% dodecyl sodium sulfate (SDS); 3 mM $MgCl_2$, 0.32 M sucrose. The homogenate was filtered through cheese cloth and centrifuged at 15,000 *g* for 20 min. One volume of water-saturated phenol was added to the supernatant. After stirring at 4°C for 30 min, the mixture was centrifuged at 4000 *g* for 30 min. One volume of buffer (100 mM Tris-HCl pH 9.0; 0.5% SDS) was added to the non-aqueous residue, and stirring and centrifuging were repeated. The aqueous phases were combined and the RNA was precipitated by addition of 0.1 vol of 1 M NaCl and 2.5 vol of ethanol. The alcoholic solution was stored at –20°C overnight. Following centrifugation, the precipitate was washed 3 times with 0.1 M NaCl–66% ethanol solution. The RNA was dissolved in distilled H_2O and washed 5–6 times with equal volumes of cold ether. The final ether traces were removed with an air stream. The total RNA fraction, diluted 20-fold, was passed through a nitrocellulose column of 40 mm length and 15 mm diameter equilibriated with the diluent buffer (500 mM KCl, 1 mM $MgCl_2$, 10 mM Tris-HCl, pH 7.6). The nitrocellulose was transferred into a centrifuge tube and the RNA eluted with high pH buffer (100 mM Tris-HCl, pH 9.0, 0.5% SDS). The RNA solution was repeatedly chilled to 4°C and centrifuged to remove traces of SDS.

The same procedure was used to prepare purified RNA from the oviducts of stilbestrol-primed, progesterone-treated pigeons that were six weeks old when hormone treatment was begun.

2.2. In vivo studies

The nitrocellulose-trapped RNA, dissolved in saline solution, was placed into the oviducts of ether-anesthetized chicks or pigeons by means of a 27 gage needle and a Hamilton syringe. The volume used was 100 μl. The recipient birds were of the same age, and had the same schedule of stilbestrol-priming as described above. The birds were sacrificed 24 hr later, and the avidin content of each treated oviduct, as well as various controls, determined by means of a sensitive and specific assay based on binding of avidin to ^{14}C-carbonyl-labelled biotin. A similar procedure was used to test chick RNA in *Xenopus laevis.* Adult females were anesthetized with Finquel (tricaine methanesulphonate) and a segment of the oviduct isolated by silk surgical sutures. Into this segment was placed the preparation of RNA in a volume of 100 μl, and the avidin content determined 24 hr later. The *Xenopus* females had intact ovaries which were pre-ovulatory, so that pre-treatment with estrogen was not required.

2.3. Avidin assay

Avidin assay was based on the method of Korenman and O'Malley [17]. Tissue was homogenized in 6 ml of 0.02 M phosphate buffer, pH 7.1 containing 0.07 M KCl, 0.004 M $MgCl_2$, and 0.07 M NaCl, and centrifuged at 5000 *g* for 30 min. To 2 ml of supernatant was added 0.5 ml of ^{14}C-biotin (0.01 μCi) in 0.02 M ammonium carbonate solution. The ^{14}C-biotin (20 mCi/mM) was obtained from The Radiochemical Centre, Amersham, England. The mixture was stirred and left at room temperature for 15 min. One milliliter of bentonite solution (10 mg of bentonite per ml of 0.2 M ammonium carbonate) was added. The mixture was centrifuged at 2000 *g* for 10 min. The precipitate was dissolved in 2 ml of 0.2 M ammonium carbonate and centrifugation and purification procedure repeated until the supernatant gave no significant count above the background when dissolved in aquasol. The precipitate was then dissolved in 10 ml of aquasol and counted. Computation of micrograms avidin per gram oviduct was based on a standard curve using avidin obtained from Sigma Chemical Company.

2.4. Electron microscopy

The 7-day-old chicks were anesthesized with a mixture of ether and 5% CO_2 in O_2. The heart was exposed and after opening the inferior cava vein at the heart level, the left ventricle was punctured with an 18 gage needle and 10 ml of diluted Ringer with 15% distilled water were perfused to clear the blood from the circulatory system. The aorta was exposed and cannulated. Five per cent glutaraldehyde in 0.1 M s-collidin buffer, pH 7.4 was perfused at room temperature. The perfusion lasted approximately 30 min. The oviduct was removed and the magnum portion sectioned into small strips. Fixation was continued by immersion in fresh fixative to complete 1 hr, followed by 1 hr wash in s-collidin buffer, 0.1 M, pH 7.4 with $CaCl_2$ and sucrose [18]. The subsequent steps were refixing during 1 hr in 1% OsO_4, 0.1 M s-collidin buffer, pH 7.4 with $CaCl_2$ and sucrose, treating for 20 min in 1%

uranyl acetate in water, dehydrating through ethyl alcohol, propylene oxide and embedding in epon aralidite mixture. Thick and thin sections were cut in a Porter Blum ultramicrotome and stained with toluidine blue borax and lead-uranyl salts respectively. Observations and micrographs were made in a Phillip 300.

Portions of the magnum of perfused oviducts were opened and the inner surface exposed and sectioned in small squares of approximately 5 mm. They were left in 1% OsO_4 0.1 M s-collidin buffer, pH 7.4. with $CaCl_2$ and sucrose in the cold during 4 hr, then dehydrated through acetones from 70% to 100% and dehydrated and dried in the critical point drying apparatus (Denton) with liquid CO_2. The blocks were mounted in colloidal graphite and covered with carbon and gold in a Denton vacuum evaporator. The specimens were studied in an Autoscan (ETEC Corp.) at 20 kV at a magnification of 1500X to 20,000X.

3. Results

3.1. Morphology

The 7-day chick oviduct is characterized by a simple, low columnar epithelium. There are occasional invaginations, but no distinct glandular development. The fine structure reveals an absence of secretory activity. More details of the fine structure are described in the legend to fig. 2. At the level of the scanning electron microscope, it is evident that some hormonally-independent differentiation of the epithelium occurs. Occasionally, single cilia are observed across the entire surface of the oviduct, and scattered clusters of cilia, from either a single cell or groups of contiguous cells can be observed (fig. 1).

Stilbestrol treatment causes an increase in cell height (fig. 4), but the most characteristic feature of the estrogen-treated oviduct is the development of secretory glands, opening into the lumen of the oviduct. The development of cilia is evident, but large areas of the surface do not have cilia (fig. 3). At the opening of glands, some of the columnar lining epithelial cells have dense or clear secretory granules in abundance. There is no evidence of secretory activity along most of the free border of the oviductal lumen.

Exposure to progesterone, after treatment with stilbestrol, causes an enhancement of secretory activity and more extensive ciliogenesis. The latter process is striking when viewed on the surface (fig. 5) or in the ultra-thin sections (fig. 6). The cell height is increased, the cytoplasm has an abundance of membrane organelles and polysomes. The nucleus is rich in particulate matter and the nucleoli are complex and elaborate. RNA-treated animals have oviducts that are almost indistinguishable from those resulting from progesterone treatment. Cilia formation is extensive and there is evident secretory activity (fig. 7). Basal bodies are seen at the free border and in deeper portions of the cells, as well. It is evident that ciliary processes are being organized even in many cells that do not have cilia protruding into the lumen (fig. 8).

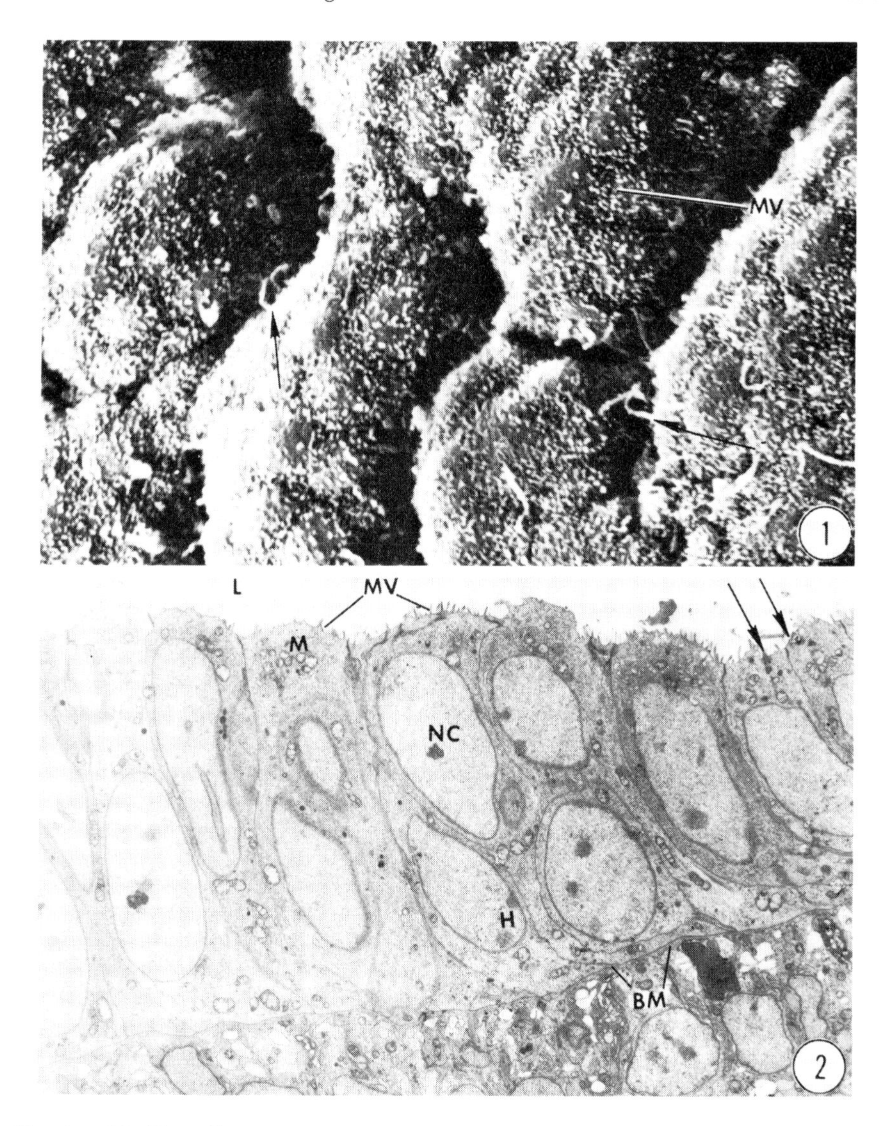

Figs. 1 and 2. These figures reveal a portion of the magnum zone of the oviduct of untreated 7-day-old chicks. Fig. 1 is a scanning electron micrograph (SEM) and fig. 2 is a transmission electron micrograph (TEM), both at a magnification of 2400X. Fig. 1 shows few ciliated processes (arrows) emerging from the oviduct inner surface which is partially covered with short microvilli (MV). Fig. 2 shows a portion of the epithelial surface. It shows the scattered short microvilli (MV); a basal body and a portion of cilia (arrows), the pale nuclei with heterochromatin clumps (H) and a few small and dense nucleoli (NC): mitochondria are pale, globular and with few cristae (M). Lumen (L) and basement membrane (BM) are indicated.

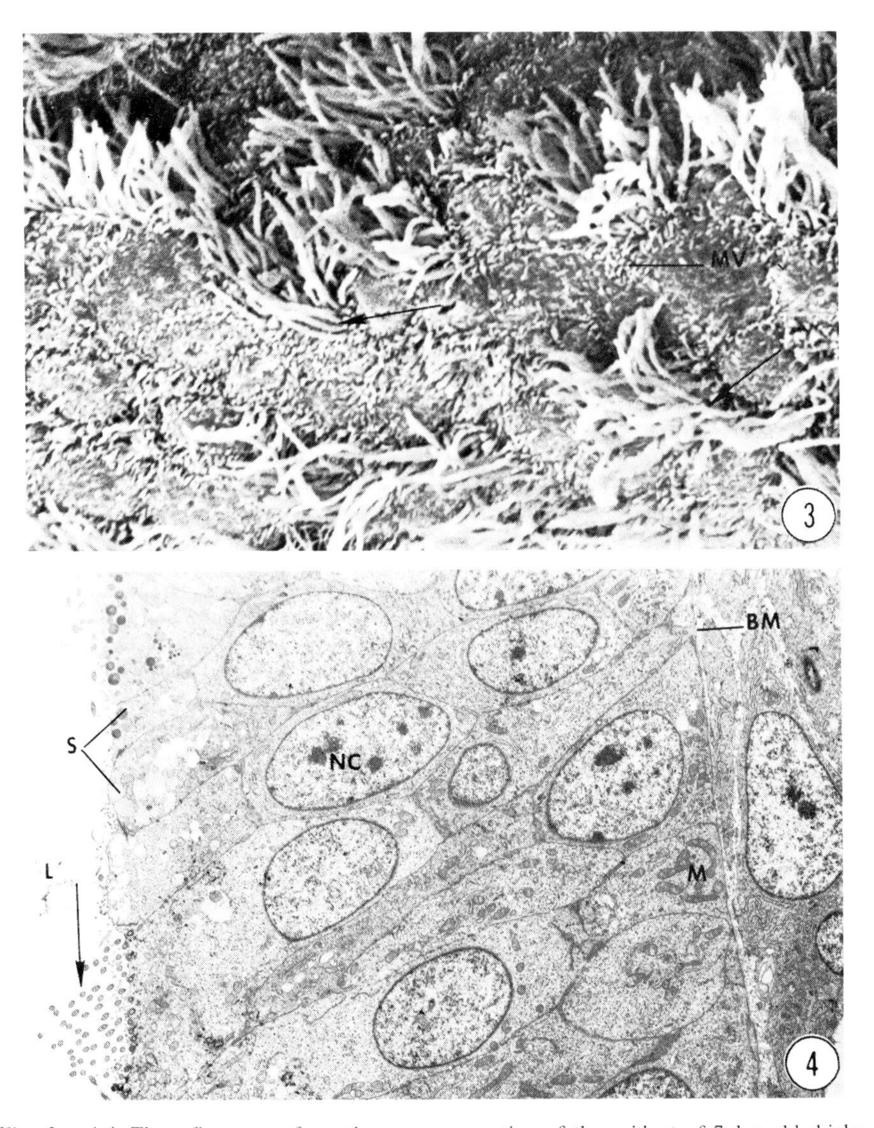

Figs. 3 and 4. These figures are from the magnum portion of the oviduct of 7-day-old chicks, treated with stilbestrol and saline. Fig. 3 is a SEM micrograph and fig. 4 is a TEM micrograph both at 2400X. Fig. 3 shows the oviduct surface partially covered by clusters of cilia (arrows) and short microvilli (MV). Fig. 4 shows a portion of the epithelial surface, lumen (L) at left and basement membrane (BM); the cells show ciliated processes (arrow) and some secretory activity (S). Nuclei are denser and with more prominent nucleoli-(NC); mitochondria are elongated, dense and more numerous (M) than those from fig. 2.

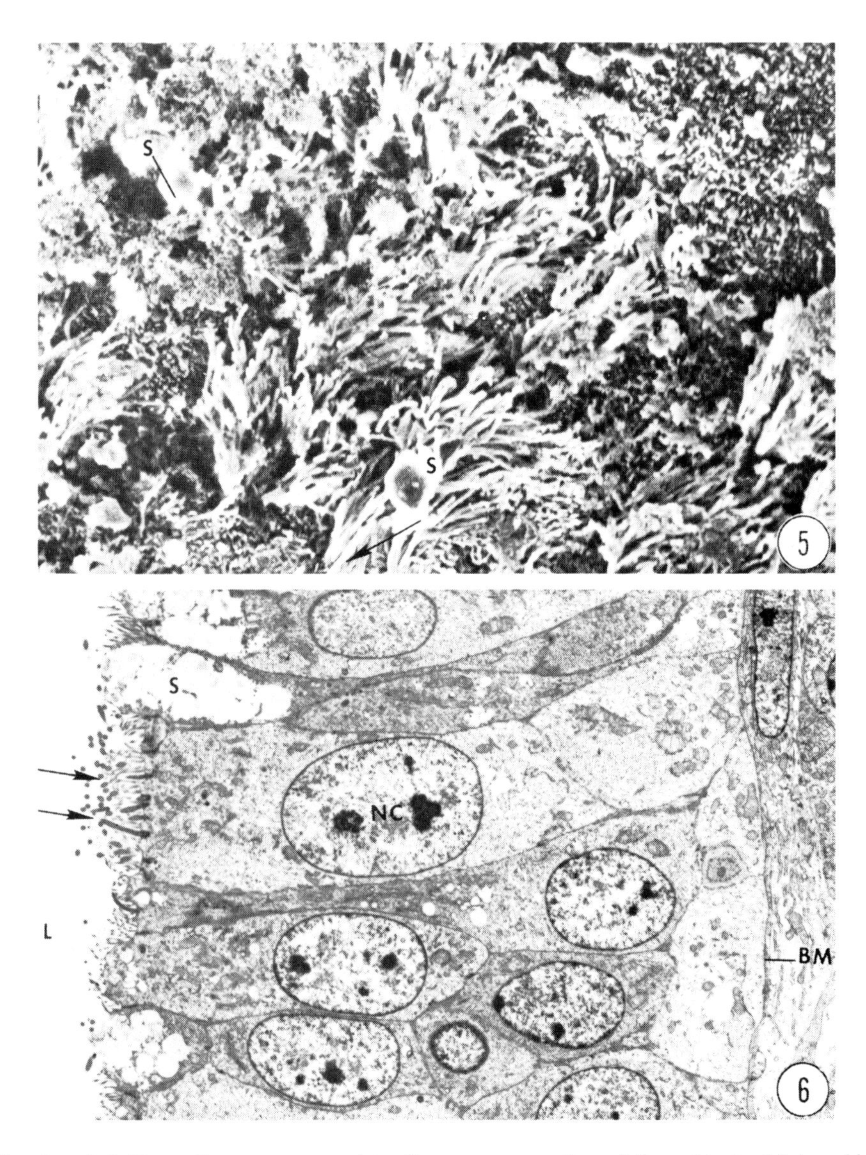

Figs. 5 and 6. These figures correspond to the magnum portion of the oviduct of 7-day-old chicks, treated with stilbestrol and progesterone. Fig. 5 is a SEM micrograph and fig. 6 a TEM micrograph, both taken at 2400X. Fig. 5 shows a surface view of the numerous clusters of ciliated processes (arrow) coated by the secretory product of the glandular cells. Numerous tightly packed microvilli are shown in the open areas (MV). Fig. 6 shows a portion of the surface epithelium with ciliated cells (arrow) and glandular cells (S). The nuclei are larger, denser with prominent nucleoli (NC) and the cells are taller than those in figs. 2 and 4. Lumen (L) and basement membrane (BM).

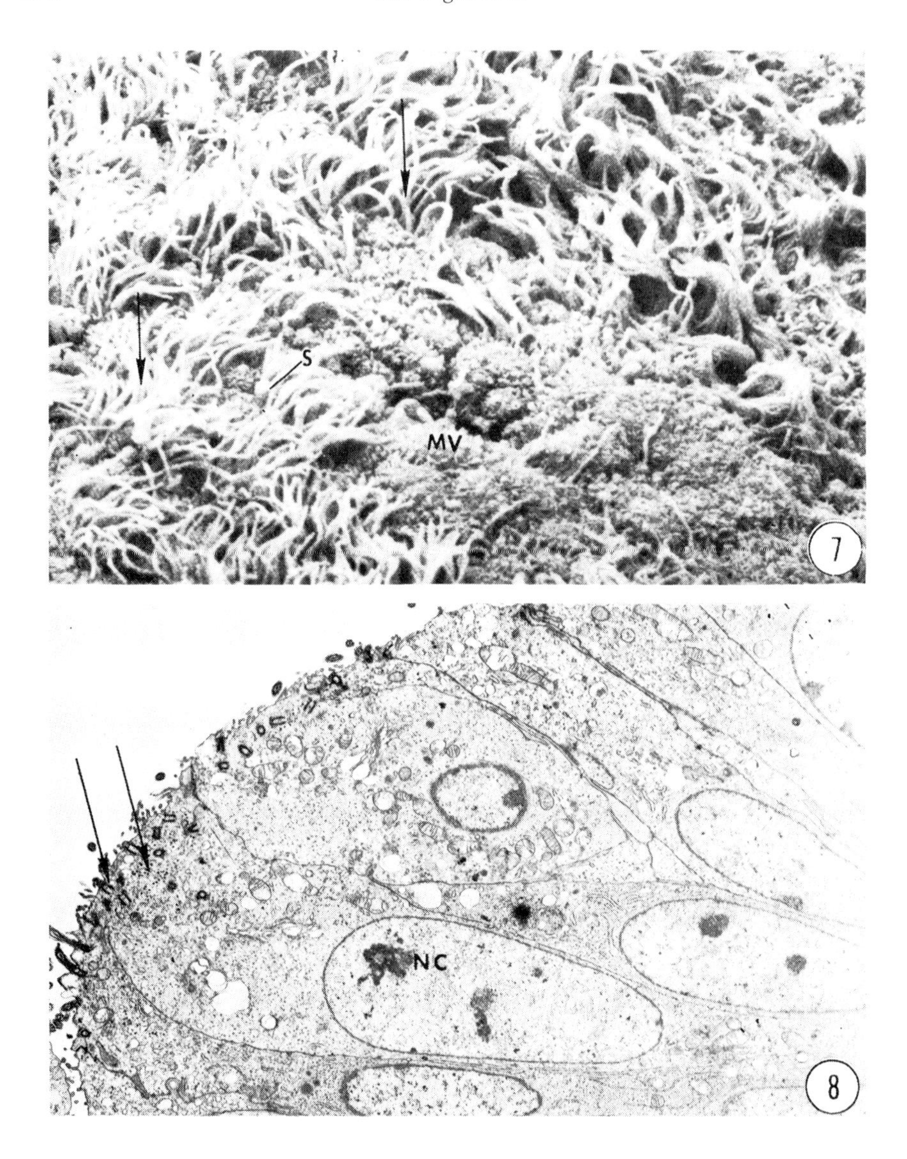

Figs. 7 and 8. These figures represent the magnum portion of the oviduct of 7-day-old chicks treated with stilbestrol and RNA. Fig. 7 is a SEM micrograph and fig. 8 a TEM micrograph, both at 2400X. Fig. 7 shows the luminal surface with numerous ciliated processes (arrows) emerging from an irregular surface with microvilli (MV) and secretory material (S). Fig. 8 shows a portion of the surface epithelium; the cells contain numerous basal bodies and dense particles related to ciliogenesis (arrows). The nuclei are large and show well-developed nucleoli (NC). The secretory activity is less marked than that shown in fig. 6. Lumen (L).

These observations can be summarized as follows. The undifferentiated oviductal cells can undergo a spontaneous specialization of ciliated cells in the absence of hormone stimulation. This results in isolated cilia across the surface of the oviduct, but there is no evidence of secretory function. With stilbestrol treatment, waves of ciliogenesis are initiated, glandular development occurs, and an abundance of dense secretory granules in the glandular epithelial cells and occasionally in the bordering oviductal lining, appear. Subsequent treatment with progesterone increases the amount of secretory activity. The secretory granules found within the glandular epithelium are frequently polymorphic, and secretory material is present in the glands and in the oviductal lumen. The luminal secretion tends to cause a matting of cilia when portions of the oviduct are prepared for scanning electron microscopy. The addition of progesterone has a pronounced effect on cilia formation as well. Within 24 hr, almost the entire oviductal surface becomes ciliated.

3.2. Avidin synthesis

Avidin synthesis by the chick oviduct is shown in table 1. It should be noted that earlier results revealed that an intra-oviductal dose of 500 μg of total RNA from stilbestrol/progesterone-treated chick oviducts was required to stimulate avidin synthesis in estrogen-primed chicks [3]. Nitrocellulose trapping effects a purification of approximately 50-fold. The RNA recovered after nitrocellulose chromatography stimulates avidin synthesis at a dose of 10 μg. This activity is lost when the poly A RNA is pre-treated with pancreatic RNase. The oviduct of the stilbestrol-primed chick responds to poly A RNA extracted from the oviducts of pigeons treated with stilbestrol and progesterone. There is no avidin present in the oviducts of stilbestrol-primed chicks that served as controls (intra-oviductal instillation of saline).

TABLE 1
Avidin synthesis by the chick oviduct in response to homologous and heterologous RNA

Number of cases	Final 24 hr treatment [a]			Avidin (μg/g)
	Material	Amount (μg)	Route	
6	None	–	–	0
6	Saline	–	i.o.	0
6	Progesterone	5000	s.c.	1.86 ± 0.76
7	Chick RNA	10	i.o.	0.83 ± 0.42
6	Chick RNA, RNase-incubated	10	i.o.	0
10	Pigeon RNA	10	i.o.	0.5 ± 0.2

[a] All chicks pre-treated with 0.5 mg stilbestrol daily for 5 days. s.c.: subcutaneous; i.o.: intra-oviductal. Bovine pancreatic RNase was used. Treatment of RNA was for 10 min at 37°C, followed by inactivation of the enzyme by heat (65°C). The chicks used were Rhode Island Reds, obtained from Hall Brothers Hatchery, Wallingford, Connecticut 06492, USA.

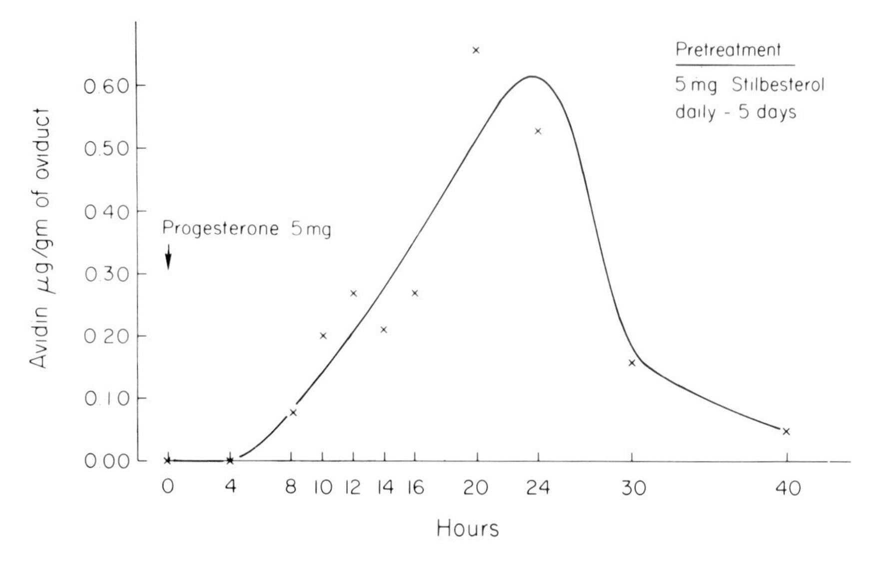

Fig. 9. Avidin synthesis by pigeon oviduct.

Like the chick, and several other species reported previously [19], the oviduct of the estrogen-primed pigeon synthesizes avidin in response to progesterone. At approximately 24 hr after the administration of progesterone, the highest levels of avidin in the pigeon oviduct are achieved (fig. 9). This response is brought about, also, by the intra-oviductal administration of homologous oviductal RNA, prepared from hormone-treated pigeons (table 2).

A dose of 10 µg of poly A RNA elicits a significant level of avidin synthesis. The pigeon oviduct responds, also, to poly A RNA recovered from hormone-treated

TABLE 2
Avidin synthesis by the pigeon oviduct in response to homologous and heterologous RNA

Number of cases	Final 24 hr treatment [a]			Avidin (µg/g)
	Material	Amount (µg)	Route	
10	None	–	–	0
12	Progesterone	5000	s.c.	0.9 ± 0.4
6	Pigeon mRNA	10	i.o.	0.12 ± 0.03
10	Chick mRNA	50	i.o.	0.53 ± 0.25
6	Chick mRNA, RNase-incubated	50	i.o.	0

[a] All pigeons pre-treated with 0.5 mg stilbestrol daily for 5 days. s.c.: subcutaneous; i.o.: intra-oviductal. Bovine pancreatic RNase was used. Treatment of RNA was for 10 min at 37°C, followed by inactivation of the enzyme by heat (65°C). The pigeons used were white carneaux, obtained from Palmetto Pigeon Plant, Sumter, South Carolina 29150, USA.

TABLE 3
Avidin synthesis by *Xenopus* oviduct in response to heterologous RNA

Number of cases	24 hr treatment [a]			Avidin (μg/g)
	Material	Amount (μg)	Route	
4	None	–	–	0
4	Progesterone	5000 mg	i.o.	0
4	Progesterone	5000 mg	d.l.	0
4	Chick RNA	50 μg	i.o.	0.13

[a] *Xenopus* were mature females with preovulatory ovaries intact (*Xenopus laevis* Daudin). i.o.: intra-oviductal; d.l.: dorsal lymph sac.

chicks. Pre-treatment of the RNA preparation with pancreatic RNase eliminates the activity.

Avidin is found, also, in the egg jelly of amphibians [19]. Presumably, it is synthesized in some portion of the oviduct, although this has not been demonstrated previously. The administration of progesterone in high dosage, either into the dorsal lymph sac, or intra-oviductal, fails to cause measurable avidin synthesis by the oviduct of *Xenopus laevis* females. Avidin is produced, however, after the instillation of 50 μg of mRNA isolated from the oviduct of progesterone-treated chicks (table 3).

4. Discussion

The heterospecific transfer of hormonal stimulation, through the mediation of hormone-induced RNA in vivo, has been described. This finding is of interest with respect to the specificity of RNA coding in the vertebrate classes, as well as in relation to the evolution of hormonal control mechanisms. Evidently, the progesterone-receptor complex [20] interacts with highly similar or identical chromatin acceptor sites in the nuclei of either the chick, the pigeon or of *Xenopus*, and does not discriminate among them. Thus, once the hormone causes the production of RNA coded for avidin synthesis in the oviductal nuclei of any of the species tested, the hormone molecule itself is no longer required and the newly synthesized RNA can complete the steps leading to avidin synthesis, interchangeably among the oviductal cells of the several species. This is illustrated in fig. 10 and may be the physiological basis for the inter-species cross-reaction of many hormones. The activity of chick RNA in *Xenopus* is of particular interest. Progesterone itself, at least under the conditions used, is unable to cause avidin synthesis in *Xenopus* oviduct. Possibly, the dose, mode of administration or timing of exposure and sacrifice were not optimal for the demonstration of the role of progesterone in the control of avidin synthesis by the amphibian oviduct. Nevertheless, the response to chick RNA

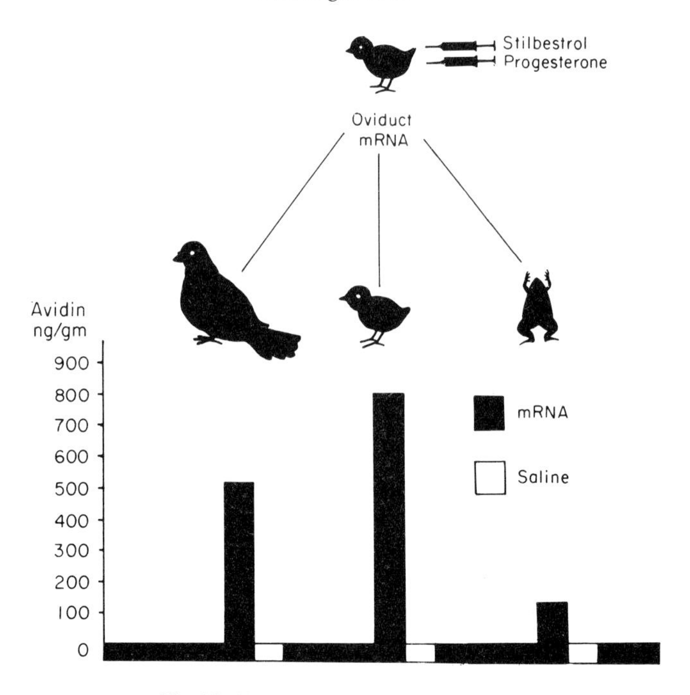

Fig. 10. Hormone effect induced by mRNA.

suggests that the synthetic pathways for avidin synthesis have already evolved in *Xenopus* oviduct, and that this evolutionary development occurred prior to the evolution of the control mechanism by progesterone found to exist in avian species. It will be of interest to determine whether progesterone itself, or a related progestin, is normally secreted by the *Xenopus* ovary.

The morphological observations clarify that a small degree of spontaneous differentiation of ciliated cells occurs in the epithelium of the chick oviduct, prior to gonadal development and sex hormone stimulation. There is no secretory gland development without estrogen influence. With the onset of estrogenic stimulation, ciliogenesis and glandular development proceed and secretory activity begins. The addition of progesterone markedly increases the secretory activity, and causes new types of secretory granules to appear. The oviductal surface becomes abundantly ciliated within 24 hr after the initiation of progesterone stimulation. All of the morphological effects of progesterone can be achieved by RNA extracted from hormone-stimulated oviducts. Thus, both the morphological and physiological effects of progesterone can be brought about by the homologous or heterologous transfer of RNA [21].

Most likely, the active component is a form of messenger RNA, since it is highly purified by nitrocellulose trapping [15]. It is clear that a contaminating residue of progesterone cannot account for the avidin synthesis noted after RNA administration [14]. This conclusion is confirmed by the present observation of avidin induction with the purified RNA preparation that is active at a dose as low as 10 μg. As in earlier experiments with either progesterone-induced RNA or estrogen-induced RNA, enzymatic degradation with pancreatic RNase eliminates the activity.

These experiments do not reveal whether the exogenous RNA is translated directly or acts in some manner at the nuclear level to influence transcription of avidin-coded RNA. A lack of absolute antigenic identity among the avidin molecules produced by different species, if found to exist, would provide an excellent experimental basis to resolve this question.

Acknowledgements

The authors are grateful to Mr. Andrew Gonzalez, Mr. Delbert Layne and Mr. Edward Tovar for their technical assistance. This work was supported by USPHS Grant Number HD 05671.

References

[1] S.J. Segal and W. Scher, In: Cellular biology of the uterus. Ed. R.M. Wynn (Appleton-Century-Crofts, New York, 1967) 114–150.
[2] S.J. Segal, In: Control mechanisms in developmental processes. Ed. M. Locke (Academic Press, New York, 1967) 264–280.
[3] C.M. Szego and J.S. Davis, Proc. Natl. Acad. Sci. U.S. 58 (1967) 1711.
[4] G.C. Mueller, In: Mechanisms of hormone action. Ed. P. Karlson (Academic Press, New York, 1965) 228–245.
[5] G.P. Talwar, S.J. Segal, A. Evans and O.W. Davidson, Proc. Natl. Acad. Sci. U.S. 52 (1964) 1059.
[6] G.P. Talwar and S.J. Segal, In: The sex steroids. Ed. K.W. McKerns (Appleton-Century-Crofts, New York, 1971) 241–272.
[7] E.V. Jensen and H.T. Jacobsen, Rec. Progr. Horm. Res. 18 (1962) 387.
[8] B.W. O'Malley, W.L. McGuire, P.O. Kohler and S.G. Korenman, Rec. Progr. Horm. Res. 25 (1969) 105.
[9] D. Trachewsky and S.J. Segal, Biochem. Biophys. Res. Commun. 27 (1967) 588.
[10] G.P. Talwar and S.J. Segal, Proc. Natl. Acad. Sci. U.S. 50 (1963) 226.
[11] S.J. Segal, O.W. Davidson and K. Wada, Proc. Natl. Acad. Sci. U.S. 54 (1965) 782.
[12] P. Tuohimaa, S.J. Segal and S.S. Koide, J. Steroid Biochem. 3 (1972) 503.
[13] R. Hertz, R.M. Fraps and W.E. Sebrell, Proc. Soc. Exp. Biol. Med. 52 (1943) 142.
[14] P. Tuohimaa, S.J. Segal and S.S. Koide, Proc. Natl. Acad. Sci. U.S. 69 (1972) 2814.
[15] G. Brawerman, J. Medecki and S.Y. Lee, Biochemistry 11 (1972) 637.
[16] G.C. Rosenfeld, J.P. Comstock, A.R. Means and B.W. O'Malley, Biochem. Biophys. Res. Commun. 47 (1972) 387.
[17] S.G. Korenman and B.W. O'Malley, Biochem. Biophys. Acta 140 (1967) 174.
[18] M.H. Burgos, R. Vitale-Calpe and M.T. Tellez de Iñón, J. Microscopie 6 (1967) 457.
[19] R. Hertz and W.E. Sebrell, Science 96 (1942) 257.
[20] B.W. O'Malley and D.O. Toft, J. Biol. Chem. 246 (1971) 1117.
[21] S.J. Segal, R. Ige, P. Tuohimaa and M. Burgos, Science (1973) in press.

Session Six

Mechanism of RNA Action

Chairman

M.C. NIU

Department of Biology, Temple University, Philadelphia, Pennsylvania

Chairman's introduction

In the previous sessions of this symposium, we have considered RNA metabolism in developing embryos and organ systems, RNA programmed protein synthesis, RNA-initiated embryonic differentiation, RNA-mediated hormone action, RNA mediation of immune response and the RNA mediated transfer of genetic traits. We are now coming to the sixth and final session dealing with the mechanism by which RNA acts on the target cells. All cellular RNAs are known to be produced in the nucleus. In living tissue, many factors from without can "turn on" or "turn off" the gene action. When it is on, new RNA is synthesized. This RNA is rapidly labelled and generally considered to be informational. We are most fortunate to have Prof. Dr. J.-E. Edström with us to start this session by discussing the use of hormones to "turn on" the genome and the immediate synthesis of new RNA. In the insect *Chironomus*, application of ecdysone result in the formation of Balbiani ring puffs in which the rapidly labelled RNA is produced. Poly A joins the 3′ end of some of the newly synthesized RNA and migrates into the cytoplasm to prime protein synthesis. Apparently, this polysome-bound poly A-attached RNA (mRNA) is what earlier speakers of this symposium had discussed in their respective studies.

In mammalian cells, the rapidly labelled RNA is huge in molecular size and heterogeneous (HnRNA). One of the recent developments in the study of RNA metabolism has revealed that HnRNA is the precusor of other nuclear RNAs. Upon cleavage, it gives rise to messenger RNA (mRNA) on the one hand and chromosomal RNA on the other. Direct support of the precursor concept comes from the functional study of HmRNA from the liver of the mouse embryo. This RNA was injected into *Xenopus* oocytes and found to program the synthesis of mouse hemoglobin. This messenger activity of HnRNA suggests the presence of 9S mRNA for hemoglobin in its long chain of nucleotides. The function of chromosomal RNA and its relation to HnRNA will be discussed by Drs. David Holmes and James Bonner. According to these authors, chromosomal RNA does not leave the nucleus, but affects the genome through histone. In recent years, however, interesting evidence has appeared showing a role of nuclear acidic protein in the gene (DNA) function. In view of this, we have asked Dr. T. Y. Wang to discuss briefly his work on nuclear acidic protein.

Soon after the discovery that liver-RNA induced albumin synthesis in mouse ascites cells, the whereabouts of exogenous-RNA was traced using H^3 or C^{14}-labelled RNA. The genotropic nature of exogenous-RNA was first reported in 1962.

Subsequent studies confirmed the association between exogenous-RNA from liver and the nuclear material of mouse ascites cells. RNA treatment was shown to be long lasting in its effect on synthesis of tryptophan oxygenase and albumin. Of particular relevance to the present theme is the observation that exogenous liver RNA promoted RNA synthesis and that this stimulatory effect was sensitive to actinomycin D and to histone. Similarly, actinomycin D inhibited the uterine RNA promoted activity of alkaline phosphatase in the uterus of the spayed mouse. Moreover, in developmental cytology chromatin diminution was found in *Ascaris* and formation of Balbiani ring puffs was observed at different loci in *Chironomus*. Both were results of differentiation, apparently initiated by mRNA. All of these data speak for the regulatory role of exogenous RNA in the function and structure of chromosomes.

In the isolated nuclei of thymocytes, addition of nuclear RNA resulted in an increase of RNA synthesis. Taking advantage of (i) that the nucleotide sequence of nuclear RNA is complementary to DNA and (ii) that one strand of DNA is functional during transcription, Dr. John Frenster will talk about his proposed mechanism of RNA action. In essence, nuclear RNA and one strand of the DNA form a hybrid, thus leaving the other DNA strand for transcription. An alternative proposal is based on the recent discovery that reverse transcriptase was also found in embryonic and adult tissues as well. With a pool of 4 deoxyribonucleoside triphosphates and mRNA, the enzyme catalyzes the synthesis of DNA. The details of the RNA-primed DNA synthesis and its subsequent incorporation into the host genome will be discussed by Drs. C.Y. Kang and Howard Temin. In contrast to the regulatory mechanism described earlier, the present one is transformative and, by and large, similar to the DNA-induced transformation of *Pneumococcus*. Accordingly, RNA-transformed cells would acquire the ability to produce new proteins characteristic of the mRNA donor tissue, while the RNA-regulated cells would synthesize increased amounts of proteins specific to the cells, and independent of the mRNA used. With these distinctions clearly defined, it is hoped that the mechanism of RNA action in differentiation will soon be clarified.

M.C. Niu

Niu and Segal (eds.). The role of RNA in reproduction and development
North-Holland Publ. Co., 1973

Appearance and decay of ribonucleic acids in the cytoplasm of salivary gland cells of *Chironomus tentans*

J.-E. EDSTRÖM
Department of Histology, Karolinska Institutet, S-104 01 Stockholm, Sweden

Four main classes of RNA with a nuclear origin appear in the cytoplasm of *Chironomus tentans* salivary gland cells, ribosomal RNA, 4 S RNA and two classes of heterogeneous RNAs with properties of messenger RNA. One of the messenger like fractions appears early in the cytoplasm and has a comparatively short life. It resembles in size the (non-specific) messenger RNA reported for other systems. The other heterogeneous RNA fraction arrives after the ribosomal RNA, has a considerably longer half-life and is specific for this tissue. This RNA migrates slowly in gels, possibly because of high molecular weight. It is a candidate for the messenger for the differentiated cell products. Ribosomal RNA after a lag period of 90–120 min first appears in the central parts of the cell and migrates slowly in a peripheral direction while the RNA fraction with properties of the non-specific messenger like RNA spreads quickly. The 4 S RNA is mainly, possibly exclusively, intranuclear during early labelling times.

1. *Introduction*

The polytene chromosomes have been a favored material in studies on gene expression since the demonstration 20 years ago that these chromosomes exhibit tissue and phase specific modifications of their banding structure [1–3]. The general conclusions of this work were that the banding pattern is constant in different tissues and during development while on the other hand the activity of the chromosome is subject to modification. A band was found to exist in either of two alternative states, compact, seemingly inactive, or puffed, with signs of RNA production.

In the chironomids certain puffs, so-called Balbiani rings are notable for two reasons, because of their size and because of their tissue specificity [1]. They are best known from the salivary gland chromosomes and there is a body of circumstantial evidence implicating them in the synthesis of the differentiated products, the secretory proteins [4–6]. At the same time the chromosomes contain a large number of puffs in the conventional size range, of the order of hundreds. The majority of non-Balbiani ring puffs do not show tissue specificity in *Drosophila* [7] and it is a reasonable guess that at least the majority of these puffs, if engaged in gene expression at all, are related to the nondifferentiated functions, in particular since the number of secretory proteins is low, in the 5–10 numbers range, in the chironomids [6, 8].

The salivary gland cells of *Chironomus tentans* provide these different types of chromosomal activities. It therefore appeared desirable to identify and describe the types of RNA that enter the cytoplasm of the salivary gland cells, with the ultimate aim of relating these to their genomic origin and to differentiated or undifferentiated functions. In this work microdissected salivary gland cells were used after fixation of the cells. In our experience fixation, in this case ethanol–acetic acid, is suitable to combine with studies of high molecular weight RNA [9–11].

Some studies of the RNA export in these cells have already been carried out. Export has been followed for ribosomal RNA [12] and transfer RNA [13] by analyses of nuclei and unfractionated cytoplasm. Furthermore granules with a highly characteristic morphology, present in the Balbiani rings [14] have been located in the cytoplasm, close to the nuclear envelope [15]. Lambert [16] has found that RNA which can be extracted from microdissected rings shows a specific hybridization to DNA in the ring from which it was taken. Such RNA is also present in the cytoplasm [17]. Daneholt [11] has found that the Balbiani ring RNA shows a characteristic distribution after gel electrophoresis, suggesting the presence of one or a few transcripts of similar size. RNA of similar migration properties can also be extracted from the cytoplasm [18].

To characterize the main classes of RNA entering the cytoplasm we have injected fourth instar larvae with tritiated uridine, sacrificed the animals after varying periods and fixed the salivary glands in ethanol-acetic acid. In a first study [19] the distribution of newly synthesized RNA in isolated nuclei and cytoplasm was investigated. The cells were liberated from the glands by micromanipulation in an oii chamber arrangement [20]. Most of the cytoplasm was then removed and used as the cytoplasmic fraction. During this work a layer of cytoplasm immediately surrounding the nucleus was, however, not included in the main cytoplasmic fraction because of the danger of nuclear contamination. After removal of remaining cytoplasm the nuclei were collected as one fraction. In each analysis cells from only one animal were used, as a rule about a dozen cells per analysis. The pooled cell components were dissolved in a pronase-sodium dodecyl sulphate solution for 3 min [11] and afterwards precipitated with ethanol and NaCl overnight together with carrier RNA. After redissolution the extract was applied to 1% agarose gels for separation of the main classes of RNA. The gels were sliced and the slices counted in a scintillation counter. The appearance and fate of four types of RNA could be followed.

2. *Main classes of RNA entering the cytoplasm*

The types of RNA that were studied, ribosomal RNA, 4 S RNA and two types of

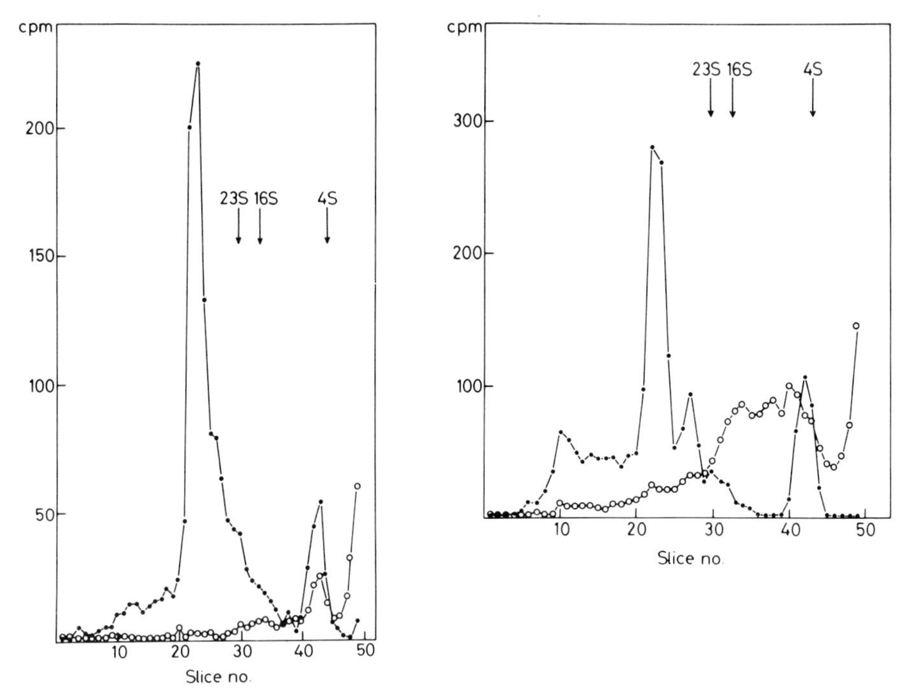

Fig. 1. Animals were injected with 50 μCi of tritiated uridine (50 Ci/mmole) and 50 μCi of tritiated cytidine (25 Ci/mmole) and sacrificed 15 min afterwards (a) or were given 25 μCi of tritiated uridine and sacrificed 45 min afterwards (b). In each case 27–28 cells were dissected into nuclei and cytoplasm, the RNA extracted and separated in 1% agarose gels. Filled circles nuclei, open circles cytoplasm. The arrows denote added *E. coli* RNA markers as described by Daneholt [11].

messenger-like RNAs appeared in the cytoplasm after distinct lag periods indicating a nuclear origin. Thus after a short incorporation time of 15 min little activity is present in the cytoplasm as compared to the nuclei (fig. 1a). After 45–60 min a ribonuclease sensitive fraction appears in the region between the 4 S and 23 S markers but with the activity extending towards the high molecular weight regions (fig. 1b). This fraction is designated the early heterogeneous (H) RNA and will be shown to possess properties of messenger RNA. Later on, after 2 hr, ribosomal RNA begins to appear (fig. 2a). After this time mainly the 18 S fraction is seen but the 28 S component is about to follow and both components are well visible after 3 hr (fig. 2b). After 3 hr a new class of heterogeneous RNA appears, previously described [18]. Its migration rate is similar to that of the Balbiani ring RNA, but is somewhat slower. This fraction is designated the late heterogeneous (H) RNA and will be shown to possess certain properties characteristic of messenger RNA. The 4 S RNA is not visible as a distinct peak during times before 3 hr and most of it is

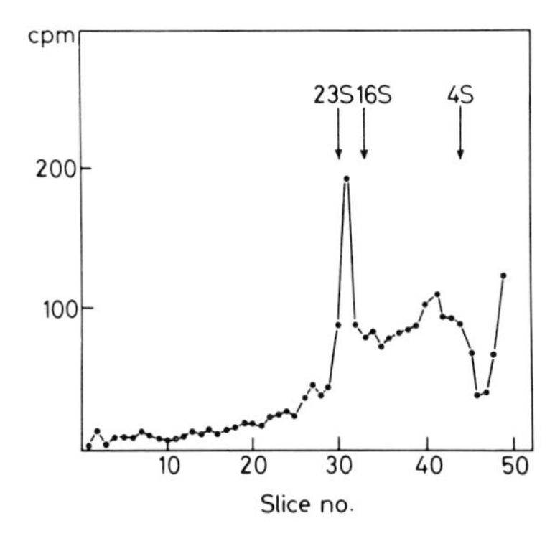

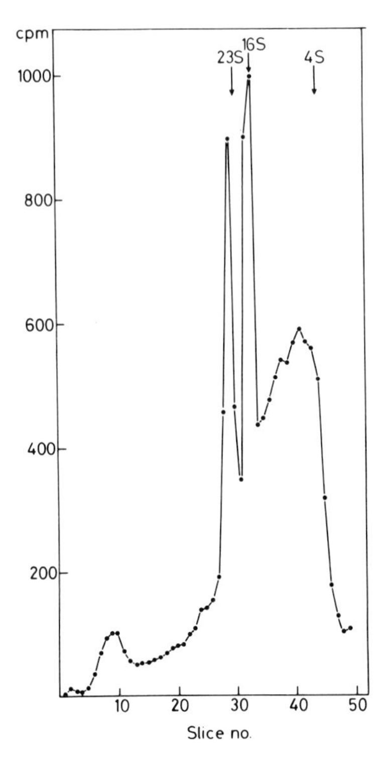

Fig. 2. Animals were injected with 25 μCi tritiated uridine and sacrificed after 120 min (a) and 180 min (b). Cytoplasm was isolated from 12 salivary gland cells, extracted for RNA and separated in 1% agarose gels. For other details see legend to fig. 3.

apparently contained in the nucleus during early times. With increasing times, after 6 hr, the early H RNA shows a relative decrease, ribosomal RNA and 4 S RNA begin to reach their definite proportions (fig. 6a).

3. Order of appearance of RNA classes in the cytoplasm

Each kind of RNA enters the cytoplasm after a distinct lag period. This has been well established in other systems for the two ribosomal components (ref. [21], for review) and also has been described for different classes of messenger RNA [22]. Less is known for the 4 S RNA since by traditional preparatory techniques it is likely to appear immediately in the cytoplasm as a result of preparatory leakage, irrespective of its actual localization. Although the present system seems considerably slower in exporting ribosomal RNA than e.g. the HeLa cell there is an overall similarity to other systems. The early H RNA is similar to the general unspecified

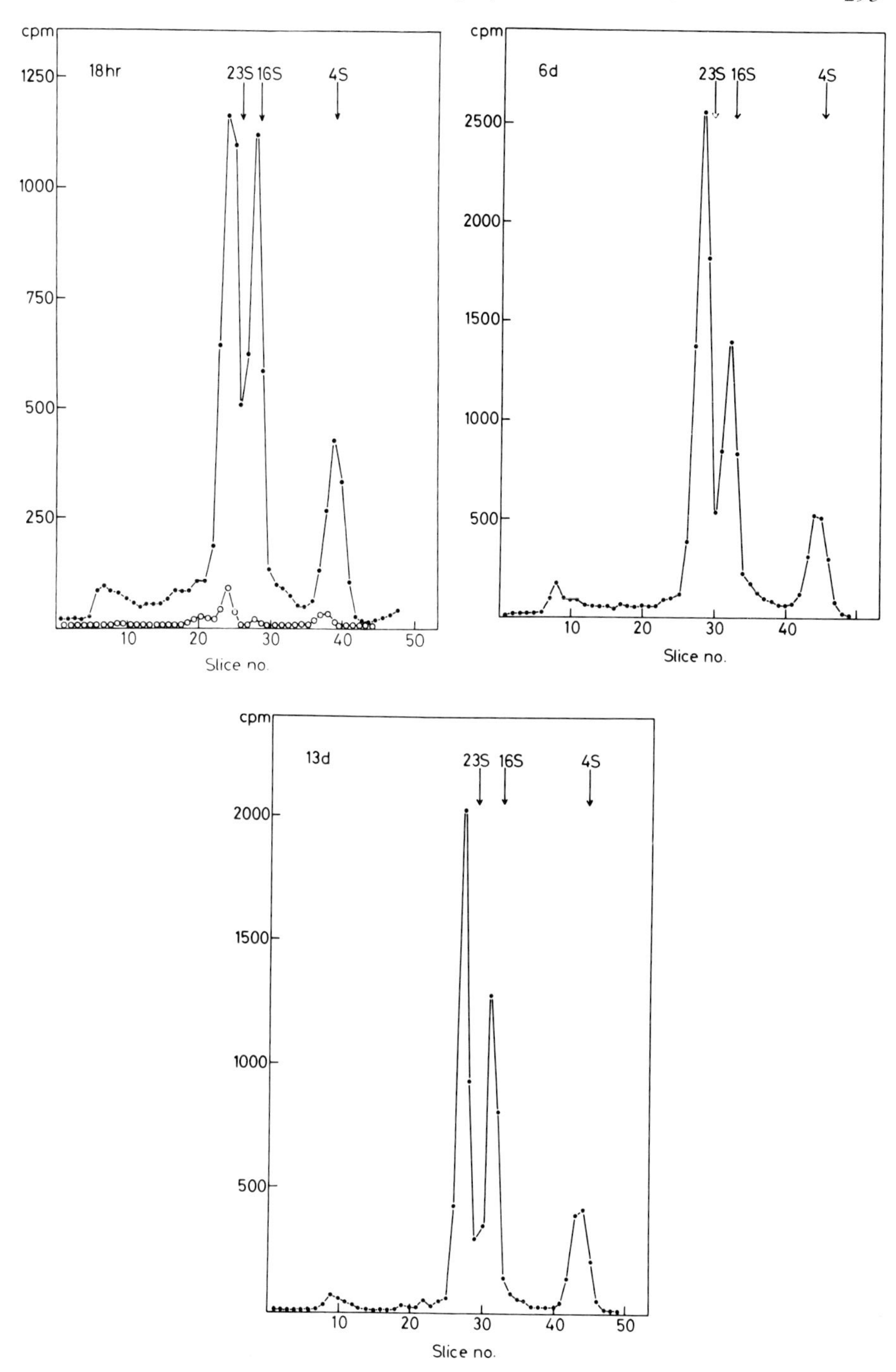

Fig. 3. Animals were injected with 25 μCi tritiated uridine and sacrificed after 18 hr (a) 6 days (b) and 13 days (c). Cytoplasm was isolated. In the 18 hr experiment the nuclear profile is also shown (open circles). For other details see legend to fig. 3. (From Edström and Tanguay [19].)

messenger RNA in its rate of exit. The late H RNA is, however, remarkable in the very long time required for appearance in the cytoplasm. This is of interest in view of the similarity of H RNA to Balbiani ring RNA. The latter kind of RNA shows a relative accumulation in the nuclear sap [11, 23–25]. A long relative nuclear residence may explain this relative accumulation.

The exact time for appearance of the 4 S RNA cannot be decided from the present analyses, the low percent agarose gel used not being capable of resolving 4 S RNA from the early H RNA. From the profiles it is possible nevertheless to deduce that at early times most of it is intranuclear. This is in good agreement with previous in vitro studies showing that maturation of 4 S RNA takes place largely in the chromosomal and nuclear sap compartments [26].

4. *Stability of the cytoplasmic RNA*

The activity in nuclear RNA drops soon after precursor injection and is hardly measurable after 18 hr (fig. 3a). In the cytoplasm the early H RNA is not easily detectable after a couple of days but the late H RNA can still be discerned after one and two weeks (fig. 3b and c). Analyses of glands 5, 12 and 19 days after precursor administration indicated similar amounts of label in RNA suggesting either a long or infinite life for ribosomal and 4 S RNA, their relative proportions remaining constant (table 1). On the assumption of an infinite life for ribosomal RNA a half-life of 9 days could be calculated for the late H RNA on the basis of a rather limited number of determinations (table 1), sufficient, however, to indicate the order of magnitude.

5. *Nature of the main RNA classes in the cytoplasm*

The fact that all classes of RNA appear after a distinct lag period in the cytoplasm

TABLE 1
Radioactivities in late H RNA and 4 S RNA relative to ribosomal RNA at different times after precursor injection in salivary gland cells of *Chironomus tentans* (ribosomal RNA = 100)

Time	Late H RNA	4S RNA
18 hr	15.4	21.3
2 days	25.8	25.4
2 days	18.0	18.3
6 days	11.3	18.4
13 days	9.5	24.7
13 days	6.5	25.4

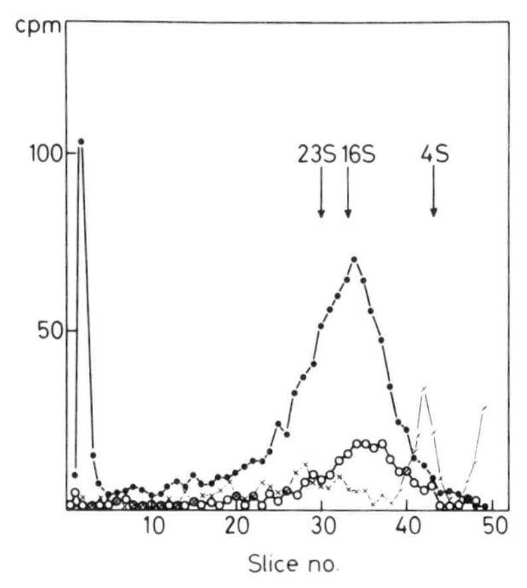

Fig. 4. Cytoplasmic RNA from salivary gland cells from an animal injected with 25 μCi of tritiated uridine 60 min before sacrifice was subjected to Millipore filtration as described by Lee et al. [28] in which technique the poly-adenylic acid containing messenger RNA is retained on the filter. The filter retained RNA (first extract of the filter) is shown by filled circles, the filtrate by crosses, a second extract of the filter is shown by open circles. Essentially all early H RNA is removed from the extract by filtration (leaving only some 4 S RNA) and RNA of a profile similar to the early H RNA can be extracted from the filter. The RNA used in this experiment was taken from cells of the same gland as used in the experiment shown in fig. 9. For the filtering experiment the whole cytoplasm was, however, used. The number of cells was lower in the filtering experiment, corresponding to $\frac{2}{3}$ of the number used in the experiment of fig. 9. (From Edström and Tanguay [19].)

indicates that they have a nuclear origin. A mitochondrial and/or microbial origin is, therefore, excluded.

The early H RNA shows some similarity to an unspecified class of messenger RNA in a rapid arrival and relatively short life in the cytoplasm. It is similar to the size distribution of rapidly lebelled cytoplasmic RNA from another insect, *Rhynchosciara* [27]. To determine its relation to messenger RNA we have employed the nitrocellulose filtering technique of Lee et al. [28]. We found [19] that the early H RNA fraction is almost completely retained by the filter in this test (fig. 4), and by this definition is messenger RNA. Compared to the filter retained fraction from HeLa cells [29] it shows maximum distribution somewhat more towards the low molecular weight range, with a peak distribution in the 14 S–16 S range.

The late H RNA with its property of long life has the characteristics of a messenger for a differentiated cell product. If it is such a messenger it should be absent in other tissues. This was also found to be the case. In analyses of midgut cells (known to lack the Balbiani rings typical of the salivary glands, ref. [1] this RNA was absent as well as in other tissue (gastric caeci), in animals where it was distinct in the salivary glands (fig. 5). In Lambert's work [17] RNA was used from animals supplied continuously with isotopes in the culturing medium. Since ribo-

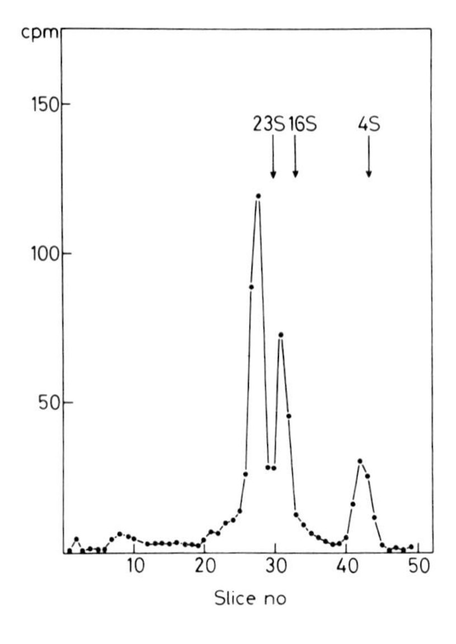

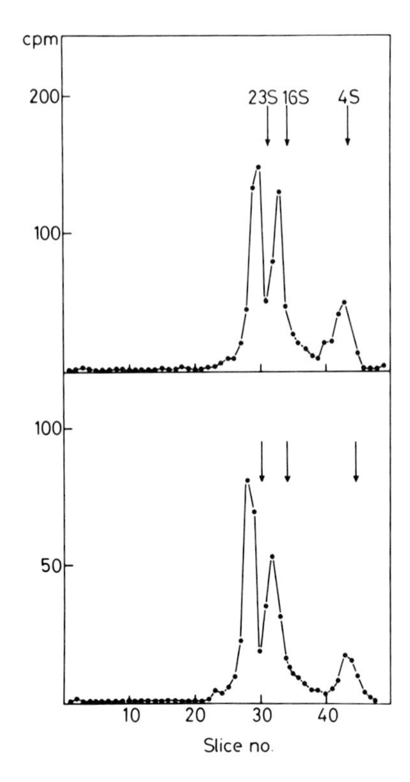

Fig. 5. Animals were given tritiated uridine and cytidine in the culturing medium for a day and sacrificed after two days. The RNA from salivary glands (a), midgut tissue (b, lower) and gastric caeci (b, upper) was analyzed by electrophoresis in agarose gels. The gastric caeci were from one animal and salivary glands and midgut from another one. The salivary gland profile corresponding to the gastric caeci was similar to the gland profile shown here.

somal RNA and 4 S RNA do not show hybridization to the Balbiani ring DNA [24] the hybridization observed was due to heterogeneous RNA. Since both late and early H RNA should be present during such labelling conditions it is difficult to state what fractions(s) gave rise to the hybridization. Late H RNA is similar to the migration properties of Balbiani ring RNA but since the agreement is not exact the similarity may be fortuitous. Nevertheless the late H RNA is an obvious candidate for a messenger for the differentiated products and the cytogenetic data suggest that the Balbiani rings are involved in its production.

Also by the results of the nitrocellulose filtering technique the late H RNA behaves like a messenger RNA. Six hours after precursor injection activity is present both in late and early H RNA and both fractions are retained on nitrocellulose filters in contrast to ribosomal RNA and 4 S RNA which pass through the filter (fig. 6).

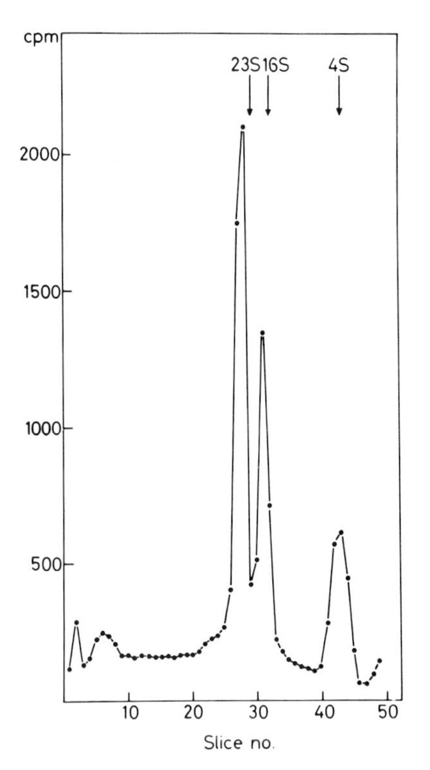

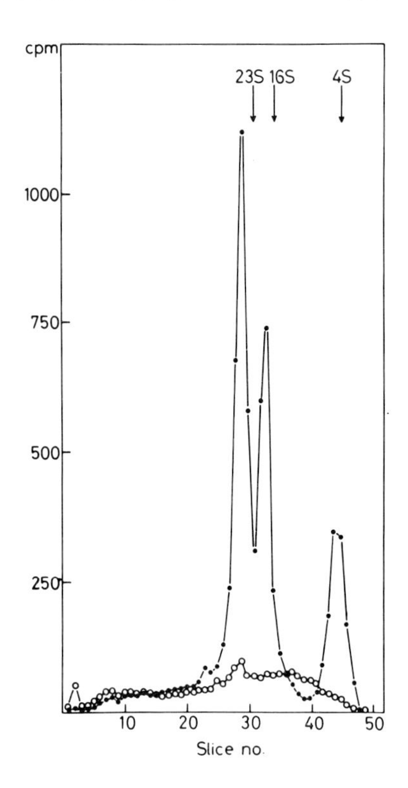

Fig. 6. Cytoplasmic RNA from salivary gland cells of an animal injected with 25 μCi of tritiated uridine 6 hr before sacrifice. Cytoplasm from 12 cells was used for extracting RNA for direct analysis in 1% agarose gels (a) and cytoplasm from another 12 cells was used for extracting RNA for filtering (b) as described in the legend to fig. 4. The filter retained RNA is shown by open circles (first extract), the filtrate by filled circles. The filter retained RNA in this experiment contains both the early and the late H RNA.

6. Migration of RNA through the cytoplasm

The fate of ribosomal RNA in the ribosomal subunits and its relation to messenger RNA after the exit to the cytoplasm has been subjected to much careful study. On the whole these studies have been related to the question of how subunits interact and combine with messenger RNA and the endoplasmic reticulum. Little is known however, regarding the movements within the cytoplasm of the newly extruded RNA components. Such information could be of considerable help in clarifying the relations between the different classes of RNA and thus their functional nature. The salivary gland cells offer particular advantages in this regard. This is because they can be dissected into zones situated at different distances from the nucleus

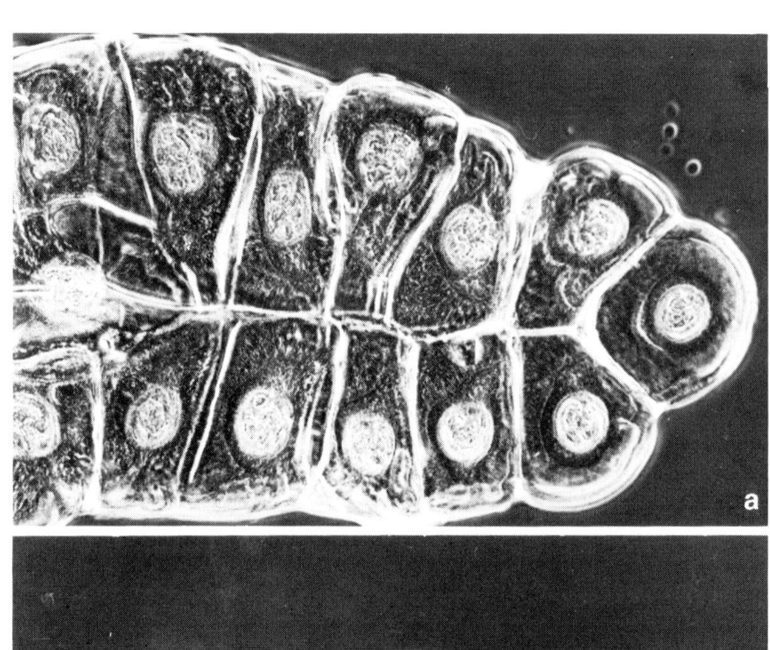
a

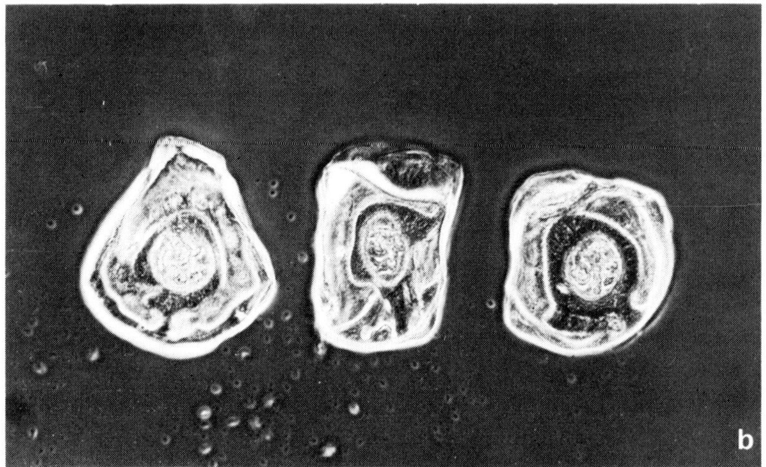
b

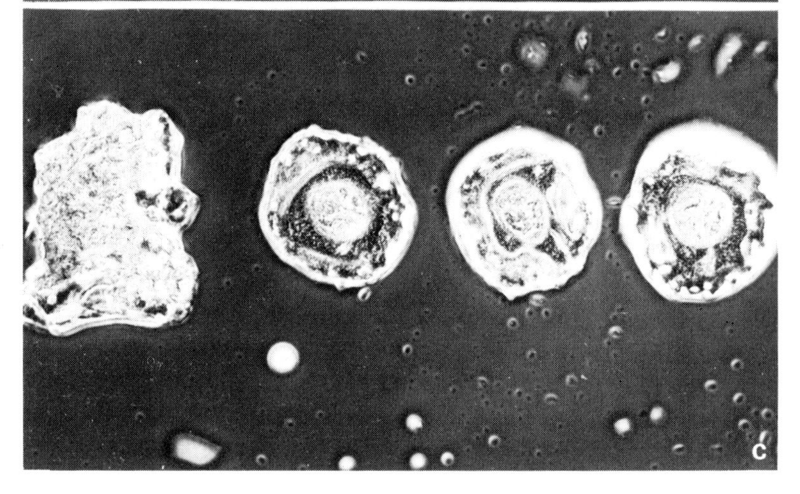
c

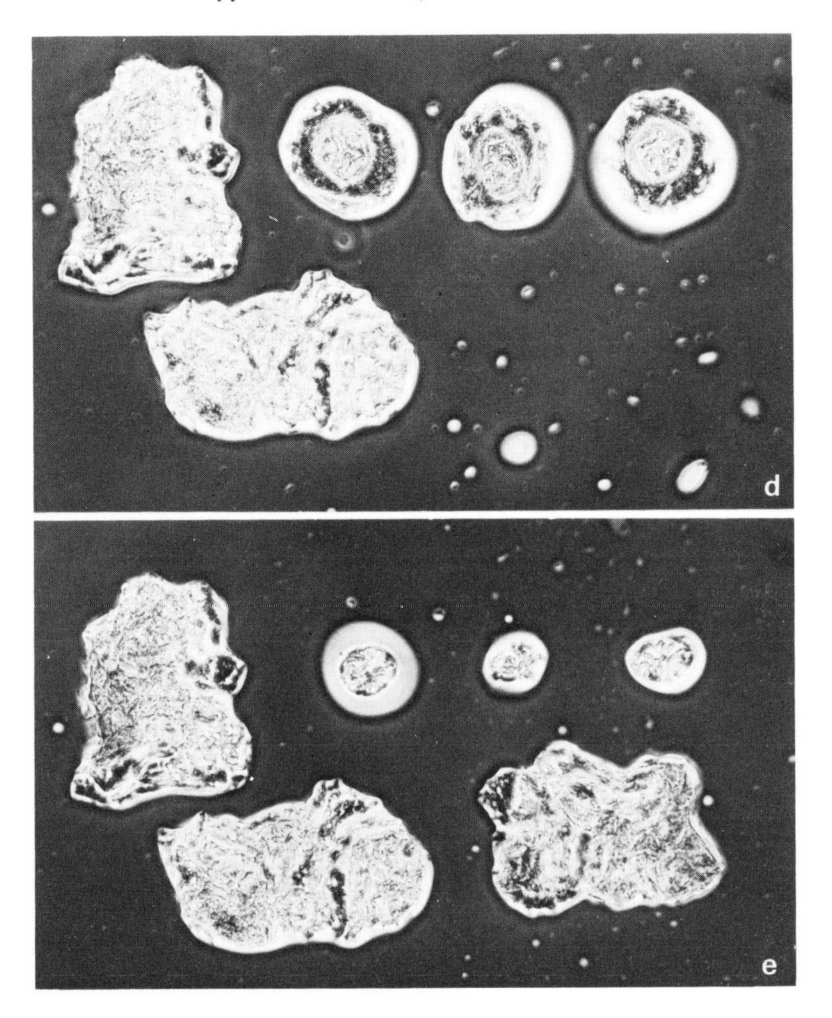

Fig. 7. An illustration of the dissection procedure for obtaining cytoplasmic zones at different distances from the nucleus, as seen in the oil chamber with the phase contrast microscope at a magnification of 320X (a) Shows part of the intact, fixed gland; (b) three isolated cells from the part shown in (a); (c) the outer zones collected to the left of the remainders of the three cells; (d) the outer and middle zones separately placed to the left and below the cell residues; and (e) the remaining nuclei and the tissue masses from the three different cytoplasmic zones.

(fig. 7) and the cytopasm within these zones can be analyzed during the exit period of RNA from the nucleus. When judging the results of such analyses it should be kept in mind that the inner zone is difficult to obtain free from nuclear contamination. In certain connexions this is irrelevant, in others it is of great importance. In

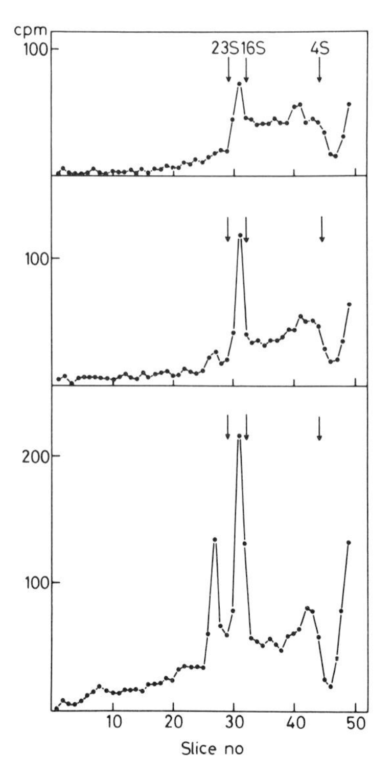

Fig. 8. An animal was injected with 25 μCi of tritiated uridine and sacrificed after 120 min. Three cytoplasmic zones were isolated from 12 cells and extracted for RNA which was separated in 1% agarose gels. The profiles represent RNA from the outer (upper panel), middle (middle panel) and inner (lower panel) zones. For other details see the legend to fig. 3.

the latter case a three-zone analysis is necessary, in the former adequate information can be obtained from a technically simpler two-zone analysis.

Two hours after precursor injection provides an interesting starting point for these investigations since it is during this period that ribosomal RNA first appears in the cytoplasm. In analysis of three-zone dissected cells (fig. 8) one finds that there is a gradient of labelled 18 S RNA extending into the outer zone and one of 28 S RNA reaching into the intermediate zone. The inner zone may be contaminated by nuclear sap but there is relatively little 18 S RNA in the nucleus and the gradient is therefore likely to exist throughout the whole radial extent of the cytoplasm. There is, however, much 30 S RNA in the nucleus which in 1% agarose gels cannot be distinguished from cytoplasmic 28 S RNA and this result therefore does not necessarily indicate the presence of a gradient throughout the whole cytoplasm. It can, however, be demonstrated for the peripheral parts.

A primary question is whether the gradients simply reflect differences in RNA content between the different zones such that peripheral parts contain less RNA

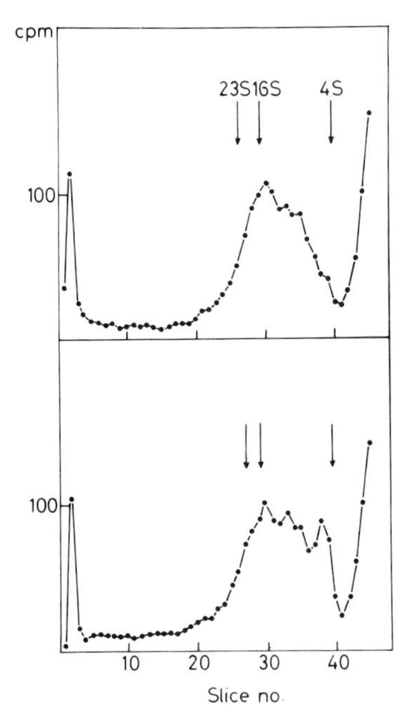

Fig. 9. An animal was injected with 25 μCi of tritiated uridine and sacrificed after 60 min. Two cytoplasmic zones were isolated from 22 cells and extracted for RNA which was separated in 1% agarose gels. The profiles represent RNA from the inner zone (upper) and outer zone (lower). For other details see the legend to fig. 3.

than central parts. Results of analyses after longer times show, however, that with increasing times the distribution of label in ribosomal RNA in different parts of the cytoplasm becomes more and more equal, with the 28 S RNA approaching the typical 2 : 1 ratio.

A second question is whether the gradients are due to slow diffusion of the components containing 18 S and 28 S RNA caused by the particular conditions of the cytoplasm. To evaluate this possibility we have investigated the distribution of the early H RNA after 1 hr of labelling, i.e. during the period of its increase in the cytoplasm. There is no evidence of any gradient at this time (fig. 9). Since the early H RNA distribution should include particles of the size of the ribosomal subunit irrespective of whether the early H RNA is bound in vivo to proteins or not one can conclude that the ribosomal RNA is in some way held back from diffusing freely. An attractive possibility is that it is bound in polysomes, possibly connected to the endoplasmic reticulum, of which these cells are full. If one assumes that both subunits are bound to the same structure as in polysomes, it should be possible to obtain a rough measure of the rate of net peripheral movement of ribosomal RNA through the cytoplasm. A difference in exit time of 25–30 min between 18 S and 28 S RNA has been observed in other systems [21] and a somewhat shorter time has been recorded for insect salivary glands [27]. The distance between the labelled gradients, claculated on a molar basis was found in two different experiments to be about 30 μm. A migration rate of the order of 1 μm per min is, therefore, indicated. It remains to be decided whether this is a rate of movement of the endoplasmic reticulum. If the ribosomal RNA is found to be associated to the reticulum it will provide a unique marker for it which should permit study of its movements in the cytoplasm and of the cellular machinery responsible for such movements.

An aspect of these studies is that at least one of the putative messengers, the early H RNA, does not comigrate with any of the labelled ribosomal components which is in good agreement with the view that messenger RNA is exported independently of ribosomal RNA. If most of the early H RNA is a messenger for non-differentiated products it might be related to free rather than membrane bound polysomes. It will consequently be of particular interest to study the late H RNA, the putative messenger for the secretory proteins. One should not neglect the possibility that it may show an entirely different relation to ribosomal RNA and it is an open possibility that it engages with ribosomal subunits already in the nucleus. Studies of cytoplasmic zones after longer labelling times have not yet revealed the presence of any gradients for the late H RNA but the question is still an open one.

The mechanisms in gene expression of the higher cell show many similarities to those of the prokaryote. The essence of the multicellular condition lies, however, in the possibility it affords of a division of labour between cells, resulting from the phenomenon of differentiation. Properties in gene expression which are unique to the eukaryote may therefore largely be connected to the expression of the differentiated state. It is in this sense that a cell like the *Chironomus* salivary gland cell is of particular importance. It will offer the possibility of following and comparing the parallel lines of gene activities for differentiated and nondifferentiated functions from the level of the genome to the sites of functional expression.

Acknowledgements

Elisabet Ericson carried out the analyses and helped me with the micromanipulations, Chana Szpiro cultured the animals, Hannele Jansson typed the manuscript, and Agneta Askendal drew the illustrations, for all of which I am greatly indebted. I also thank Robert Tanguay for permission to use unpublished material. The work described in this paper has been carried out with the support of the Swedish Cancer Society.

References

[1] W. Beermann, Chromosoma 5 (1952) 139.
[2] F. Mechelke, Chromosoma 5 (1953) 511.
[3] M.E. Breuer and C. Pavan, Chromosoma 7 (1955) 371.
[4] W. Beermann, Chromosoma 12 (1961) 1.
[5] W. Baudisch and R. Panitz, Exptl. Cell Res. 49 (1968) 470.
[6] U. Grossbach, Chromosoma 28 (1969) 136.
[7] H.D. Berendes, J. Exptl. Zool. 162 (1966) 209.
[8] U. Wobus, R. Panitz and E. Serfling, Molec. Gen. Genetics 107 (1970) 215.
[9] J.-E. Edstrom and B. Daneholt, J. Mol. Biol. 28 (1967) 331.
[10] C. Pelling, Cold Spring Harbor Symp. Quant. Biol. 35 (1970) 521.
[11] B. Daneholt, Nature New Biol. 240 (1972) 229.
[12] U. Ringborg and L. Rydlander, J. Cell Biol. 51 (1971) 355.
[13] E. Egyházi and J.-E. Edström, Biochem. Biophys. Res. Commun. 46 (1972) 1551.
[14] W. Beermann and G.F. Bahr, Exptl. Cell Res. 6 (1954) 195.
[15] B.J. Stevens and H. Swift, J. Cell Biol. 31 (1966) 55.
[16] B. Lambert, J. Mol. Biol. 72 (1972) 65.
[17] B. Lambert, Nature 242 (1973) 65.
[18] B. Daneholt and H. Hosick, Proc. Natl. Acad. Sci. U.S. 70 (1973) 442.
[19] J.-E. Edström and R. Tanguay, manuscript in preparation (1973).
[20] J.-E. Edström, In: Methods in Cell Physiol., Vol. 1, Ed. D. M. Prescott (Academic Press, New York, 1964) pp. 417–447.
[21] J.E. Darnell, Bacteriol. Rev. 32 (1968) 262.
[22] G. Schochetman and R.P. Perry, J. Mol. Biol. 63 (1972) 591.
[23] B. Daneholt and L. Svedhem, Exptl. Cell Res. 67 (1971) 263.
[24] B. Lambert, L. Wieslander, B. Daneholt, E. Egyházi and U. Ringborg, J. Cell Biol. 53 (1973) 381,
[25] B. Lambert, B. Daneholt, J.-E. Edström, E. Egyházi and U. Ringborg, Exptl. Cell Res. 76 (1973) 381.
[26] E. Egyházi, B. Daneholt, J.-E. Edström, B. Lambert and U. Ringborg, J. Mol. Biol. 44 (1969) 517.
[27] H.A. Armelin and N. Marques, Biochemistry 11 (1972) 3663.
[28] S.Y. Lee, J. Mendecki and G. Brawerman, Proc. Natl. Acad. Sci. U.S. 68 (1971) 1331.
[29] R.H. Singer and S. Penman, Nature 240 (1972) 100.

Niu and Segal (eds.). The role of RNA in reproduction and development
North-Holland Publ. Co., 1973

Sequence composition and organization of the genome and of the nuclear RNA of higher organisms: an approach to understanding gene regulation

David S. HOLMES and James BONNER
California Institute of Technology, Division of Biology
Pasadena, Calif. 91109, USA

Whereas considerable progress has been made in our understanding of transcriptional control of gene expression in bacteria, comparatively little is known about this matter in higher organisms. It is evident, however, that insight into the structural basis of gene regulation in higher organisms requires an understanding of the types of sequences present, and their organization, in the genomic DNA and its RNA transcripts.

Based on techniques involving nucleic acid hybridization and electron microscopy of DNA and RNA, evidence is presented that: (i) Under specific conditions of duplex stability *Drosophila* DNA consists of sequences some of which are present only per genome (single copy), some are repeated tens of times (middle repetitive) and some are repeated thousands of times (highly repetitive). (ii) Middle repetitive sequences are short, about 100–300 bases long, and are interspersed throughout the genome between single copy sequences with an average distance of 700–1000 bases between repetitive sequences. (iii) In rat, a population of chromosomally-associated RNA molecules (cRNA) consists almost exclusively, of middle repetitive transcripts. (iv) Rat nuclear RNA (HnRNA) consists of both repetitive and single copy sequences. (v) Many of the repetitive sequences of HnRNA are similar to those of cRNA. (vi) Circumstantial evidence is presented which implicates cRNA in the control of gene activity. These results are discussed in the context of the Britten–Davidson model of gene regulation [8].

1. Introduction

During the past few years it has been fashionable to argue that a large part of the differentiation and development of higher organisms is due to the regulation of gene expression at the level of transcription. This tenet does not mitigate the role of other levels of control. Indeed, it is most probable that differentiation proceeds only by the careful integration of a number of controls at various levels of cellular and intercellular organization.

The evidence for specificity of transcriptional control comes from three main categories of experiments: (i) The observation that localized RNA synthesis is both associated with puffing at specific chromosomal loci in diptera and correlated with the occurrence of characteristic cells products [1]. (ii) The demonstration that populations of single copy RNA transcripts in different tissues [2] and at various developmental stages [3] are different. It remains, however, to establish that the differences are due to specificity of transcription and not to differential viability of the RNA trancripts. (iii) The observation that chromatin from different tissues of the same organism supports the in vitro synthesis of different repetitive RNA sequences using RNA polymerase from a bacterial source to transcribe the chromatin [4, 5]. In addition to questions relating to the validity of using heterologous RNA polymerase, the interpretation of hybridization experiments that seek to test differences in repetitive RNA sequences has been challenged [6]. We are thus led to the view that whereas the evidence for specificity of transcription is good it is still insufficient to prove it rigorously. Nevertheless, we will proceed, like many others, on the assumption that additional evidence will be forthcoming.

In the history of scientific endeavor it is evident that a reciprocal relationship exists between the extent of our knowledge on a particular subject and the number of theories pertaining to that subject; the regulation of transcription during development is no exception to this rule. Of the numerous theories on transcriptional regulation, several are based on the Jacob and Monod model of bacterial regulation with extensions designed to accommodate the additional complexity of higher organisms [7–12]. One of these theories, that of Britten and Davidson [8], has been the focus of considerable attention. In this paper we will discuss some recent experiments that explore the nature and distribution of sequences present in genomic DNA and in nuclear RNA. While our results can be explained within the conceptual framework of the Britten–Davidson model they do not prove it.

2. The Britten–Davidson model of gene regulation

Fig. 1 summarizes graphically the important elements of the Britten–Davidson model. The salient features are: (i) Producer genes which code for RNAs other than those directly involved in gene regulation. For example they code for tRNA, rRNA, 5 S RNA and mRNA. The producer genes are organized in batteries; the genes of a battery are activated simultaneously. (ii) Contiguous with each producer gene is a receptor gene which is recognized by a diffusable activator molecule such as RNA. It is the specific recognition and interaction of activator RNA with the receptor gene that control the activity of the associated battery. (iii) There are integrator genes whose transcriptional product is activator RNA. (iv) The genes of an integrator set are activated in concert in response to inducing agents, such as hormones, which recognize and act on a sensor gene contiguous with the integrator set.

The key features of this model are that it can account for the control of dispersed genes, and can provide a means whereby their activation can result from a single initiatory event leading to a hierarchial and integrated program of gene regu-

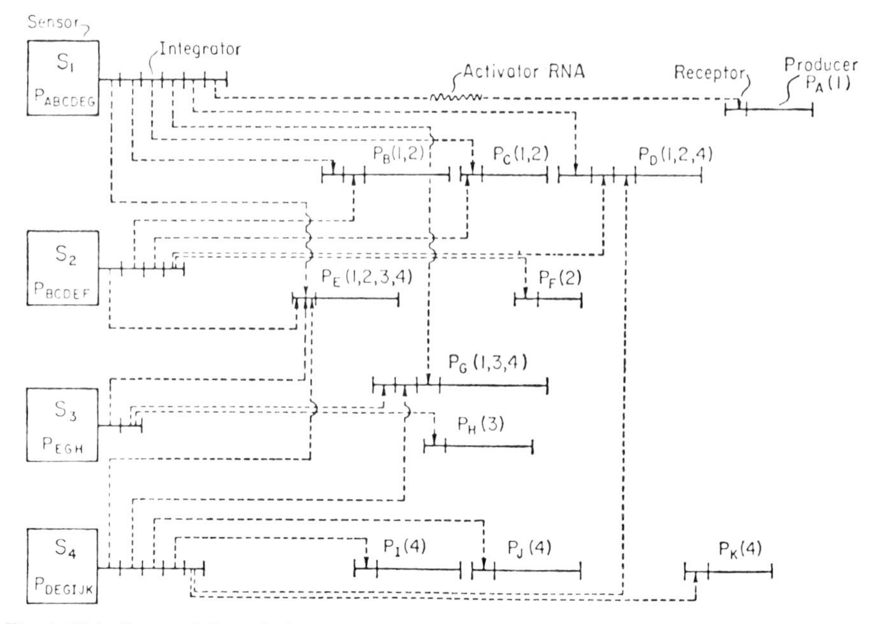

Fig. 1. This diagram is intended to suggest the existence of overlapping batteries of genes and to show how, according to the model, control of their transcription might occur. The dotted lines symbolize the diffusion of activator RNA from its sites of synthesis, the integrator genes, to the receptor genes. The numbers in parentheses show which sensor genes control the transcription of the producer genes. At each sensor the battery of producer genes activated by that sensor is listed. In reality many batteries will be much larger than those shown and some genes will be part of hundreds of batteries. (Taken from [8] with permission of the editor.)

lation. It also suggests a function for some of the nuclear RNA and some of the "excess" DNA of higher organisms, particularly the repetitive sequences.

Although in this scheme the activator molecule is an RNA, it could equally well be a protein without altering the fundamental design of the model. The idea that proteins are involved in the specific recognition and control of genes is currently being examined but this paper will not discuss this area [13].

Fig. 2 illustrates a few of the possible ways in which the basic design of the model can be varied. In fig. 2a an integrator set is shown in two modes of transcription: (i) each activator RNA is transcribed independently, or (ii) the whole set is transcribed into a polycistronic product with post-transcriptional processing to yield activator RNA. In either mode the organization of the integrator genes within a set invokes contiguity or close linkage of repetitive sequences, and requires the presence of a sensor site adjacent to, or close to, one end of the set. Other categories of sequences may be present between integrator genes involved in the mechanics of transcription or coding for regions of the RNA transcripts involved in their post-transcriptional processing.

The organization and processing of the transcripts in the two modes differ more

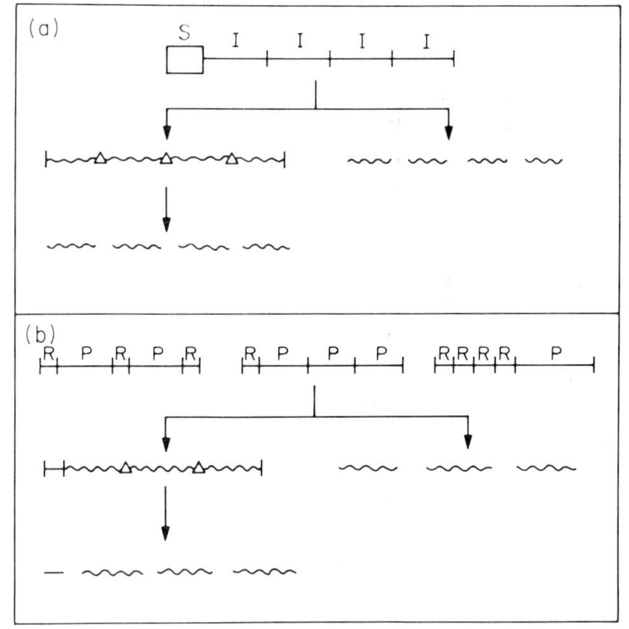

Fig. 2. Various modifications of the Britten–Davidson model as exemplified in fig. 1. (a) Two possible modes of transcription of an integrator set. (b) Three different organizations of receptor and producer genes. One organization is shown in two modes of transcription. S = sensor gene, I = integrator gene, R = receptor gene, P = producer gene. ——— DNA, ~~~ RNA, △ RNA sequence, other than a "messenger" sequence, e.g. a region involved in post-transcriptional processing of the RNA.

markedly, because in one instance a precursor–product relationship between the immediate product of an integrator set and activator RNA in its functional form is called for and in the other instance activator RNA is the immediate transcript of an integrator gene.

Fig. 2b illustrates three arrangements of receptor and producer genes and two possible ways in which one of these variants could be transcribed into RNA. The figure is largely self-explanatory. We would like to emphasize that we have not illustrated all possible combinations of the basic elements nor do we wish to imply that the incidence of any one or more of the variants necessarily excludes the occurrence of the others in the same genome. The object of the exercise is to take the basic elements of the Britten–Davidson model and to generate a variety of different arrangements in the sequences of both the DNA and the RNA transcripts. These operations are, hopefully, consistent with the fundamental purpose of the model and are open to experimental verification.

3. Sequence composition and organization of eukaryotic DNA

The point of departure in these studies is an investigation into the classes of sequences present and their organization within the genome of *Drosophila*. We have chosen *Drosophila* because its genome size is considerably smaller than, for example, that of mammals. There is also available for *Drosophila* a range of genetic tools and structural features such as polytene chromosomes which increase the possibility of manipulating the genome to suit our experimental purposes.

Fig. 3 shows the reassociation profile of *Drosophila* DNA in the form of a Cot curve as described by Britten and Kohne [14]. At the criterion chosen for reassociation there appears to be three major reassociating components. About 7% of the DNA reassociates more rapidly than can be measured in this experiment. This corresponds, at least in part, to the centromeric DNA some of which is probably not transcribed into RNA [15, 16]. About 15% of the DNA reassociates with kinetics indicating that the major component of this fraction is repeated approximately 35 times per genome. This has been termed the middle repetitive DNA [17]. The remainder of the DNA reassociates with kinetics indicating that each sequence is present only once per genome [17].

Although the proportions and repetition frequencies of the reassociating compo-

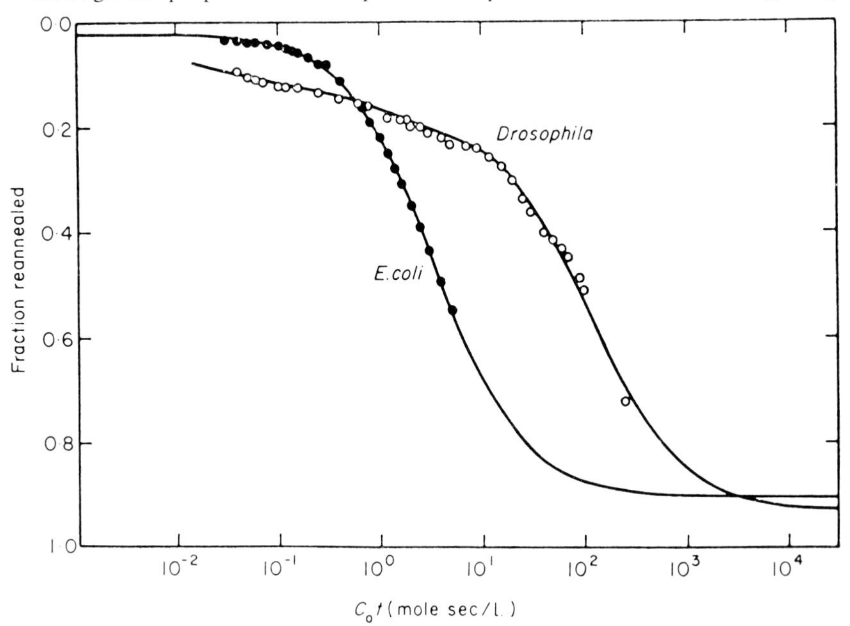

Fig. 3. Optical reassociation profiles for *Drosophila melanogaster* and *E. coli* DNA (both sheared to circa 350–400 base pairs in length). Reannealing in 0.12 M phosphate buffer pH 6.8 at 66°C. Concentration of DNA: for *Drosophila* 0.2 O.D. 260 (1 cm path-length cell) for upper portion of profile and 2.0 O.D. 260 (1 mm path-length cell) for lower portion; for *E. coli* 0.2 O.D. 260 (1 cm path-length cell). (Taken from [17] with permission of the editor.)

nents may vary between organisms, it is apparent that all eukaryotic genomes examined so far, at least above the fungi, exhibit both repeated and single copy sequences [18].

By analogy with bacterial systems the single copy sequences of eukaryotes are thought to include genes encoding proteins. In the terminology of Britten and Davidson [8] these would correspond to producer genes. Several recent experiments confirm the view that at least some of the single copy sequences, or sequences of very low reiteration, are producer genes [19–21]. The middle repetitive DNA of eukaryotes includes the genes coding for rRNA, 5 S RNA, tRNA and histone mRNA [22–24]. However, these genes probably account for only a very small portion of the middle repetitive DNA under specific conditions of duplex stability. The remainder of the middle repetitive DNA is of unknown function although it is known that a considerable proportion of it is transcribed into RNA as shown below and in ref. [25].

The middle repetitive DNA, by virtue of its repetitiousness, qualifies as a candidate for the integrator and activator genes of the Britten–Davidson model. To explore further this possibility the size and distribution of the middle repetitive sequences along the genomic DNA were investigated. Fig. 4 illustrates the methods used and the results of these experiments. The majority (94%) of the middle repetitive sequences are between 100 and 400 bases in length (fig. 4b) as visualized by the aqueous Kleinschmidt technique. When the distance between middle repetitive tails is measured, using the formamide Kleinschmidt technique, it is apparent that most of the tails are separated by about 100 to 500 nucleotides (fig. 4d). This corresponds closely to the estimated average length of the middle repetitive sequences. More interestingly there appears to be a second peak in the histogram with a spacing of about 700–1000 nucleotides between repetitive sequences. This corresponds to the expected size of a producer gene coding for an average sized protein [26]. Some repetitive sequences are separated by as much as 3000 nucleotides. In some instances two pairs of tails were visualized very close together on the DNA (< 500 bases) [17]. This might indicate that two repeated sequences are either very close or are contiguous. In addition, it is worth noting that some of the repeated sequences are separated by much more DNA than is expected for one producer gene. We therefore advance the view that: (i) the middle repetitive sequences are short (about 100–300 bases), (ii) a large portion of individual repeated sequences are interspersed with short non-repeated DNA segments (circa 1000 bases), (iii) there are other regions in which two repeated sequences are found relatively close together, (iv) no segments containing many contiguous repeated sequences which might correspond to integrator sets (see fig. 2a) were detected. However, if such segments were present as a very small percentage of the total DNA they could easily have been overlooked in this type of experiment. We are, therefore, proceeding with further experiments to investigate this matter [27]. (v) There are some long stretches of non-repeated DNA between repeated sequences.

The picture of the *Drosophila* genome which emerges is essentially a synthesis of

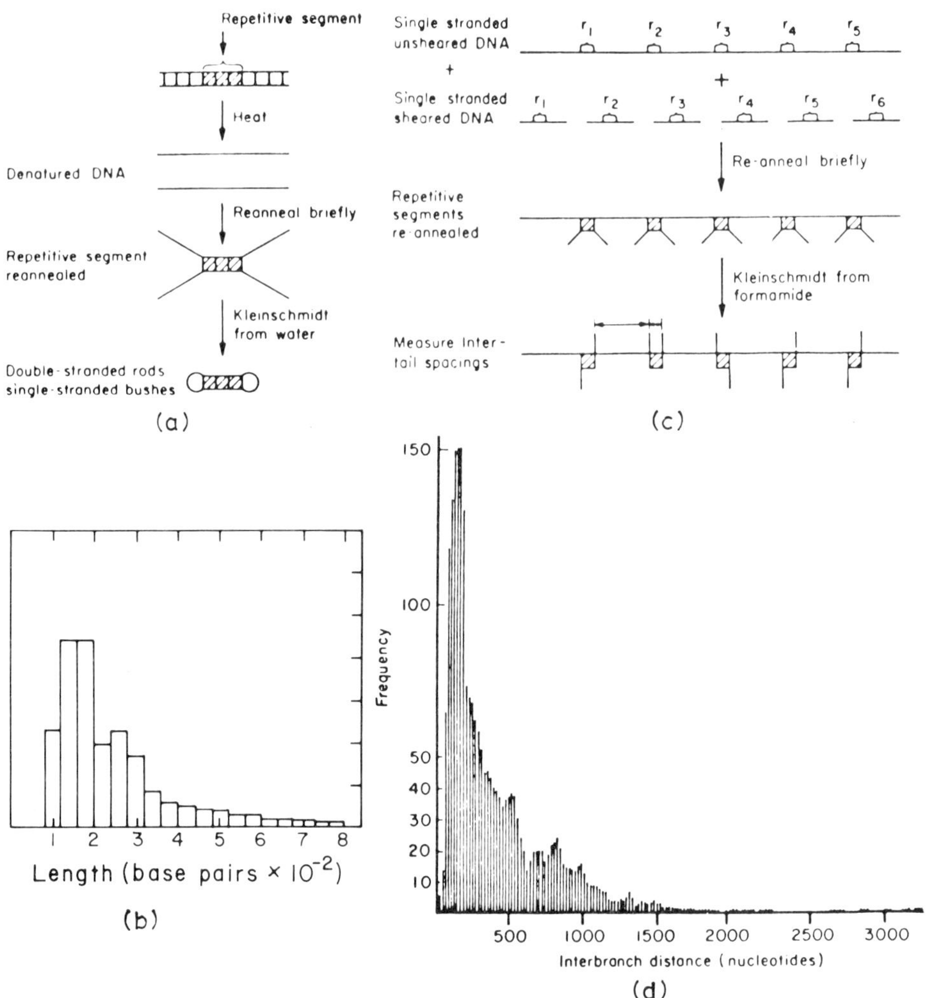

Fig. 4. Length and distribution of the repetitive DNA segments of *Drosophila melanogaster*. (a) Schematic representation of the procedure used for the determination of the length of the middle repetitive DNA segments. DNA (sheared to circa 800 bases) which reassociated between a Cot of 0.05 and 3 was separated by chromatography on hydroxyapatite from the remainder of the DNA and defined as middle repetitive [17]. The experimental strategy for length determinations is based on the observation that single stranded but not double stranded DNA collapses to form bushes when spread onto grids by the aqueous Kleinschmidt technique. The length of the duplex structure can then be measured. 100 bases of duplex is about the limit of resolution that can be achieved by this method. (b) Frequency distribution of the lengths of reannealed duplexes of repetitive DNA prepared as described above. Reannealing was to a Cot of 3.0. (c) Scheme for determination of the distribution of the middle repetitive segments. DNA was denatured and reassociated as described in the legend to fig. 2. Both double and single stranded DNA remain extended when spread on grids by the formamide Kleinschmidt procedure. Two parameters can be measured (i) the distance between tails of a single repetitive segment and (ii) the distance between tails of two or more separate repetitive segments. (d) Frequency distribution of the interbranch distances of repetitive DNA segments prepared as described above. (Redrawn from [17].)

the three types of genomic organization depicted in fig. 2b. However, this data cannot be used to support, unconditionally, the Britten–Davidson model because there is no experimental evidence attributing a function to the sequence examined, i.e. there is no experimental correlation between repetitive sequences and receptor genes or between any of the single copy sequences and producer genes. The isolation of pure producer DNA and its utilization as a probe in the types of experiments outlined above will, no doubt, greatly expedite our understanding of the functional organization of the genome. A second thrust in this direction might come from the use of variants of genomic organization such as the maze systems used by McClintock in the study of controlling elements [28]. Variants, such as these, might be amenable to analysis by the types of experiments just described.

4. The role of RNA in gene regulation

The second category of experiments that we will describe concerns the properties of and the relationship between two classes of nuclear RNA: giant nuclear RNA (HnRNA) and a population of small RNA molecules associated with chromatin, termed chromosomal RNA (cRNA). To orientate these experiments in the context of the present discussion requires a slight digression. Chromatin (interphase chromosomes), isolated by standard procedures, contains DNA, histones, non-histone proteins and RNA [28]. It is widely held, but still open to argument, that histones bind relatively non-specifically to DNA rendering it unavailable for transcription, and that some other class of molecules is necessary to recognize and activate specific genes [29]. Molecules with these properties presumably interact, if only in a transitory fashion, with the DNA, at some stage when the genome is to be programmed for transcription. Therefore the RNA and non-histone protein components of the chromatin have become candidates for specific control elements [13].

Chromosomal RNA (cRNA), one of several categories of RNA molecules associated with chromatin, has been extensively investigated over the past decade. A considerable amount of circumstantial evidence exists correlating cRNA with gene activation. We will mention some of this evidence below. During the same period another novel class of RNA termed nuclear RNA (HnRNA) was being studied. Early experiments showed that both cRNA and HnRNA hybridize to a considerable proportion of the repetitive DNA [30, 31]. Therefore it became expedient to determine whether the repetitive sequences might be similar in both classes of RNA. With the publication of the Britten–Davidson model an analogy could be drawn between cRNA and activator RNA, and less completely between some of the HnRNA and the polycistronic product of the integrator sets (see fig. 2a). It was conceivable that a portion of the HnRNA was a long chain precursor to cRNA. In this way, a single event at a sensor site could stimulate an integrator set to transcribe polycistronic RNA which could be post-transcriptionally processed to yield a number of activator RNAs. A precedent for the production of functional RNA by post-transcriptional processing of a long chain precursor exists in the case of rRNA [32]. It was, therefore, considered reasonable to examine the properties of cRNA and HnRNA and the relationship between them in the light of the Britten–Davidson model.

We turn now to our results in this context and approach the matter by asking a series of questions: (i) Is cRNA an activator RNA? (ii) Is HnRNA the transcriptional product of integrator sets? (iii) Is HnRNA the precursor to cRNA?

5. A candidate for activator RNA

What properties might one predict for activator RNA? Since activator RNA is a product of the integrator genes, it should be produced in response to stimuli such as hormones, and different sequences of activator RNA should be produced in response to different stimuli. Assuming that differential gene activity reflects the presence of different activator sequences then different tissues of the same organism should contain different sequences of activator RNA. Sequences of activator RNA should be complementary to sequences of DNA corresponding to integrator and receptor genes, and since the model requires considerable repetition among the integrator and receptor genes, the activator RNA in turn should contain transcripts of repeated sequences.

Several properties of cRNA lend credence to the view that it is an activator RNA. New sequences of cRNA are produced during rat liver regeneration [33], different sequences are found in different tissues of the same organism [30, 34], and the kinetics of hybridization of cRNA to homologous DNA immobilized on filters suggest that cRNA contains repetitive transcripts [35, 36]. In order to further substantiate the parallel between cRNA and activator RNA it is necessary (i) to show that the occurrence of different cRNA sequences in different tissues and during liver regeneration is due directly to physiological stimulation and not to differential degradation of RNA sequences mediated by the stimulus, and (ii) to examine more definitively the sequence composition of cRNA under conditions which will assay single copy as well as repetitive transcripts.

Two types of hybridization experiments have been carried out to investigate more definitively the sequences represented in cRNA. In the first experiment, shown in fig. 5, uniformly labeled ^{3}H-cRNA from rat ascites was hybridized to unlabeled sheared rat DNA present in great excess. Under these conditions the rate of hybrid formation is dependent predominantly on the concentration of complementary DNA sequences [37, 38]. Using a computer-fitted single second order reaction curve to describe the rate of hybridization it is clear that cRNA hybridizes, as far as detected, at a rate ($Cot_{1/2}$ = 1.92) similar to that of the major reassociating component of the middle repetitive DNA ($Cot_{1/2}$ = 1.06). The single second order curve displayed in fig. 5, only partially describes the hybridization reaction and it is likely that there is more than one kinetic component in the cRNA population [40].

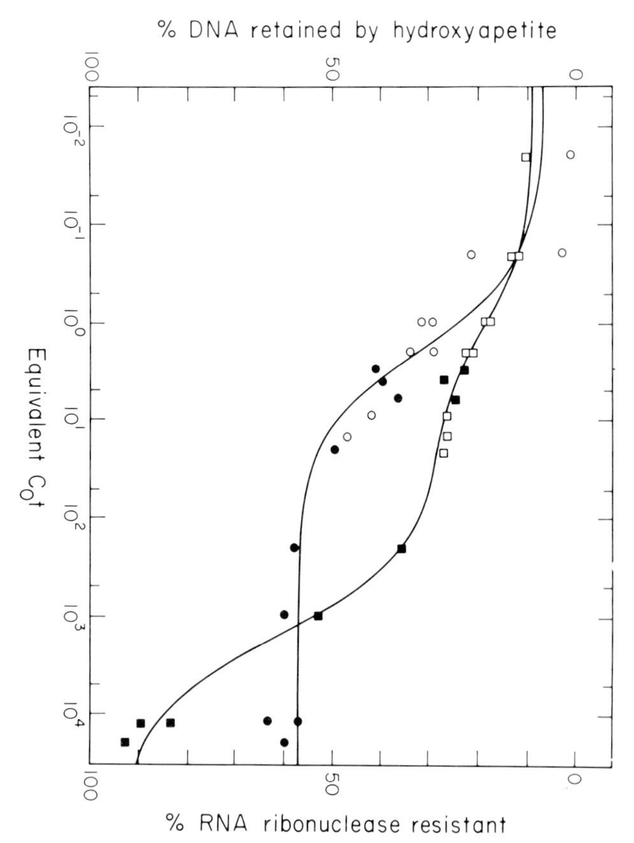

Fig. 5. Computer analysis of the hybridization of dimethyl sulfate labeled ^3H cRNA (specific activity about 82,000 cpm/μg) from rat ascites in the presence of an excess of unlabeled sheared (circa 350 nucleotides) rat nuclear DNA. The ratio of ^3H cRNA : DNA was about 1 : 111,000. The hybridization was carried out in either (open symbols) 0.12 M phosphate buffer pH 6.8 or (filled symbols) 0.48 M phosphate buffer pH 6.8 at 62°C or 66°C respectively. The reassociation of DNA (squares) was followed by monitoring the A260 of an aliquot of the reaction mixture after passage through a hydroxyapatite column. After mild ribonuclease treatment (20 μg/ml ribonuclease A and 20 units ribonuclease T_1 in 0.24 phosphate buffer for 15 min at 37°C) the RNA : DNA hybrids (circles) were monitored by the TCA precipitation method of Melli et al. [38]. (Taken from [47].)

Fig. 6 shows the ability of ascites cRNA to hybridize to purified single copy and middle repetitive rat DNA. About 16% of the middle repetitive and 1% of the single copy DNA are complementary to cRNA.

cRNA is thus complementary to a considerable proportion of the rat DNA under relatively stringent conditions of hybridization and most, if not all, of this DNA belongs to the middle repetitive components of the genome. This is consistent with the expected properties of activator RNA.

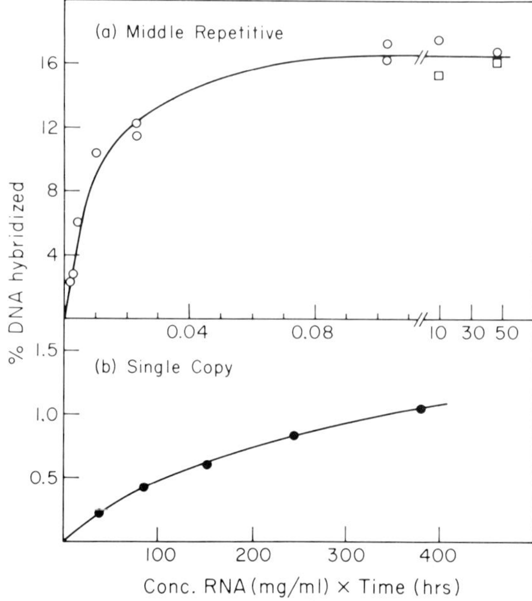

Fig. 6. Hybridization of dimethyl sulfate labeled ^{3}H cRNA from rat ascites (specific activity about 82,000 cpm/μg) to either unlabeled (a) middle repetitive or (b) single copy rat nuclear DNA (about 350 nucleotides long). After hybridization in 0.12 M phosphate buffer pH 6.8 at 62°C (open symbols) or in 0.48 M phosphate buffer pH 6.8 at 66°C (closed symbols) the reaction mix was treated with ribonuclease A (50 μg/ml) and ribonuclease T_1 (50 units/ml) in 0.24 M phosphate buffer pH 6.8 at 37°C for 10 min. The reaction mix was chromatographed on Sephadex G-100. Radioactivity associated with the DNA in the void volume of the G-100 column was scored as ribonuclease resistant RNA. (Taken from [47].)

It is appropriate, therefore, to look at the distribution of cRNA sites along the nuclear DNA to see if the pattern in any way approaches that expected for the receptor genes. One approach to this question might be to visualize cRNA–DNA hybrids under the electron microscope by a method analogous to that described above for the mapping of DNA sequences. However cRNA–DNA hybrids are difficult to distinguish from single-stranded DNA in formamide spreadings under the electron microscope. To circumvent this problem, ferritin, an electron-opaque marker, was covalently attached to the 3′ end of rat ascites cRNA and the conjugates hybridized to total unsheared rat nuclear DNA. Unhybridized RNA ferritin was removed and the DNA–RNA ferritin hybrids mounted on grids for visualization under the electron microscope.

The results of a preliminary experiment of this type are shown in figs. 7 and 8. The interferritin distance estimates the spacing between hybrids. Some ferritins are separated by only 100 to 200 nucleotides, suggesting that some cRNA sites are either very close or are contiguous. The remainder of the cRNA ferritin sites are distributed in a fashion characteristic of the organization of middle repetitive DNA

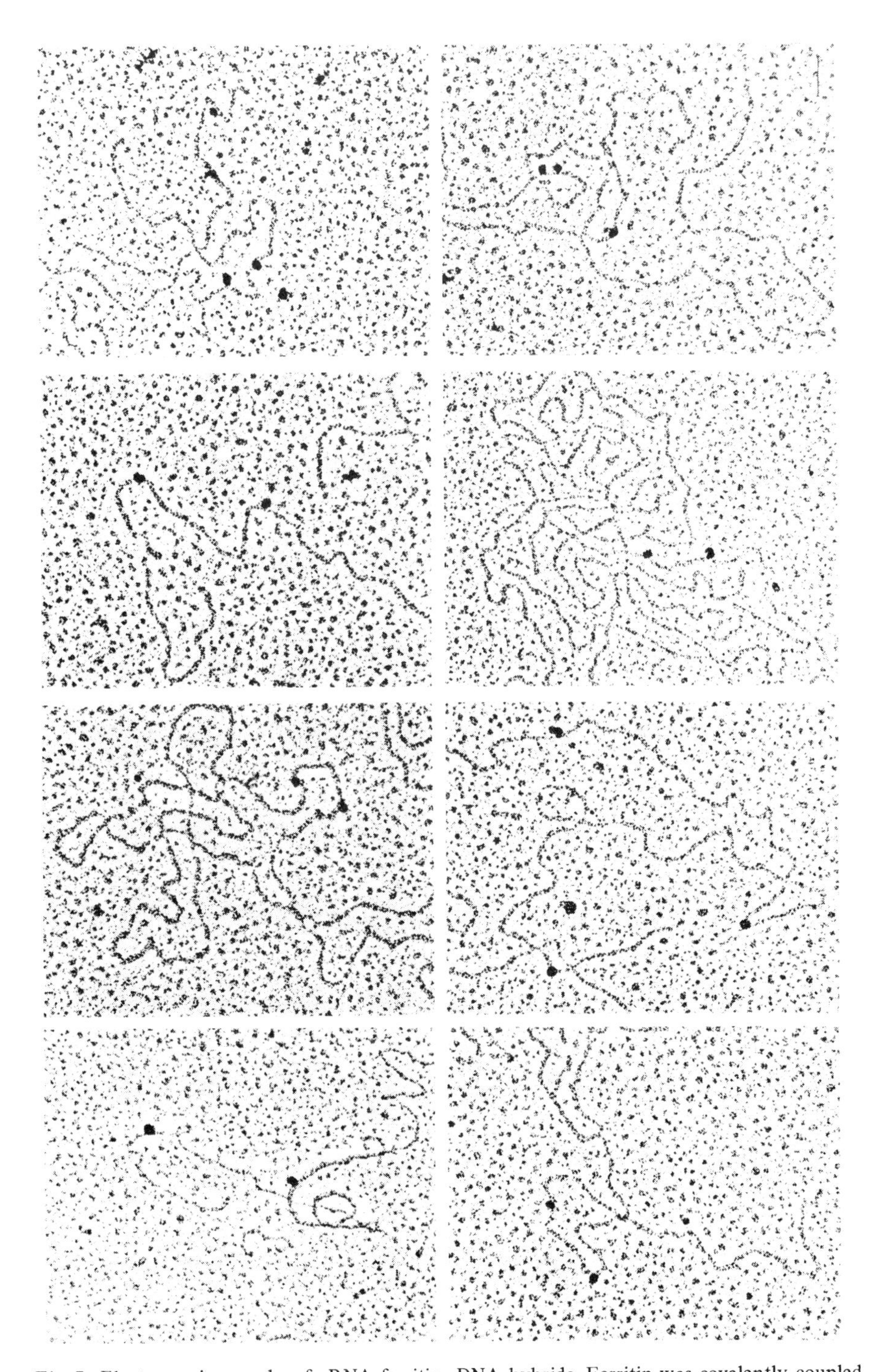

Fig. 7. Electron micrographs of cRNA ferritin–DNA hybrids. Ferritin was covalently coupled to rat ascites cRNA by a method described in [39]. Hybridization of the cRNA ferritin conjugates to total unsheared denatured rat nuclear DNA was carried out as described in the legend to fig. 6. Unhybridized cRNA ferritin conjugates were removed by mild rebonuclease treatment, followed by phenol extraction. The hybrids were spread on grids for visualization under the electron microscope by the Kleinschmidt formamide procedure. Control experiments indicate that ferritin coupling does not interfere with the fidelity of hybridization of the RNA and ferritin by itself does not bind to DNA.

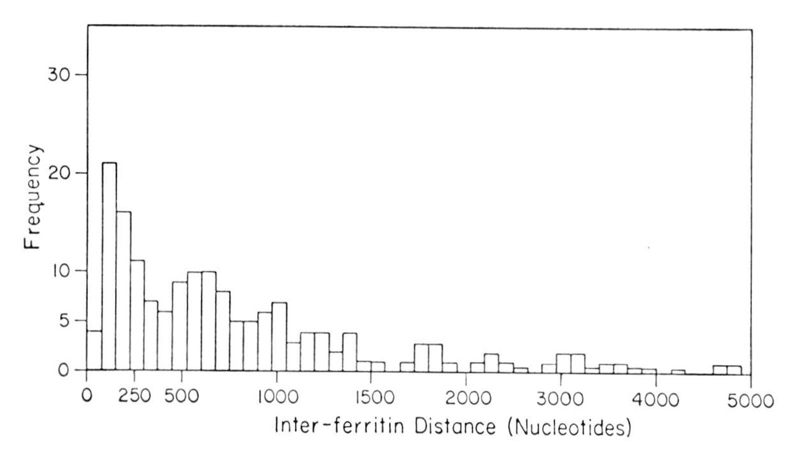

Fig. 8. Frequency distribution of the cRNA interferritin distances observed on electron micrographs such as those shown in fig. 7.

[40]. In this particular experiment 30 pairs of adjacent (< 200 nucleotides separation) cRNA ferritins were scored, but no more than two adjacent ferritins were detected. We cannot conclude that there are no regions with more than two contiguous cRNA sites, regions which might correspond to integrator genes, because in this experiment the cRNA was added at a concentration below that necessary to saturate all the available DNA sites. Thus, it is unlikely that regions of multiple contiguous cRNA sites would be detected if they exist as a small percent of all cRNA sites. Since the experiment has proved to be technically feasible this experimental deficiency is being rectified [40].

6. *Does HnRNA contain transcripts of integrator sites?*

HnRNA is a class of rapidly labeled, low G + C, nuclear RNA which sediments heterogenously in sucrose gradients [41]. Indirect evidence suggests that whereas a portion of this RNA is precursor to polysomal mRNA, much of the remainder of the RNA contains sequences not found in mRNA, some of which are repetitive and are confined to the nucleus [41]. It is known that new repetitive sequences of HnRNA are produced in response to hormone stimulation [42], and liver regeneration [43].

It is of interest, therefore, to examine the sequences of HnRNA and to determine by competition hybridization their relationship to cRNA. Fig. 9 shows the result of hybridizing sonicated uniformly-labeled, ^{3}H-HnRNA prepared from rat ascites cells to sheared total rat nuclear DNA present in a vast excess. Under the conditions employed about 22% of the reacting HnRNA hybridizes with a rate ($Cot_{1/2}$ = 5.65)

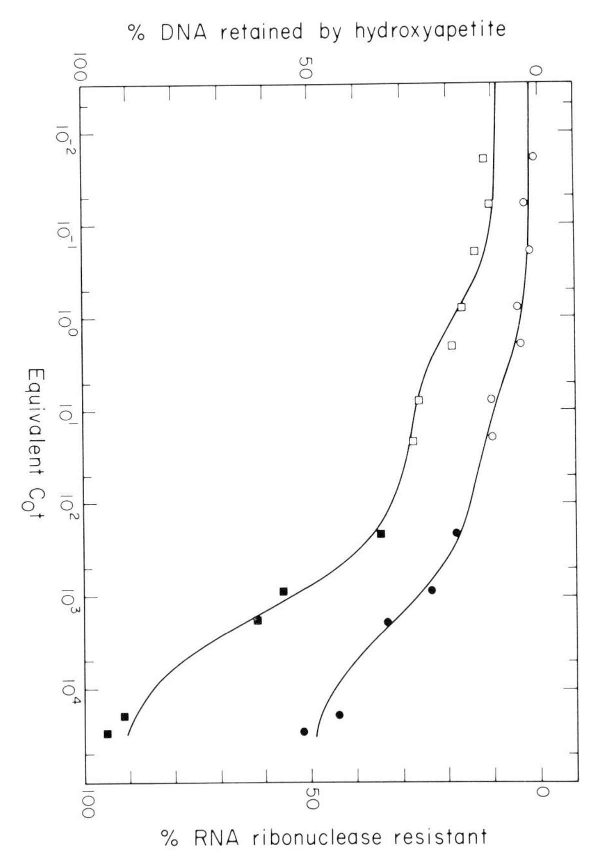

Fig. 9. Computer analysis of the hybridization of rat ascites dimethyl sulfate labeled ^{3}H giant HnRNA (sonicated to about 4–8S) (circles) to an excess of unlabeled sheared (circa 350 nucleotides long) rat nuclear DNA (squares). The ratio of RNA : DNA was about 1 : 170,000. The reaction was carried out and the hybridization of RNA : DNA and the reassociation of DNA were monitored as described in the legend to fig. 5. Open symbols indicate reaction in 0.12 M phosphate buffer pH 6.8 at 62°C and filled symbols indicate reaction in 0.48 M phosphate buffer pH 6.8 at 66°C. (Taken from [50].)

similar to that characterizing the major reassociating component of the middle repetitive DNA, the remainder hybridizes more slowly, with kinetics ($Cot_{1/2}$ = 2.08×10^3) similar to those describing the single copy DNA sequences.

Fig. 10 shows the result of an experiment in which sonicated HnRNA was hybridized to purified ^{125}I middle repetitive (fig. 10a) or single copy (fig. 10b) rat DNA. HnRNA hybridizes to about 5% of the single copy and 10% of the middle repetitive DNA.

These experiments demonstrate that most of the sequences of HnRNA are transcripts of the single copy DNA. However, the 22% or so of the sequences that are

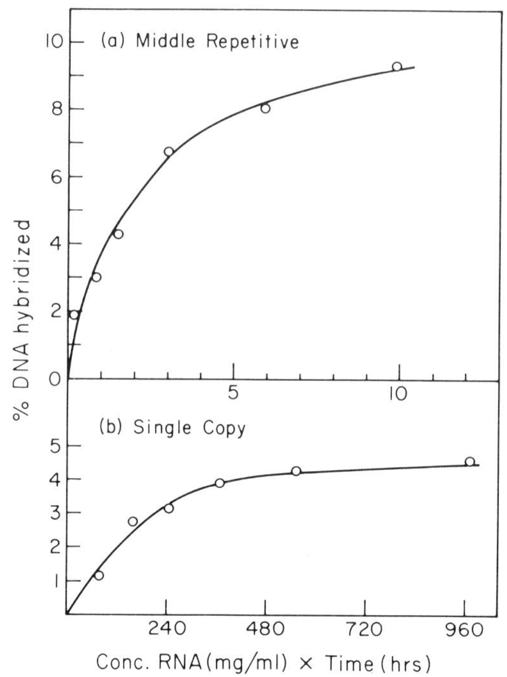

Fig. 10. Hybridization of unlabeled giant HnRNA from rat ascites (sonicated to about 4–8 S) to either (a) ^{125}I middle repetitive or (b) ^{115}I single copy rat nuclear DNA (about 350 nucleotides long). Hybridization was carried out as described in the legend to fig. 6. Hybrids were assayed by passage through a hydroxyapatite column as described in [37]. The single copy DNA–RNA hybrids melted with a T_m about 3°C below native DNA and middle repetitive–RNA hybrids melted with T_m or about 11°C below native DNA [50]. (Taken from [50].)

transcripts of the repetitive DNA hybridize to a very large percentage of the DNA (10%), close to, although slightly below the value recorded for cRNA (16%). In the next section we describe competition hybridization experiments designed to detect similarity between repetitive sequences of HnRNA and cRNA.

7. Is part of HnRNA a precursor to cRNA?

Two findings are consistent with, but do not prove, the idea that there may exist a precursor-product relationship between HnRNA and cRNA. First, new sequences of cRNA are produced during liver regeneration after the production of new sequences of HnRNA [33]. Second, preliminary evidence indicates that the production of both HnRNA and cRNA is inhibited by α-amanitin, whereas the synthesis of rRNA is not [44].

To investigate further the possibility of a precursor-product relationship, the

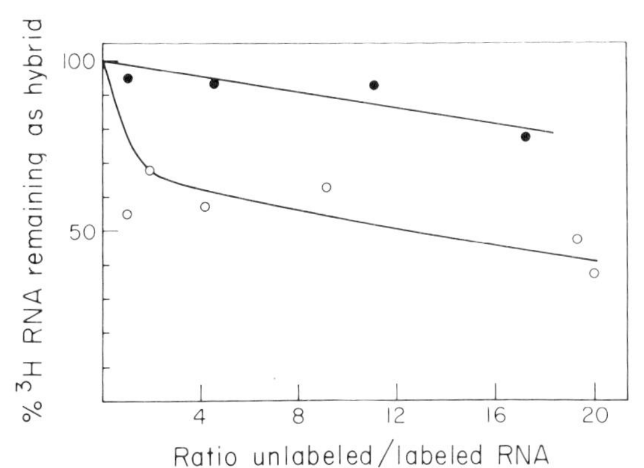

Fig. 11. Hybridization competition between rat ascites cRNA and giant HnRNA (sonicated to about 4–8 S in size). cRNA or HnRNA was labeled in vitro with ^{3}H dimethyl sulfate. Hybridization to total unlabeled rat nuclear DNA (containing trace amount of ^{14}C DNA) immobilized on filters [45] was carried out in 0.12 M phosphate buffer pH 6.8–50% (v/v) formamide. Sufficient quantity of labeled RNA was added to nearly saturate all repetitive complementary sites on the DNA and unlabeled and labeled RNA were added simultaneously. After incubation at 37°C for 18 hr unreacted RNA was removed by washing the filters and by mild ribonuclease treatment as described by Dahmus and McConnell [35]. •——• ^{3}H cRNA × unlabeled giant HnRNA; ○——○ ^{3}H giant HnRNA × unlabeled cRNA. (Taken from [47].)

sequences present in HnRNA were compared to those in cRNA by competition hybridization (see fig. 11) using the Gillespie and Spiegleman filter technique which assays essentially only repetitive transcripts [45, 46]. When uniformly labeled ^{3}H-cRNA is added to the reaction in sufficient quantities to nearly saturate all the complementary DNA sites, the extent of competition with unlabeled HnRNA increases only very slowly with increasing additions of competitor. This indicates that only a few of the sequences present in HnRNA are similar to those of cRNA. This is expected because as a rough approximation only 22% of the input HnRNA is represented by repetitive transcripts. In the reciprocal competition, uniformly labeled HnRNA added at near saturating amounts is competed to an extent indicating that at least 50% of the sequences of cRNA are found in HnRNA. Due to some base-pair mismatch in the hybrids as judged by their melting profile [47] it is possible to infer only that there is a similarity and not necessarily an identity between the competing repetitive sequences in cRNA and HnRNA.

It is improbable that contamination of HnRNA preparations with cRNA can explain the competition experiments because HnRNA has been shown by sedimentation velocity and electron microscopy under stringent denaturing conditions to consist predominantly of molecules in the weight range $5–10 \times 10^6$ daltons [48], whereas cRNA is approximately 1.6×10^4 daltons [35]. Serious consideration should be given to the possibility that cRNA is an artifactual degradation product of HnRNA. It is unlikely that cRNA consists of random breakdown products of total HnRNA because it contains few detectable single copy transcripts whereas HnRNA consists mainly of single copy transcripts. However, it is possible that cRNA is the breakdown product of either (i) a portion of HnRNA which might be organized predominantly as multiple contiguous repeated sequences, or (ii) repeated sequences of HnRNA which are interspersed between single copy sequences. We will return to this point shortly. It is unlikely that cRNA from rat ascites cells is degraded tRNA or rRNA although this possibility cannot be ruled out for all tissues [35, 36].

The observations that rat ascites cRNA becomes labeled considerably later than HnRNA after a short labeling pulse and that new sequences of cRNA occur some time after the production of new sequences of repetitive HnRNA in rat liver [33] are consistent with the idea that cRNA is not a breakdown product of HnRNA resulting from some isolation procedure. On the other hand even if cRNA is a "biological" product of repetitive segments of HnRNA, the possibility cannot be ruled out that it undergoes further degradation during isolation.

8. Alternative interpretations

The hybridization competition experiments suggest that part of giant HnRNA has repetitive sequences in common with cRNA. It is important at this juncture to determine the organization of the repetitive sequences in HnRNA. The RNA excess and the DNA excess hybridization experiments described above strongly argue for the presence of both repetitive and single copy sequences in HnRNA but tell us nothing about their organization. One possibility is that HnRNA could contain two populations of molecules, one of which is exclusively composed of repetitive sequences, the other exclusively single copy sequences. Alternatively both types of sequences might be found on the same molecules perhaps distributed in a way analogous to the sequence organization of the nuclear DNA. A third possibility is that all these categories of organization could be represented. If it transpires that cRNA is related to a small percentage of HnRNA organized as multiple contiguous repetitive sequences then this would provide additional circumstantial evidence in favor of the Britten–Davidson model. On the other hand if the repetitive sequences of HnRNA are interspersed between single copy sequences then a number of additional speculations become more realistic centering around the idea that the repetitive sequences represent (i) transcriptional products of operator or RNA polymerase sites, etc. [10] or (ii) regions of the RNA involved in post-transcriptional processing and packaging. In these instances cRNA could represent the residue after post-transcriptional modification of HnRNA.

9. Conclusion

It is evident that insight into the structural basis of gene regulation requires an understanding of the sequence organization of genomic DNA and of the RNA transcripts. In this context our main conclusions are (i) under specific conditions of duplex stability the genome of higher organisms consists of single copy and repetitive sequences. The latter are of varying repetitiousness and includes a middle repetitive component. (ii) The middle repetitive sequences of *Drosophila* DNA are short, about 100–300 base pairs long and are interspersed throughout the genome between single copy sequences. These observations are being extended to the rat genome [49]. (iii) In rat, a population of chromosomally-associated RNA molecules (cRNA) consist almost exclusively of middle repetitive transcripts. (iv) Rat giant nuclear RNA (HnRNA) consists of both repetitive and single copy sequences. (v) Many of the repetitive sequences of cRNA are similar to some of the repetitive sequences of HnRNA.

As a second approach we have collated the circumstantial evidence pertaining to the function of repetitive RNA in gene regulation in the light of the Britten–Davidson model. This approach is less incisive than the first, suffering chiefly from the difficulty of demonstrating that a class of RNA is directly responsible for differential gene activation and is not merely a result of it.

We foresee considerable progress in understanding the organization of sequences in DNA and RNA, coming from a combination of technical advancements such as the use of electron microscopy, and the isolation of defined genomic components such as pure producer genes, together with new approaches such as the use of genetic tools to generate interesting organizational variants of the genome.

Acknowledgements

We thank our colleague Dr. Jerry Johnson for his critical evaluation of the manuscript. Work is supported in part by USPHS Grant GW 13762.

References

[1] H.D. Berendes, Chromosoma 17 (1965) 35.
[2] L. Crouse, M.-D. Chilton and B.J. McCarthy, Biochemistry 11 (1972) 798.
[3] R.A. Firtel, J. Mol. Biol. 66 (1972) 363.
[4] J. Paul and R. Gilmour, Nature 210 (1966) 992.
[5] C.H. Tan and M. Miyagi, J. Mol. Biol. 50 (1970) 641.
[6] H.C. Birnboim, J.J. Pène and J.E. Darnell, Proc. Natl. Acad. Sci. U.S. 58 (1967) 320.
[7] K. Scherrer and L. Marcaud, J. Cell Physiol. 72 (1968) 181.
[8] R.J. Britten and E.H. Davidson, Science 165 (1969) 349.
[9] W.J. Dreyer, In: Developmental aspects of antibody formation and structure, Vol. II, Eds. J. Sterzl and I. Riha (Academic Press, New York, 1970) 919–933.

[10] G.P. Georgiev, A.P. Ryskov, C. Coutelle, V.L. Mantiera and E.R. Arakyan, Biochim. Biophys. Acta 259 (1972) 259.
[11] G.D. Wasserman, Molecular control of cell differentiation and morphogenesis (M. Dekker Inc., New York, 1972).
[12] J. Monohan and R. Hall, A new model for gene regulation in eukaryotic cells. J. Theor. Biol. (1973) in press.
[13] S.E. Elgin and J. Bonner, In: The biochemistry of gene expression in higher organisms, Eds. J.W. Lee and J.K. Pollak (D. Reidel, Dordrecht, Holland, 1972).
[14] R.J. Britten and D.E. Kohne, Carnegie Institution of Washington Year Book 65 (1966) 78.
[15] M. Botchan, R. Kram, C.W. Schmid and J.E. Hearst, Proc. Natl. Acad. Sci. U.S. 68 (1971) 1125.
[16] R. Kram, M. Botchan and J.E. Hearst, J. Mol. Biol. 64 (1972) 103.
[17] J.-R. Wu, J. Hurn and J. Bonner, J. Mol. Biol. 64 (1972) 211.
[18] R.J. Britten D. Kohne (1967). Repeated nucleotide sequences. Carnegie Institution of Washington Year Book 66 (1967) 73-88.
[19] J.O. Bishop, R. Pemberton and C. Baglioni, Nature New Biology 235 (1972) 231.
[20] Y. Suzuki, L.P. Gage and D.D. Brown, J. Mol. Biol. 70 (1972) 637.
[21] P.R. Harrison, A. Hell, G.D. Birnie and J. Paul, Nature 239 (1972) 219.
[22] K.D. Tartoff and R.P. Perry, J. Mol. Biol. 51 (1970) 171.
[23] L. Hatlen and G. Attardi, J. Mol. Biol. 56 (1971) 535.
[24] L.H. Kedes and M.C. Birnstiel, Nature New Biology 230 (1971) 165.
[25] B.R. Hough and E.H. Davidson, J. Mol. Biol. 70 (1972) 491.
[26] M.O. Dayhoff (Ed.), Atlas of protein sequence and structure, Vol. 5 (The National Biomedical Research Foundation, 1972).
[27] D. Holmes, M. Wilkes and J. Bonner (1973) in preparation.
[28] B. McClintock, Carnegie Institution of Washington Year Book 70 (1971) 5-17.
[29] S.E. Elgin, S.C. Froehner, J.E. Smart and J. Bonner, In: Advances in cell and mocular biology 1 (1971) 1–57.
[30] J. Bonner and J. Widholm, Proc. Natl. Acad. Sci. U.S. 57 (1967) 1379.
[31] R. Shearer and C. McCarthy, Biochemistry 6 (1967) 283.
[32] G. Attardi and F. Amaldi, Ann. Rev. Biochem. 39 (1970) 183.
[33] J.E. Mayfield and J. Bonner, Proc. Natl. Acad. Sci. U.S. 69 (1972) 7.
[34] J.E. Mayfield and J. Bonner, Proc. Natl. Acad. Sci. U.S. 68 (1971) 2652.
[35] M.E. Dahmus and D.J. McConnell, Biochemistry 8 (1969) 1529.
[36] D.S. Holmes, J.E. Mayfield, G. Sander and J. Bonner, Science 177 (1972) 72.
[37] A.H. Gelderman, A.V. Rake and R.J. Britten, Proc. Natl. Acad. Sci. U.S. 68 (1971) 172.
[38] M. Melli, C. Whitfield, K.V. Rao, M. Richardson and J.O. Bishop, Nature New Biology 231 (1971) 8.
[39] M. Wu, N. Davidson, G. Attardi and Y. Aloni, J. Mol. Biol. 71 (1972) 81.
[40] In preparation (1973).
[41] J.E. Darnell, B.E.H. Maden, R. Soeiro and G. Pagoulatos, In: Problems in development. RNA in development, Ed. E.W. Hanley (1971) 315–324.
[42] R.B. Church and B.J. McCarthy, Biochem. Biophys. Acta 199 (1970) 103.
[43] R.B. Church and B.J. McCarthy, J. Mol. Biol. 23 (1967) 459.
[44] G. Montecuccoli, F. Novello and F. Stirpe, FEBS Letters 25 (1972) 305.
[45] D. Gillespie S. Spiegelman, J. Mol. Biol. 12 (1965) 829.
[46] M. Birnstiel, B.H. Sells and I.F. Purdom, J. Mol. Biol. 63 (1972) 21.
[47] D. Holmes, J. Mayfield, L. Murthy and J. Bonner, III Hybridization properties of chromosomal RNA. Biochemistry (1973) submitted for publication.
[48] D. Holmes and J. Bonner, The preparation and properties of giant nuclear RNA. I. Preparation, molecular weight, base composition and secondary structure. Biochemistry 12 (1973) 2330.

[49] J.-R. Wu, M. Wilkes, D. Holmes and J. Bonner, Sequence organization of the rat genome. J. Mol. Biol. (1973) submitted for publication.
[50] D. Holmes and J. Bonner, The preparation and properties of giant nuclear RNA. II. Properties of hybridizable sequences. Biochemistry (1973) in press.

Niu and Segal (eds.). The role of RNA in reproduction and development
North-Holland Publ. Co., 1973

Nonhistone proteins as gene derepressor molecules*

T.Y. WANG and N.C. KOSTRABA
Biology Department, State University of New York, Buffalo, N.Y., USA

In eukaryotic organisms, the DNA is complexed with proteins and RNA. As a result, most of the genetic loci of the DNA are masked and not transcribed. It has been generally believed that the specific transcription of chromatin is, in part, determined by the nonhistone chromosomal proteins. These nonhistone proteins have been shown to stimulate transcription of DNA [1] and chromatin [2–4] in vitro. The RNA transcribed from chromatin in the presence of nonhistone proteins is characterized by the appearance of new RNA species not found in the unstimulated chromatin transcript. An example of this phenomenon, using Walker tumor, is illustrated in fig. 1. As can be seen, when tumor chromatin transcript is hybridized with DNA that had been saturated with activated chromatin transcript, there is no further annealing. However, when DNA saturated with tumor chromatin transcript was further annealed with RNA transcribed from stimulated chromatin, there was additional hybrid formation. The data indicate that activation of chromatin by nonhistone proteins results in transcription of additional genome sequences. Moreover, this activation is more enhanced by homologous than by heterologous nonhistone proteins, as shown in fig. 2. Further, homologous and heterologous nonhistone proteins activate chromatin differentially, resulting in synthesis of different RNAs [5–7]. Fig. 3 represents one such study. These data suggest that the nonhistone proteins are involved in gene derepression. If this is indeed the case, the activation of transcription by nonhistone proteins should manifest specificity, characteristic of the tissue origin of the nonhistone proteins. The following study, using the Walker tumor and rat liver systems, demonstrates such specificity and suggests a regulatory function for the nonhistone proteins.

The nonhistone proteins were prepared by a method described previously [4]. Chromatin was isolated from cell nuclei according to the procedure of Tan and Miyagi [8]. Molecular hybridization was performed according to the procedure of Gillespie and Spiegelman [9] with slight modifications [8], and synthesis of RNA in vitro from chromatin was as described elsewhere [7].

* Discussant's paper to the contribution of D.S. Holmes and J. Bonner.

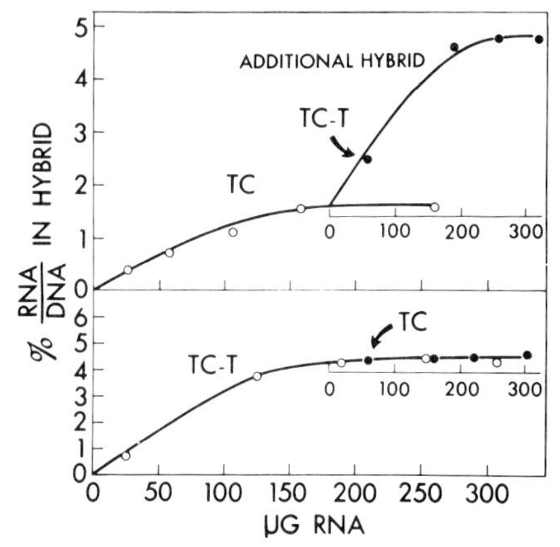

Fig. 1. Reciprocal double saturation hybridization of Walker tumor DNA with RNA transcribed from Walker tumor chromatin and with RNA synthesized from tumor chromatin stimulated by tumor nonhistone proteins. Upper curves: tumor DNA was annealed with tumor chromatin transcript to saturation level (○). This saturated DNA was further annealed with transcript of tumor chromatin stimulated by tumor nonhistone proteins (TC-T), yielding additional hybrid (●). Lower curve: tumor DNA was saturated with TC-T RNA (○) and further annealed with TC RNA (●).

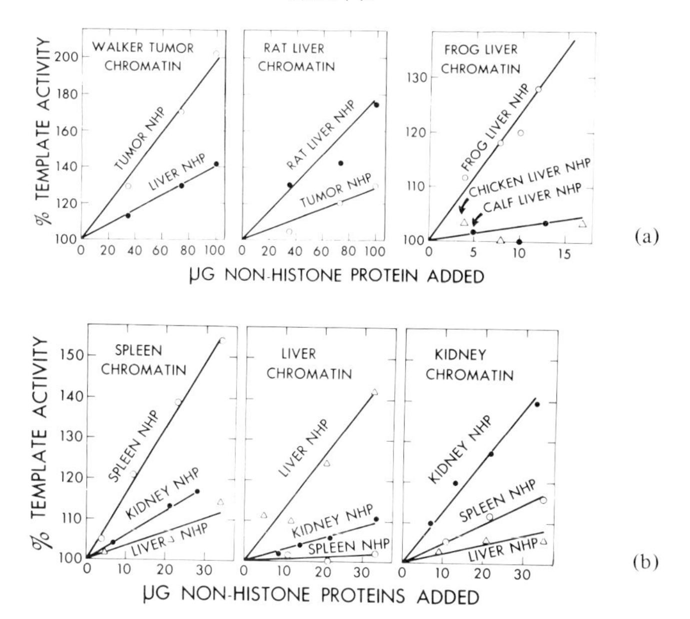

Fig. 2. (a) Template activities of chromatins from Walker tumor, rat liver and frog liver stimulated by nonhistone proteins (NHP) prepared from homologous and heterologous tissues. (b) Template activities of chromatins from rat spleen, liver and kidney stimulated by nonhistone proteins (NHP) prepared from homologous and heterologous tissues.

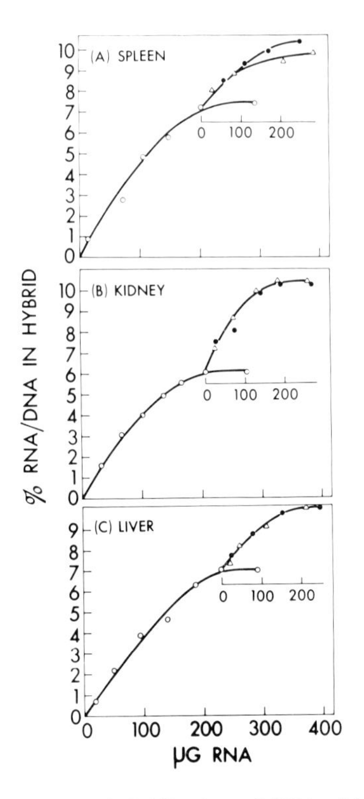

Fig. 3. Reciprocal double saturation hybridization of DNA with RNAs synthesized from chromatins of rat spleen, kidney and liver activated by homologous and heterologous nonhistone proteins (NHP). (A)○, annealing of rat spleen DNA with RNA transcribed from spleen chromatin activated by spleen NHP (S-S). The DNA filters saturated with S-S RNA were further annealed with RNAs synthesized from spleen chromatin activated by NHP prepared from liver (●) and kidney (△). (B)○, annealing of rat kidney DNA with RNA transcribed from kidney chromatin activated by kidney NHP (K-K). The DNA filters saturated with K-K RNA were further annealed with RNAs synthesized from kidney chromatin activated by NHP prepared from liver (●) and spleen (△). (C)○, annealing of rat liver DNA with RNA transcribed from liver chromatin activated by liver NHP (L-L). The DNA filters saturated with L-L RNA were further annealed with RNAs synthesized from liver chromatin activated by NHP prepared from kidney (●) and spleen (△).

When rat liver DNA was saturated with liver chromatin transcript, further annealing with RNAs transcribed from tumor chromatin or from liver chromatin that had been activated by tumor nonhistone proteins yielded 2% and 1% additional DNA–RNA hybrid, respectively, as shown in fig. 4. If the activation of chromatin transcription by nonhistone proteins is tissue-specific, the RNA transcribed from liver chromatin activated by tumor nonhistone proteins should manifest characteristics of tumor transcript and should then compete with RNA transcribed from

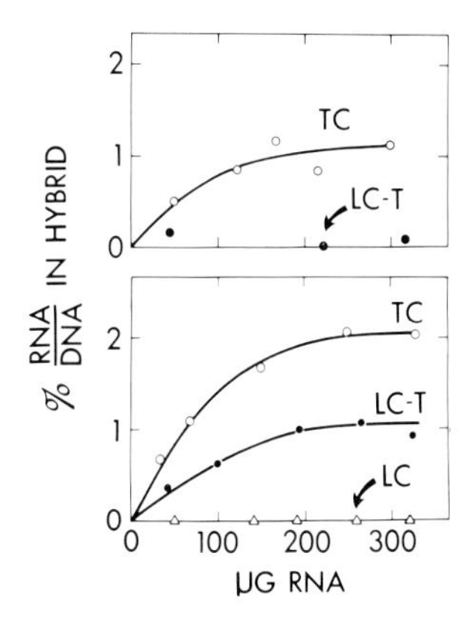

Fig. 4. Specificity of nonhistone proteins (NHP) in activation of transcription as illustrated by hybridization of rat liver DNA with RNAs transcribed from Walker tumor chromatin (TC), from liver chromatin (LC), and from liver chromatin stimulated by tumor NHP (LC-T). Lower curves: liver DNA saturated with unlabeled LC RNA was further annealed with ^{3}H-labeled TC RNA (○) and LC-T RNA (•), resulting in additional 2% and 1% hybrid, respectively. Control showing that the DNA was saturated with unlabeled LC RNA is indicated by lack of additional hybrid formation when annealed with ^{3}H-LC RNA (△). Upper curve: liver DNA saturated with LC-T RNA was further annealed with ^{3}H-TC RNA, resulting in only 1% additional hybrid. Control showing that the DNA was saturated with unlabeled LC-T RNA is indicated by further annealing with ^{3}H-LC-T RNA (•), which yielded no additional hybrid.

tumor chromatin. Since the activation by tumor nonhistone proteins is partial, the liver chromatin should transcribe normal liver RNA as well as tumor-specific RNA. Therefore, the RNA transcribed from liver chromatin activated by tumor nonhistone proteins should compete partially with tumor chromatin transcript. As can be seen in fig. 4, when DNA saturated with transcript of activated liver chromatin (by tumor nonhistone proteins) was further annealed with transcript of tumor chromatin, there was only 1% DNA–RNA hybrid formation. This indicates that approximately 50% of the tumor chromatin transcript is similar to RNA transcribed from liver chromatin that had been activated by tumor nonhistone proteins. Thus, the tumor nonhistone proteins activate liver chromatin so that transcription is characteristic of the tumor.

Similar results can also be obtained using Walker tumor chromatin and rat liver nonhistone proteins, as shown in fig. 5. In this experiment, the tumor DNA was saturated with RNA synthesized from tumor chromatin. Further annealing of these saturated DNA filters with liver chromatin transcript and with RNA transcribed from tumor chromatin activated by liver nonhistone proteins yielded 6% and 2% DNA–RNA hybrid, respectively. If the result depicted in fig. 4 was not an isolated case, the 2% DNA–RNA hybrid obtained with RNA transcribed from activated tumor chromatin (by liver non-histone proteins) should contain RNA species characteristic of liver chromatin transcript. Consequently, it should then compete with

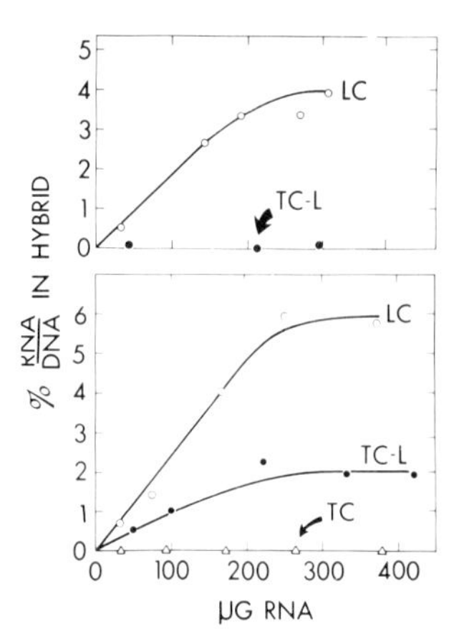

Fig. 5. Tissue-specific activation of transcription by nonhistone proteins (NHP) as illustrated by hybridization of Walker tumor DNA with RNAs transcribed from rat liver chromatin (LC), from tumor chromatin activated by liver NHP (TC-L) and from tumor chromatin (TC). Lower curves: tumor DNA saturated with unlabeled TC RNA was further annealed with ^{3}H-labeled LC RNA (○) and TC-L RNA (•), yielding 6% and 2% additional hybrid, respectively. Control showing that the DNA was saturated with unlabeled TC RNA is indicated by further annealing with ^{3}H-TC RNA (△), which resulted in no further hybrid formation. Upper curve: tumor DNA saturated with unlabeled TC-L RNA was further annealed with ^{3}H-LC RNA (○), yielding only 4% additional hybrid. Control showing that the DNA was saturated with unlabeled TC-L RNA is indicated by further annealing with ^{3}H-TC-L RNA (•), which resulted in no additional hybrid.

RNA transcribed from liver chromatin, resulting in less than 6% hybrid formation. The data in fig. 5 show that as a result of activation by liver nonhistone proteins, 40% of the transcribed product of tumor chromatin was indistinguishable from the liver chromatin transcript, as judged by DNA–RNA hybridization.

The results described indicate that nonhistone proteins specifically activate the transcription of additional DNA sequences of chromatin, and thus suggest that nonhistone proteins are regulatory molecules in gene derepression.

Acknowledgement

The studies reported here have been supported by a U.S. Public Health Service grant, GM-11698.

References

[1] C.S. Teng, C.T. Teng and V.G. Allfrey, J. Biol. Chem. 246 (1971) 3597.
[2] C.S. Teng and T.H. Hamilton, Proc. Natl. Acad. Sci. U.S. 63 (1969) 465.
[3] T.Y. Wang, Exptl. Cell Res. 61 (1970) 455.
[4] M. Kamiyama and T.Y. Wang, Biochim. Biophys. Acta 228 (1971) 563.
[5] T.Y. Wang, Exptl. Cell. Res. 69 (1971) 217.
[6] N.C. Kostraba and T.Y. Wang, Cancer Res. 32 (1972) 2348.
[7] N.C. Kostraba and T.Y. Wang, Biochim. Biophys. Acta 262 (1972) 169.
[8] C.H. Tan and M. Miyagi, J. Mol. Biol. 50 (1970) 641.
[9] D. Gillespie and S. Spiegelman, J. Mol. Biol. 12 (1965) 829.

Niu and Segal (eds.). The role of RNA in reproduction and development
North-Holland Publ. Co., 1973

RNA in gene de-repression*

John H. FRENSTER and Paul R. HERSTEIN
Division of Oncology, Stanford University School of Medicine, Stanford, Calif. 94305, USA

Normal embryogenesis, organ regeneration, neoplastic transformation, immune lymphocyte activation, and the cellular response to steroid hormones all involve activation of DNA template sites for RNA synthesis. Such gene de-repression is mediated by a variety of polyanionic nuclear ligands to DNA, the most significant of which appear to be species of de-repressor RNA. The removal of de-repressor RNA from specific gene sites offers a mechanism for selective feedback inhibition after excessive transcription at such gene sites.

1. Introduction

One of the central problems of cell physiology is the mechanism whereby the individual animal develops and maintains the diversity of cell types characteristic of the adult state out of the original single-cell fertilized ovum. Although particular adult cell types are often stable for the lifetime of the individual, such phenotypic diversity and stability is not based on a corresponding genetic diversity among these cell types. Rather, cellular and molecular techniques of ever-increasing sensitivity and resolution have revealed that most if not all diploid cells within a normal adult have an identical complement of DNA [1], while at the same time these cells display a diverse expression of such DNA in tissue-specific patterns of RNA synthesis [1, 2]. Such variable gene expression among cells possessing identical gene contents indicates the importance of gene control mechanisms, which permit significant tissue specialization and yet maximize tissue renewal and coordination for the benefit of the parent organism.

Theoretically, variable gene expression could be achieved either by selective gene repression within an otherwise fully expressed genome, or conversely, by selective gene de-repression within an otherwise fully repressed genome. Detailed ultrastructural, biochemical and metabolic studies have revealed that selective gene de-repression is physiologically the more significant gene control mechanism in animals [3].

* Supported in part by Grants CA-10174 and AM-01006 from the National Institutes of Health, and by a Research Scholar Award from the Leukemia Society to Dr. Frenster.

TABLE 1
Natural occurrence of gene de-repression

(1) Activation of sperm genome in early embryo
(2) Organ regeneration and hypertrophy
(3) Chemical and viral oncogenesis
(4) Steroid hormone activation of target cells
(5) Activation of immune lymphocytes

As a consequence, particular attention is being paid to the study of those circumstances in animal life during which de-repression of previously-repressed genes has been shown to occur (table 1).

2. *Occurrence of gene de-repression*

During the course of normal embryonic development, a progressive restriction is noted in the diversity of RNA molecules being synthesized, ranging from an extensive transcription of the oocyte genome during the lampbrush stage of oogenesis in the unfertilized oocyte [4] to the very marked restriction of RNA synthesis in the mature nucleate erythrocyte [2] or to the absence of any RNA synthesis at all in the mature sperm [4]. However, when such a non-transcribing mature sperm hybridizes with the mature ovum during fertilization, a very rapid de-repression of the sperm genome occurs, in what is probably the most crucial example of gene de-repression in animal systems [4]. Such de-repression of the paternal genome converts the haploid ovum into a functional diploid cell, and permits an equal paternal contribution to gene expression in all resulting cells of the embryo and the adult [4].

During the course of embryogenesis in the developing liver, a progressive restriction in the diversity of RNA types is observed in the adult liver as compared to the embryonic liver [5], but when the adult liver is induced to regenerate following partial hepatectomy de-repression of previously-repressed genes again allows reappearance of those RNA species characteristic of the embryonic state [5]. A similar de-repression of previously-repressed genes also is noted if neoplastic cells are assayed for the diversity of RNA species as the neoplasm progresses from a

TABLE 2
Human neoplasms displaying specific fetal antigens

Adenocarcinoma of colon
Primary hepatoma of liver
Hodgkin's lymphoma
Chronic lymphocytic leukemia

benign nodule to a spontaneous neoplasm to a transplantable neoplasm [6]. Such de-repression of normally-repressed RNA species has also been demonstrated within human leukemic lymphocytes [7]. A similar de-repression of fetal genes may account for the re-appearence within adult human neoplasms (table 2) of cell surface antigens characteristic of normal fetal life [8], and quite similar de-repression of fetal antigens has been observed in both chemical-induced [9] and viral-induced [10] animal neoplasms. In this regard, de-repression of normally-repressed viral oncogenes is the central feature of the oncogene hypothesis of neoplastic transformation [11].

Detailed studies of the mechanism of action of steroid hormones indicates that each steroid hormone, including estrogen, testosterone, progesterone, aldosterone, and cortisol can be shown to bind to a corresponding specific receptor protein

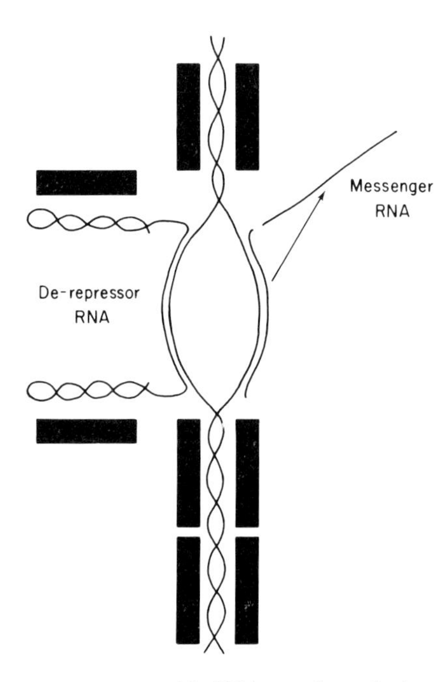

Fig. 1. Interaction of histone repressors with DNA templates during gene repression and de-repression. Under conditions of gene repression, polycationic histone repressors (dark bars) form strong electrostatic bonds with the phosphate groups on the exterior of the closed DNA helix [3]. Experimentally, hydrolysis of such histone repressors by proteolytic enzymes such as trypsin, or physical displacement of such histone repressors by strong synthetic polyanions such as polyethylene sulfonate, results in de-repression of the underlying DNA template and increased rates of RNA synthesis [3]. Under physiologic conditions, nuclear polyanions such as phosphoproteins, acidic proteins, and nuclear RNA are capable of similarly displacing histone repressors and increasing the rates of RNA synthesis [3]. Certain species of nuclear RNA appear capable of recognizing specific base sequences on the DNA template during such gene de-repression [3], and thus are strong candidates as the molecular species which effects locus-specific and strand-specific gene de-repression in animals [19].

found in the cytoplasm of cells responsive to that steroid [12]. Following such binding, the hormone-receptor complex is transported into the nucleus, interacting there with the chromatin complex [13] and increasing the synthesis of tissue-specific messenger RNA species [13].

Finally, within immune lymphocytes undergoing activation by specific antigens or by defined mitogens, marked stimulation of the synthesis of heterogenous nuclear RNA is noted [14]. If these stimulated RNA species are found to be absent from non-stimulated cells, then the process of lymphocyte activation can be viewed as involving gene de-repression, and that is already suggested by the character of ultrastructural changes observed during lymphocyte activation [15].

3. Mechanism of gene de-repression

Because gene de-repression occurs so widely under natural conditions (table 1), a variety of physiological systems are now available for defining the molecular mechanisms of gene de-repression. These studies have taken advantage of the fact that repressed DNA sequences, found in condensed heterochromatin [16], can be separated from de-repressed DNA sequences, found in extended euchromatin [16], and these can be separately analyzed within their native chromatin complexes by standard biochemical, biophysical, and metabolic methods [3, 17]. Such separated chromatin complexes can also be used as assay systems for those molecular species suspected of possessing repressor or de-repressor activity [3]. These studies have uniformly revealed that polycationic histone proteins serve as nonspecific repressors of template activity within associated DNA molecules (fig. 1), and must be removed from DNA before such DNA can be active in RNA synthesis [3]. Under experimental conditions, histone repressors can be removed from DNA templates by proteolytic digestion or by physical displacement with strong synthetic polyanions [3], while under physiologic conditions, nuclear polyanions such as phosphoproteins, acidic proteins, and nuclear RNA can effect such histone displacement and gene de-repression [3].

Certain species of nuclear RNA appear to possess the ability to select particular portions of the genome for specific-gene locus de-repression [3], perhaps by virtue of their ability to hybridize [18] to complementary base sequences on the anticoding strand of the DNA template [19], thereby freeing the coding strand of DNA for messenger RNA synthesis that is both locus-specific and strand-specific [19]. During such gene de-repression, only a small length of DNA, 4–5 base pairs in length, is in the strand-separated state [20]. Recent evidence indicates that de-repressor RNA species can be isolated and purified for further characterization and analysis in assay systems [21], and may form double-stranded RNA–RNA duplexes with the RNA of the immediate transcription product [22].

Because the displacement of histones from underlying DNA templates during gene de-repression results in a conformational change within the chromatin complex [16], it has become possible to develop probes of such complexes with

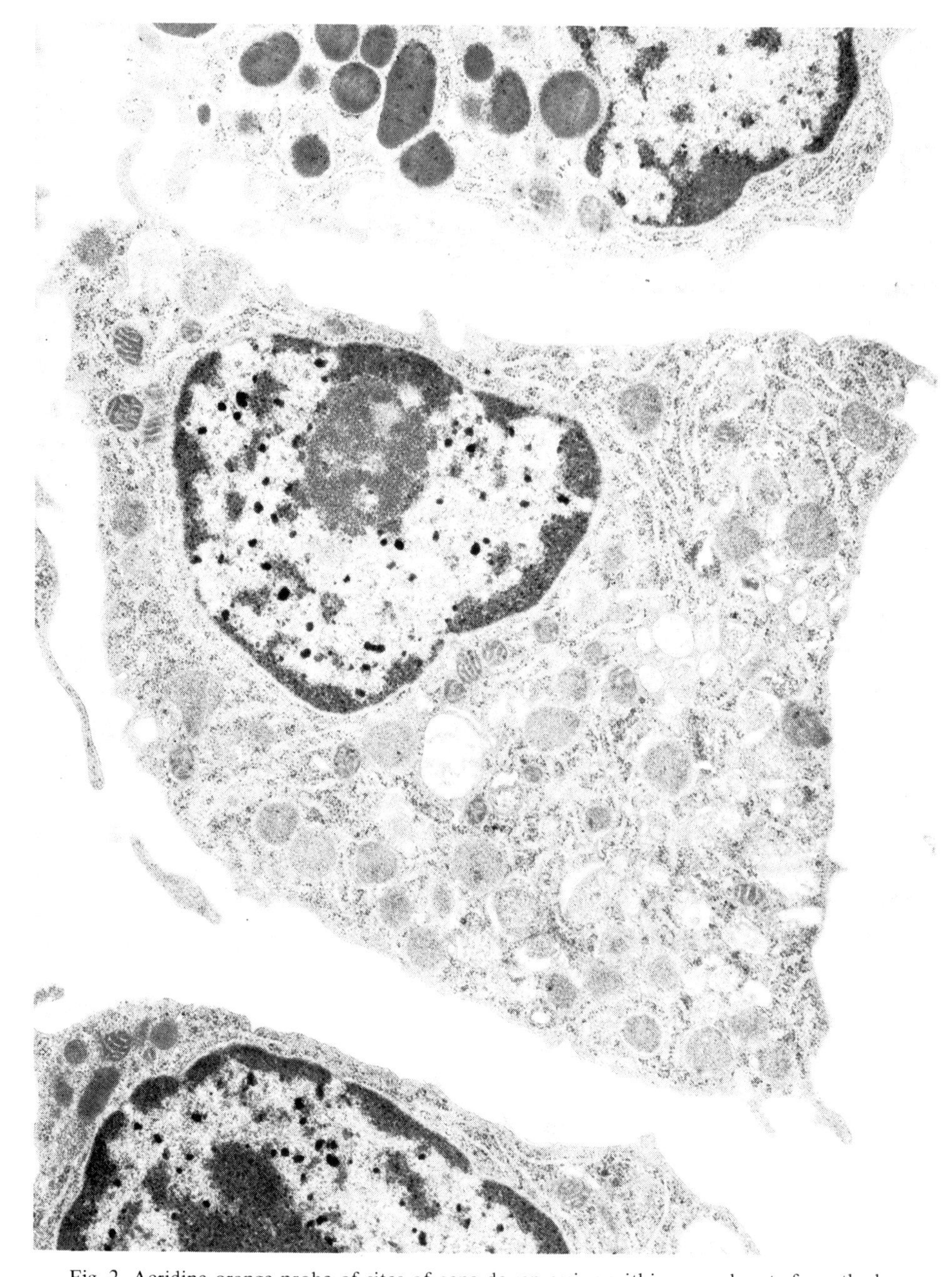

Fig. 2. Acridine orange probe of sites of gene de-repression within a myelocyte from the bone marrow of an untreated patient with chronic myelocytic leukemia [23]. The electron-dense reaction product localizes over DNA template sites that are free of histone repressors [23]. These sites are found to be confined to the active extended euchromatin portion of the cell nucleus [23] (light-staining nuclear areas in the above electron micrograph). No reaction product is observed in the cytoplasm, the nucleolus, or in the repressed condensed heterochromatin portion of the cell nucleus. Quantitative counts of individual reaction product grains per cell permit an estimation of the degree of gene de-repression for each cell within a cell population. × 9,000.

defined molecular species which bind to DNA only if histones are first displaced [23]. Acridine orange is a useful probe in this regard (fig. 2) because it binds specifically to that DNA which is free of repressor histones [23], and because it can be visualized at high-resolution by electron microscopy to reveal the exact points of gene de-repression within the nuclei of intact human leukemic bone marrow cells [23] and of living human lymphocytes [24]. By this means it is also possible to quantitate the degree of gene de-repression within each cell of a given cell population (fig. 2), providing useful data regarding variation of states of gene de-repression within such histologically-complex organs such as regenerating liver, bone marrow, and lymph nodes of patients in various states of disease or recovery.

The recognition that steroid hormones penetrate into the cell nucleus where they may effect gene de-repression in sensitive tissues [13] indicates to some degree the complexity of molecular interactions in the process of gene de-repression. This analysis can be extended by comparing those nuclear ligands which increase RNA synthesis with those which inhibits such synthesis (table 3). Such diverse nuclear ligands as estrogen, testosterone, methylcholanthrene, RNA polymerase, de-repressor RNA, and polyoma DNA all increase RNA synthesis, and all bind preferentially to single-stranded DNA [25]. Conversely such diverse nuclear ligands as histones, protamines, actinomycin D, acridine orange, chloroquine, and lac repressor all decrease RNA synthesis, and all bind preferentially to double-stranded DNA [25]. These correlations strongly support the model of gene de-repression [19] which predicts that nuclear ligands preferring single-stranded DNA will increase RNA synthesis by stabilizing DNA in the open loop conformation favorable to gene transcription [25], while nuclear ligands preferring double-stranded DNA will decrease RNA synthesis by stabilizing DNA in the closed helix conformation unfavorable to gene transcription [25].

TABLE 3
Correlation of nuclear ligand binding and RNA synthesis

Nuclear ligand	Preferred form of DNA	Effect on RNA synthesis
Histones	Double-stranded	Decreased
Protamines	Double-stranded	Decreased
Actinomycin D	Double-stranded	Decreased
Acridine orange	Double-stranded	Decreased
Chloroquine	Double-stranded	Decreased
Lac repressor	Double-stranded	Decreased
Testosterone	Single-stranded	Increased
Estradiol	Single-stranded	Increased
Methylcholanthrene	Single-stranded	Increased
RNA polymerase	Single-stranded	Increased
De-repressor RNA	Single-stranded	Increased
Polyoma viral DNA	Single-stranded	Increased

4. Control of gene de-repression

The constraints posed by both the gene-locus specificity and the long-term stability of states of gene de-repression in adult animals provide insight into the nature of the physiological systems which control gene de-repression. Somatic cell genetic analysis has revealed that all but one of the X chromosomes in adult mammalian cells are repressed [26], that the decision which X chromosome will remain active is made early in embryonic life and is then maintained with high fidelity through more than 50 generations of daughter cells [27], and that the control mechanisms are exquisitely sensitive not only to the number of X chromosomes in the cell, but perhaps also to the gene content of each X chromosome [26]. These analyses strongly suggest that feedback mechanisms reflecting messenger RNA accumulation are impinging upon and probably controlling RNA production during selective gene de-repression [26].

Recent studies on the molecular features of messenger RNA synthesis strongly indicate that a significant portion of the immediate transcription product is cleaved (fig. 3) in forming the final messenger RNA molecule [28], and that the non-messenger fragment may correspond to operator RNA with base sequences complementary to de-repressor RNA of the same gene locus [29]. Excess production of messenger RNA may lead to an accumulation of operator RNA and to the formation of double-stranded RNA–RNA duplexes between the operator RNA and the de-repressor RNA of the particular gene locus, serving to remove de-repressor RNA from the DNA template, and by this means decreasing the rate of gene transcription at the particular gene locus [29]. Both heterometric and homometric RNA–RNA duplexes (fig. 3) have recently been isolated from a variety of normal animal tissues [22].

In this regard, the ability of oncogenic viruses to de-repress fetal genes in the animal cell genome [10] may reflect either the ability of viral RNA to act as a fraudulent type of de-repressor RNA (fig. 3), or the ability of viral DNA to insert in or near the operator locus of the host genome (fig. 3), a condition postulated in the protovirus hypothesis of neoplastic transformation [30].

5. Clinical implications

A review of the circumstances in which gene de-repression normally occurs (table 1) gives some suggestion of the possible clinical significance of the emerging data concerning gene de-repression in man [31].

An understanding of gene de-repression in the early developing embryo may well give us insight into the relationships of viral infections and chemical teratogens in producing anomalies in the course of normal organ development, while an understanding of the control of gene de-repression during organ regeneration and hypertrophy opens the possibility of organ replacement via regeneration in the adult patient.

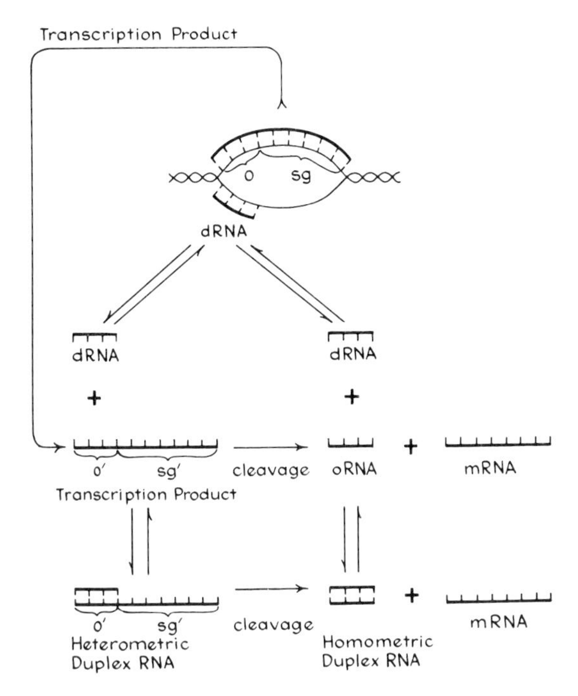

Fig. 3. Postulated feedback control of gene transcription in animals. Recent studies of the molecular mechanisms of messenger RNA synthesis in animals indicate that a significant portion of the immediate transcription product is cleaved in forming the final messenger RNA (mRNA) molecule [28]. The non-messenger portion of the transcription product includes operator RNA (oRNA), transcribed from the operator locus (o) on the DNA template [29] just prior to transcription of the structural gene (sg) which codes for messenger RNA. De-repressor RNA (dRNA) is thought to bind to the anti-coding strand of the DNA template during gene de-repression [19], and therefore to be complementary in base sequence to operator RNA [29]. Such base sequence complementarity would allow RNA–RNA duplex formation between the oRNA and dRNA of a specific gene locus during conditions of oRNA excess, that is, after excessive gene transcription, and would diminish such gene transcription by removing d RNA from the operator site on the DNA template [29]. RNA–RNA duplexes could include the messenger RNA portion (heterometric duplex RNA) or could exclude the messenger RNA portion (homometric duplex RNA). Both types of RNA–RNA duplexes have recently been isolated from a variety of normal animal tissues [22].

The increasing data linking de-repression of fetal genes (table 2) with the natural history of human neoplasms [8], as well as the well-established data in which de-repression of fetal genes can be demonstrated during chemical [9] and viral [10] oncogenesis in animals, offers both insight into the neoplastic process as well as clinical opportunities to approach the therapy [31], diagnosis [8], and prevention [8] of human neoplasms by immunologic means. The recent recognition that immunotherapeutic human lymphocytes [31, 32] may be activated via gene de-repression mechanisms [31] while the neoplastic target cells may themselves have already undergone a separate gene de-repression [31] offers some indication of the scope and complexity of biological interactions soon to be encountered at the clinical level, as does the intrinsic molecular complexity of the human genome as revealed by recent studies [33–35].

References

[1] B.J. McCarthy and B.H. Hoyer, Proc. Natl. Acad. Sci. U.S. 52 (1964) 915.
[2] L. Grouse, M.D. Chilton and B.J. McCarthy, Biochemistry 11 (1972) 798.
[3] J.H. Frenster, Nature 206 (1965) 680.
[4] E.H. Davidson, Gene activity in early development (Academic Press, New York, 1968).
[5] R.B. Church and B.J. McCarthy, J. Mol. Biol. 23 (1967) 477.
[6] R.W. Turkington, Cancer Res. 31 (1971) 427.
[7] P.E. Neiman and P.H. Henry, Biochemistry 8 (1969) 275.
[8] P. Gold, Ann. Rev. Med. 22 (1971) 85.
[9] E.H. Stonehill and A. Bendich, Nature 228 (1970) 370.
[10] J.H. Coggin, K.R. Ambrose and N.G. Anderson, J. Immunol. 105 (1970) 524.
[11] G.J. Todaro and R.J. Huebner, Proc. Natl. Acad. Sci. U.S. 69 (1972) 1009.
[12] T.C. Spelsberg, A.W. Steggles and B.W. O'Malley, J. Biol. Chem. 246 (1971) 4188.
[13] S. Liao, T. Liang and J.L. Tymoczko, Nature New Biol. 241 (1973) 211.
[14] H.L. Cooper, J. Biol. Chem. 243 (1968) 34.
[15] K. Tokuyasu, S.C. Madden and L.J. Zeldis, J. Cell Biol. 39 (1968) 630.
[16] J.H.Frenster, V.G. Allfrey and A.E. Mirsky, Proc. Natl. Acad. Sci. U.S. 50 (1963) 1026.
[17] B.L. McConaughy and B.J. McCarthy, Biochemistry 11 (1972) 998.
[18] M.E. Dahmus and J. Bonner, Fed. Proc. 29 (1970) 1255.
[19] J.H. Frenster, Nature 106 (1965) 1269.
[20] J.M. Saucier and J.C. Wang, Nature New Biol. 239 (1972) 167.
[21] D.S. Holmes, J.E. Mayfield, G. Sander and J. Bonner, Science 177 (1972) 72.
[22] L.H. Kronenberg and T. Humphreys, Biochemistry 11 (1972) 2020.
[23] J.H. Frenster, Cancer Res. 31 (1971) 1128.
[24] J.H. Frenster, Nature New Biol. 236 (1972) 175.
[25] J.H. Frenster, Nature 208 (1965) 1093.
[26] M.M. Grumbach, A. Morishima and J.H. Taylor, Proc. Natl. Acad. Sci. U.S. 49 (1963) 581.
[27] R.G. Davidson, H.M. Nitowsky and B. Childs, Proc. Natl. Acad. Sci. U.S. 50 (1963) 481.
[28] M. Melli and R.C. Pemberton, Nature New Biol. 236 (1972) 172.
[29] P.R. Herstein and J.H. Frenster, In: Proc. Sec. Conf. on Embryonic and fetal antigens in cancer, Eds. N.G. Anderson and J.H. Coggin (National Technical Information Service, U.S. Dept. of Commerce, 1972, Springfield, Virginia) 3–5.
[30] H.M. Temin, Natl. Acad. Sci. U.S. 69 (1972) 1016.
[31] J.H. Frenster and W.M. Rogoway, In: Proc. Fifth Annual Leukocyte Culture Conf. Ed. J.E. Harris (New York, Academic Press, 1970) 359-371.
[32] A.R. Cheema and E.M. Hersh, Cancer 29 (1972) 982.
[33] G.F. Saunders, S. Shirakawa, P.P. Saunders, F.E. Arrighi and T.C. Hsu, J. Mol. Biol. 63 (1972) 323.
[34] J.H. Frenster, In: The cell nucleus. Ed. H. Busch (Academic Press, New York, 1973) in press.
[35] J.H. Frenster, S.L. Nakatsu and M.A. Masek, Adv. Cell Mol. Biol. 3 (1973) in press.

Niu and Segal (eds.). The role of RNA in reproduction and development
North-Holland Publ. Co., 1973

RNA-directed DNA synthesis in viruses and normal cells: a possible mechanism in differentiation

C.-Y. KANG and H.M. TEMIN
McArdle Laboratory, University of Wisconsin, Madison, Wis. 53706, USA

Virions of RNA tumor viruses contain endogenous RNA-directed DNA polymerase activity. This endogenous RNA-directed DNA polymerase activity appears to be responsible for the synthesis of provirus DNA.

Uninfected chicken cells and normal chicken embryos contain endogenous DNA polymerase activity which is sensitive to ribonuclease, but resistant to actinomycin D and deoxyribonuclease. The early product of this chicken endogenous DNA polymerase activity is a complex of an RNA template and the newly synthesized DNA product. This complex is disrupted by ribonuclease, alkali, and heat. The DNA product of the endogenous DNA polymerase activity from chic'.en embryos hybridizes to the RNA from the same chicken cell fraction but it does not hybridize to RNA of Rous sarcoma or reticuloendotheliosis viruses.

These results demonstrate that uninfected chicken embryos contain endogenous RNA-directed DNA polymerase activity which is not related to avian RNA viruses. Therefore, RNA-directed DNA polymerase activity is not a unique property of RNA tumor viruses.

The chicken endogenous RNA-directed DNA polymerase activity may have a role in normal development.

1. Introduction

This symposium is to discuss the role of RNA in development. One of the major problems of development is how cells stably differentiate into new states. The solution to this problem must include an understanding of the mechanisms of induction and of maintenance of differentiation. Previous contributions to this volume have discussed the possibility that RNA might have a role in these processes other than as a messenger RNA.

Members of the leukovirus or RNA tumor virus group of viruses contain RNA and can stably change infected cells. The mechanism of this change is the introduction of new genes into cells by means of RNA-directed DNA synthesis. In this paper, we shall briefly discuss strongly transforming RNA tumor viruses as a model for development and then discuss the existence in normal cells of RNA-directed DNA polymerase activity. This activity might play a role in normal differentiation analogous to the role in neoplastic transformation of RNA-directed DNA synthesis by RNA tumor viruses.

2. RNA tumor viruses

RNA tumor viruses is the name of a group of animal viruses discovered about 60 yr ago (see [1] for review). RNA tumor viruses are widely distributed in nature. They are sometimes associated with tumors in animals, but more frequently are found in normal cells. They have been classified as strongly transforming, weakly transforming, and non-transforming depending upon the efficiency with which they produce tumors [2].

Members of the tumor virus group have a virion of about 100 nm diameter with a glycoprotein and lipid-containing envelope and an internal core of unknown symmetry. The core contains a 70 S RNA, which after heating dissociates into 35 S subunits, and a DNA polymerase. The replication of RNA tumor viruses is sensitive to actinomycin D treatment and requires DNA synthesis soon after infection. Virus replication does not lead to cell death.

Infection of normal chicken cells with a strongly transforming RNA tumor virus, like Rous sarcoma virus, alters the properties of the infected cells: (1) The cells are altered in morphology and orientation; (2) the cells multiply more under conditions of limited serum; (3) the cells multiply in soft agar; (4) the cells contain virion protein components like the internal group-specific antigens and the envelope glycoproteins; (5) the cells contain tumor specific transplantation antigens; (6) the cells have increased hyaluronic acid synthesis as a result of an increased activity of the enzyme hyaluronic acid synthetase; (7) the cells have an increased rate of glucose transport; and (8) the cells form a tumor if injected into chickens.

The genetic mechanism of these changes in cell properties is the introduction of Rous sarcoma virus genes into the infected chicken cells. There are three pieces of evidence for the presence of genes affecting cell properties in the Rous sarcoma virus virion: (1) virus mutants which cause different morphologies of infected cells; (2) non-transforming virus mutants which do not transform fibroblasts; and (3) virus mutants which are temperature-sensitive for transformation.

These new genes are introduced into infected cells by RNA-directed DNA synthesis. The infecting virus RNA acts as a template for the virion DNA polymerase to make a DNA provirus. This DNA provirus is a template for formation of RNA for progeny virus and for formation of RNA which causes cell transformation.

The evidence for this form of replication is (1) the regular inheritance of virus genomes in infected cells; (2) the sensitivity of virus RNA synthesis to actinomycin D; (3) the absence of evidence for RNA–RNA replication in virus producing cells; (4) the requirement for early DNA synthesis for virus infection; (5) the labeling of this early DNA with 5-bromodeoxyuridine and inactivation of infection by exposure to light; (6) the presence of a DNA polymerase in virions; (7) the isolation of infectious DNA with virus information from virus-infected cells; and (8) the Mendelian inheritance of endogenous virus-related information.

In addition to strongly transforming RNA tumor viruses like Rous sarcoma virus, there are weakly transforming RNA tumor viruses. These viruses do not cause transformation of fibroblasts in culture and cause tumors in animals only after long periods of time and extensive virus replication. There are also non-transforming RNA tumor viruses. These viruses do not cause transformation of fibroblasts in culture and do not cause tumors in animals.

Both weakly transforming and non-transforming RNA tumor viruses replicate in the same fashion as strongly transforming RNA tumor viruses, that is, their replication involves information transfer from RNA to DNA. The weakly transforming and the non-transforming RNA tumor viruses probably contain in their genomes only information for new virions, that is, they do not contain information for neoplastic transformation. The only stable differentiated change introduced into an infected cell by these viruses is the formation of progeny virus.

3. RNA-directed DNA synthesis in normal cells

In considering the origin of non-transforming RNA tumor viruses, we hypothesized that they originated from a system of RNA-directed DNA synthesis in normal cells [3]. Such a system in normal cells could provide a means for stable changes in cells. In order to secure evidence for the possible existence of RNA-directed DNA synthesis in normal cells, we looked in normal cells for endogenous RNA-directed DNA polymerase activity like that found in cores of RNA tumor viruses.

All infectious preparations of RNA tumor viruses contain endogenous RNA-directed DNA polymerase activity (see [4] for review). This activity is ribonuclease-sensitive and is partially resistant to actinomycin D. The product DNA hybridizes to the template RNA, and an early product DNA-RNA complex can be isolated.

To look for similar endogenous RNA-directed DNA polymerase activity in normal cells, we prepared a microsome fraction from normal uninfected chicken embryos [5]. When this microsome fraction was centrifuged on an isopycnic sucrose density gradient, we found that exogenous DNA-directed DNA polymerase activity was partially separated from endogenous DNA polymerase activity. The endogenous DNA polymerase activity could be further purified by velocity gradient centrifugation.

The chicken endogenous DNA polymerase activity required a divalent cation and all four deoxyribonucleoside triphosphates for full activity. It was stimulated by exogenous DNA. The activity was completely destroyed by preincubation with ribonuclease, was resistant to deoxyribonuclease, and was partially resistant to actinomycin D (fig. 1).

If a reaction was run in the presence of actinomycin D, and a 5 min product was isolated with phenol–cresol extraction followed by cetyltrimethyl-ammonium bromide precipitation, an RNA–DNA complex was found [6]. When this early product was analyzed on neutral sucrose gradients, it was found in two peaks, one at about 6 S and one at about 30 S (fig. 2). The faster sedimenting peak disappeared

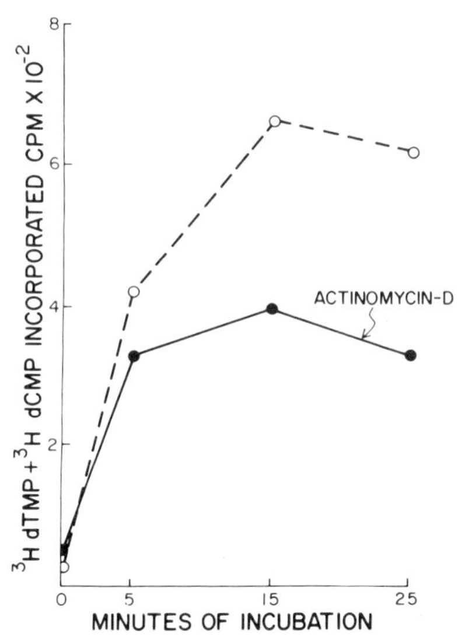

Fig. 1. Effects of actinomycin D on chicken endogenous DNA polymerase activity. 5-day-old chicken embryos were washed four times with ice-chilled phosphate-buffered saline (pH 7.2) and were centrifuged at 1000 rpm for 5 min in an International centrifuge. The pelleted embryos were resuspended in an equal volume of 0.01 M Tris-HCl (pH 7.4) containing 0.25 M ribonuclease-free sucrose, 0.001 M EDTA, and 0.03 M dithiothreitol and were homogenized with 10 strokes of a tight-fitting glass Dounce homogenizer. The homogenate was then centrifuged at 8000 rpm for 10 min in a Sorvall SS34 rotor, and the supernatant was further centrifuged at 45,000 rpm for 1 hr in a Spinco SW50.1 rotor. The pellet from the 45,000 rpm centrifugation was suspended in 0.01 M Tris-HCl (pH 8.0) containing 0.1% Nonidet P-40 (Shell Co.), 0.03 M dithiothreitol, and 0.25 M ribonuclease-free sucrose. The final protein concentration was made 20 mg/ml. Endogenous DNA polymerase activity was determined in a complete reaction mixture containing 10 nmoles dATP, 10 nmoles dGTP, 7.5 nmoles ATP, 1.875 μmoles $MgCl_2$, 2.5 μmoles KCl, 2.4 μg phosphoenol pyruvate, 10 μg pyruvate kinase, 2.5 μCi (about 0.1 nmole) [^{3}H]TTP, 2.5 μCi (about 0.15 nmoles) [^{3}H]dCTP in 100 μl of 20 mM Tris HCl (pH 8.0), 0.4 mM EDTA, 10 mM dithiothreitol. 25 μl of the chicken fraction, containing 500 μg protein, was added to 100 μl of the reaction mixture and the whole reaction mixture was incubated at 40°C. A parallel reaction was carried out in the presence of 100 μg/ml of actinomycin D. 25 μl samples were withdrawn at the indicated times, and trichloroacetic acid-insoluble radioactivity was determined. ○——○ control reaction; ●——● actinomycin D-treated reaction.

after treatment with ribonuclease, alkali, or heat. When the same early product was centrifuged in equilibrium cesium sulfate density gradients, it was found in two positions, one in the DNA region of the gradient and one in the RNA region of the gradient (fig. 3). The peak in the RNA region of the gradient completely disappeared after treatment with alkali or ribonuclease. Most of the peak at the RNA region of the gradient disappeared after heating, but a small amount was left. This small amount remaining may represent DNA product covalently attached to an RNA primer.

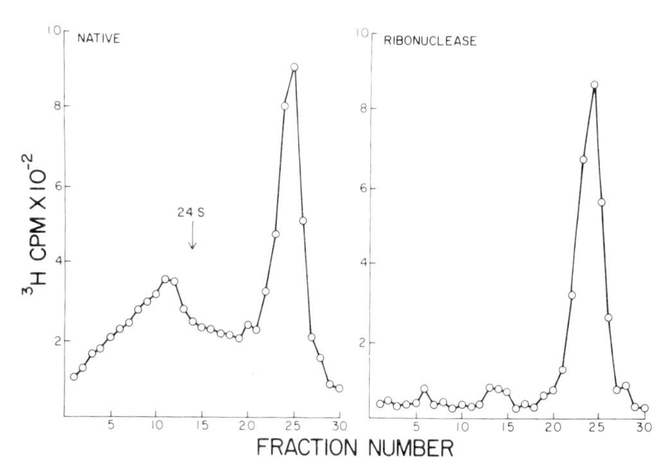

Fig. 2. Zone sedimentation of early product of chicken endogenous DNA polymerase activity. Tritium-labeled DNA product was purified from a 5-min reaction (in 5 ml) of the chicken endogenous DNA polymerase activity in the presence of 100 μg/ml actinomycin D (fig. 1). The reaction was terminated by the addition of a final concentration of 1% SDS and 50 mM EDTA. The reaction product was twice extracted with a half volume of phenol : cresol (50 g phenol, 7 ml m-cresol, 5 ml H_2O, 0.1% 8-hydroxy-quinoline), once with an equal volume of chloroform, and twice with two volumes of diethyl ether. The residual ether was removed by bubbling air through the final aqueous phase. 100 μg boiled calf thymus DNA (Worthington) and 200 μg yeast tRNA (Worthington) were added to 1 ml of product, followed by the slow addition of 50 μl/ml of 0.1 M cetyltrimethylammonium bromide (Eastman Organic Chemicals), precipitation on ice for 20 min, and collection by centrifugation at 15,000 rpm for 10 min in a Sorvall SS34 rotor. The pellet was suspended in the original volume of 1 M NaCl and was reprecipitated with four volumes of 100% ethanol at $-20°$C overnight. The precipitate was collected by centrifugation and resuspended in 0.01 M Tris-HCl (pH 7.2) containing 0.014 M NaCl. The purified early product was then divided into equal portions. A portion was preincubated at 37°C for 1 hr with 50 μg/ml of boiled ribonuclease A (Worthington) and 40 units of ribonuclease T_1 (Calbiochem). The control was preincubated with water. The samples were analyzed on 10–30% linear sucrose gradients made in 0.02 M Tris-HCl (pH 7.4) containing 0.1 M NaCl and 0.001 M EDTA. The gradients were centrifuged at 50,000 rpm in an SW50.1 rotor for 3 hr, and fractions were collected from a hole pierced in the bottom of the tube. 0.1 ml of each fraction was acid precipitated, and the radioactivity was determined. The recovery of radioactivity in the ribonuclease-treated sample was about 60%. The arrow with 24 S marker represents the position of M 13 DNA in a parallel gradient. Other portions of the purified early product were treated with ribonucleases in 0.4 M NaCl, 1 M NaOH at 70°C for 1 hr, or incubation in a boiling water bath for 10 min and were analyzed on linear sucrose gradients. The patterns were similar to those of the sample shown treated with ribonucleases in 0.014 M NaCl. This experiment is taken from [6].

These results indicate that part of the early product was an RNA–DNA complex and part was free DNA. The free DNA was of small size as shown by its slow sedimentation in sucrose gradients (fig. 2), and it was partially double-stranded as shown by its resistance to S1 nuclease [6]. The complexed DNA was attached to RNA as shown by the break-up of the complex after treatment with ribonuclease or

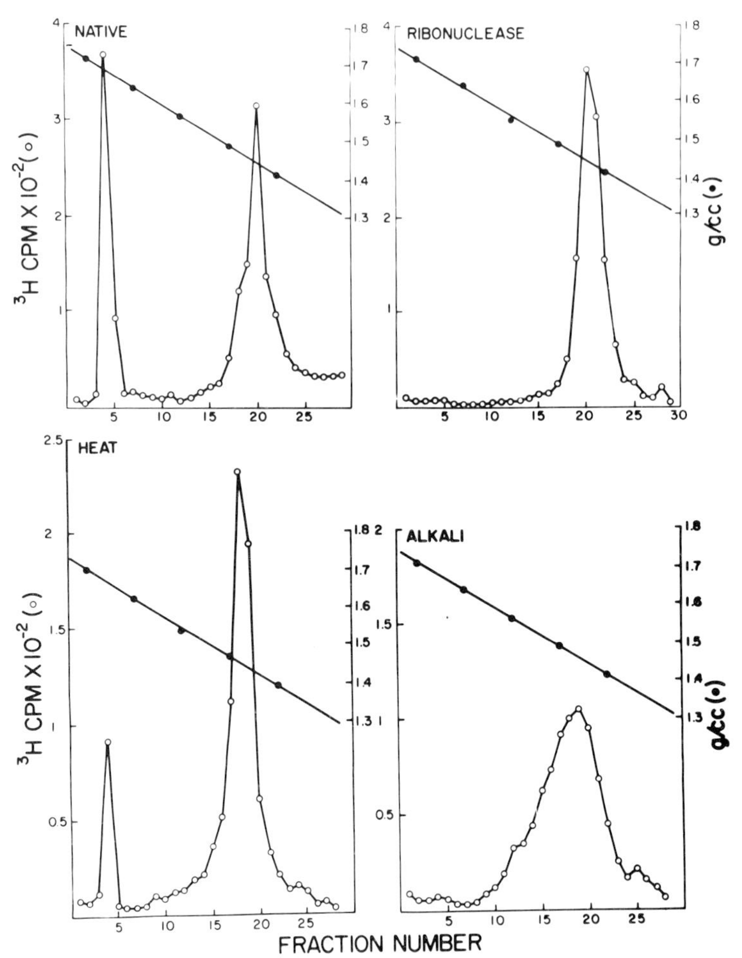

Fig. 3. Equilibrium centrifugation of early product of chicken endogenous DNA polymerase activity. The early product of a 5 min reaction of the chicken endogenous DNA polymerase activity in the presence of 100 μg/ml actinomycin D was purified as described in the legend to fig. 2. Aliquots in 0.014 M NaCl, 0.01 M Tris-HCl (pH 7.2) were treated with water (native), 50 μg/ml of boiled ribonuclease A and 40 units of ribonuclease T_1 at 37°C for 1 hr (ribonuclease), 1 M NaOH at 70°C for 1 hr (alkali), or were incubated in a boiling water bath for 10 min and quickly cooled (heat). Samples were then diluted to 1.3 ml with 0.3 M NaCl-0.03 M sodium citrate, pH 7.0, and were mixed with 1.75 ml of Cs_2SO_4 (refractive index 1.4015) and 0.1 ml of 10% Sarkosyl. The solutions were centrifuged in polyallomer tubes in an SW50.1 rotor for 65 hr at 31,500 rpm, and fractions were collected from each gradient through a hole in the bottom of the tube. The refractive index was measured, and trichloroacetic acid-insoluble radioactivity was determined. The recovery of radioactivity relative to the native sample was 73% in the sample treated with ribonucleases in 0.014 M NaCl, 51% in the sample treated with alkali, and 58% in the sample which was heated. (Others have found poor recoveries of the early product of the RNA tumor virus endogenous DNA polymerase activity after treatment with alkali or heat [7, 8].) The scale of the ordinate in the lower two gradients is different than that in the upper two gradients. This experiment is taken from [6].

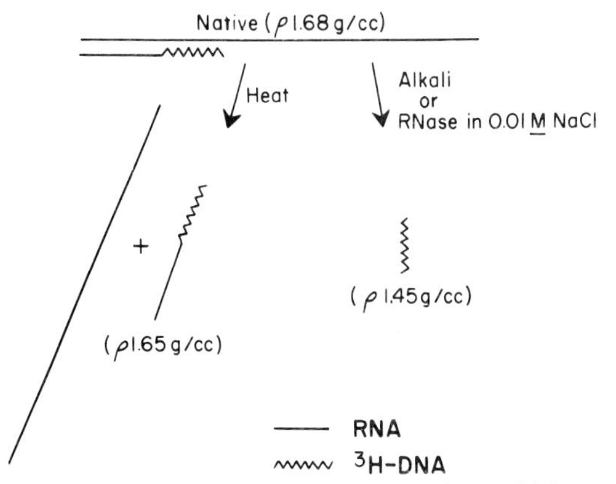

Fig. 4. A model of the early DNA product-RNA complex of the chicken endogenous DNA polymerase activity.

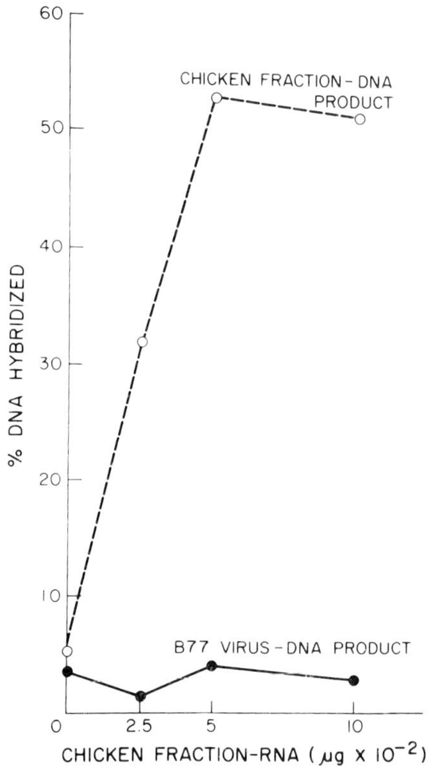

Fig. 5. Hybridization of DNA products of chicken and Rous sarcoma virus endogenous DNA polymerase activities to RNA isolated from the chicken microsome fraction. DNA products were made in the presence of 100 μg/ml actinomycin D as described in the legends to figs. 1 and 2 and were annealed for 15 hr at 68°C to the indicated amount of chicken RNA in a volume of 0.2 ml. Hybridization was monitored by resistance to S1 nuclease. 100% was approximately 2000 cpm for B77 virus DNA product and 600 cpm for chicken fraction DNA product. B77 virus is an avian sarcoma virus like Rous sarcoma virus.

alkali (fig. 4). The RNA was larger than the DNA as shown by the sedimentation of the RNA–DNA complex at approximately 30 S in sucrose and its banding in the RNA region of equilibrium cesium sulfate density gradients. The DNA was complexed to the RNA primarily by noncovalent bonds as shown by the dissociation of the DNA after heating.

To determine further whether the DNA product of the chicken endogenous DNA polymerase activity was made from an RNA template, nucleic acid hybridization experiments were performed. Product was isolated from a 5 min reaction in the presence of actinomycin D, and RNA was destroyed by treatment with alkali. The product DNA was then annealed to RNA isolated from the same chicken microsome fraction. DNA product from a Rous sarcoma virus endogenous DNA polymerase reaction was used as a control. The extent of hybridization was determined by resistance to S1 nuclease digestion. Fig. 5 shows that the DNA product of the chicken activity annealed to the RNA from the chicken microsome fraction, while Rous sarcoma virus DNA did not.

These experiments demonstrate that an entity with all of the properties of RNA-directed DNA polymerase activity is present in uninfected chicken embryos. This activity is ribonuclease-sensitive and is partially resistant to actinomycin D. The early DNA product of this activity is a complex with RNA, and the product DNA hybridizes to RNA from the same fraction.

4. Origin and function of cellular RNA-directed DNA polymerase activity

What is the origin of this chicken endogenous RNA-directed DNA polymerase activity? Could this activity be from a latent virus? To answer this question, we have asked whether the RNA template of the chicken endogenous RNA-directed DNA polymerase activity is related to the RNAs of avian leukosis viruses or reticuloendotheliosis viruses [5]. Nucleic acid hybridization experiments were carried out as described in fig. 5. It was found that the DNA product of the chicken endogenous DNA polymerase activity did not hybridize to RNA of reticuloendotheliosis virus, B77 virus, or Rous associated virus-0.

The relation of the DNA polymerase of the chicken activity to the DNA polymerases of avian leukosis virus and reticuloendotheliosis virus was also studied [5]. An antibody which neutralized the DNA polymerase activity of all avian leukosis viruses did not neutralize the chicken endogenous DNA polymerase activity. The DNA polymerase of reticuloendotheliosis virus was shown to be larger in size than the DNA polymerase of the chicken endogenous DNA polymerase activity (Mizutani, personal communication).

These experiments established that the chicken endogenous RNA-directed DNA polymerase activity was not related to any known avian RNA viruses with a virion DNA polymerase, that is, to the avian leukosis or reticuloendotheliosis viruses.

Could there be an unknown latent virus in chicken embryos? The answer to this question is mostly a matter of definition. Cells containing chicken endogenous RNA-directed DNA polymerase activity do not produce any virions and do not show any signs of transformation or of cytopathic effects. Therefore, any hypothetical latent virus would not have any observable effect on cells and would not make progeny virions. Such a latent virus would be indistinguishable from a normal cellular element. Therefore, we have called such an element a protovirus [9].

We can determine whether or not the DNA template for the RNA template of the chicken endogenous DNA polymerase activity is integrated with cellular chromosomal DNA, that is, whether the DNA belongs to a plasmid. To answer this question, we have begun hybridization experiments with the DNA product of the chicken endogenous DNA polymerase activity and chicken red blood cell DNA. Initial experiments suggest that there is hybridization and, therefore, that the RNA template of the chicken endogenous DNA polymerase activity is derived from normal chicken cell DNA. This DNA could have arisen from previous virus infection. However, at present, it appears to be cellular and not virus DNA.

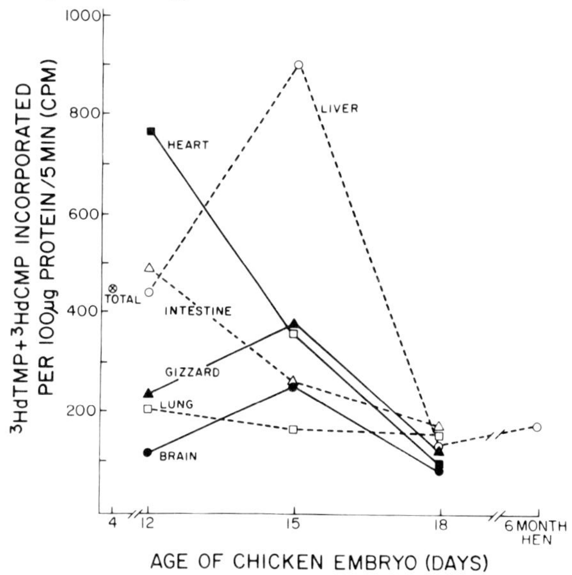

Fig. 6. Endogenous DNA polymerase activities in different organs of chicken embryos at different times. Approximately 15 chicken embryos at different ages were dissected, and the brain, gizzard, heart, intestine, liver, and lung were separately extracted. The organs were chopped with a scissors, washed in phosphate buffered saline (pH 7.0), and centrifuged at 1000 rpm for 5 min in an International centrifuge. The pellets were homogenized, and microsome fractions were prepared as described in the legend to fig. 1. Microsome fractions were also prepared from whole four day old chicken embryos and from the liver of a six month old hen. The protein concentration in each microsome fraction was determined by the Lowry method after the removal of dithiothreitol. The endogenous DNA polymerase activity was determined in a complete reaction mixture (fig. 1) containing 100 μg/ml of actinomycin D. Trichloroactic acid-precipitable counts were determined at 5 min. The incorporation of ^{3}H-dTMP and ^{3}H-dCMP is presented per 100 μg protein of each microsome fraction.

Then we can ask, is the normal cellular endogenous RNA-directed DNA polymerase activity active in normal development? As an initial approach to this question, we have looked at the amount of this activity in different organs at different times (fig. 6). We found that the activity increased and decreased in different organs at different times. We realize that these experiments cannot be accepted simply, because there could be nucleases which affect the amount of activity found. However, the results suggest that the amount of activity might vary during embryologic development.

What would the role of this activity be in development? We have suggested that this activity provides a means for cells to change stably their genomes [9]. They could move genes around and change them from inactive to active regions of the chromosomes, they could amplify genes, and they could send new genes to neighbouring cells. It will require other kinds of experiments to get evidence for these processes. We are looking at the nature of the template and of the DNA in different cells to see if we can find evidence for some such variation. Crippa and Tocchini-Valentini have presented evidence for RNA-directed DNA synthesis in gene amplification in *Xenopus* [10].

This activity may also be responsible for the induction of neoplastic transformation when RNA tumor viruses, chemical carcinogens, or radiation cause the production of new genes for neoplastic transformation.

Acknowledgements

This work was supported by Public Health Service Research Grant CA-07175 from the National Cancer Institute. C.-Y. Kang is a research fellow of the National Cancer Institute of Canada. H.M. Temin holds Research Career Development Award 10K3-CA-8182 from the National Cancer Institute.

References

[1] H.M. Temin, Ann. Rev. Microbiol. 25 (1971) 610.
[2] H.M. Temin, In: RNA viruses and host genomes in oncogenesis, Eds. P.M. Emmelot and P. Bentvelzen (North-Holland, Amsterdam, London, 1972) 351–363.
[3] H.M. Temin, Persp. Biol. Med. 14 (1970) 11.
[4] H.M. Temin and D. Baltimore, Advan. Virus Res. 17 (1972) 129.
[5] C.-Y. Kang and H.M. Temin, Proc. Natl. Acad. Sci. U.S. 69 (1972) 1550.
[6] C.-Y. Kang and H.M. Temin, Nature New Biol. 242 (1973) 206.
[7] I.M. Verma, N.L. Meuth, E. Bromfeld, K.F. Manly and D. Baltimore, Nature New Biology 233 (1971) 131.
[8] J.M. Taylor, A.J. Faras, H.E. Varmus, W.E. Levinson and J.M. Bishop, Biochem. 11 (1972) 2343.
[9] H.M. Temin, J. Natl. Cancer Inst. 46 (1971) III.
[10] M. Crippa and G.P. Tocchini-Valentini, Proc. Natl. Acad. Sci. U.S. 68 (1971) 2769.

Subject index

Acidic protein, 168
Acridine orange, 335
Actinomycin D, binding to DNA, 169, 170
Actinomycin D, influence on, 5, 271, 335
  albumin synthesis, 49
  BME treated PE, 184, 195
  informational RNA, 132
  ovalbumin synthesis, 29, 36, 39
  reverse transcriptase, 214, 221, 339
  RNA-antigen complex, 129
  testis RNA, 140
  uterine RNA, 90, 107, 156, 164
  uterine translational capacity, 167, 175
  viral RNA synthesis, 340, 346
Activator RNA, 305–323
Adrenal explant, 74
  RNA effect, 74
Adrenal gland, 74
Adrenal RNA, 82
  effect on steroid metabolism, 81–84
  in vitro effect, 81–84
  in vivo effect, 84
Albumin–antiserum, 46, 105, 107
  cell replication, 49
  mRNA stability, 51
  polysomes (membrane bound), 51
  synthesis, 46–52
  synthesis regulation, 51
Aldosterone, 332
Alkaline phosphatase, 92–100, 156, 170
Allo-DNA, 261
  effect on genetic reversion, 262
  specific effect, 265
Allo-RNA, 261
  effect on genetic reversion, 264. *See* Reversion frequency
  specific effect, 265
α-Amanitin, 318
Ameloblasts, 250
Amphibian eggs. *See Xenopus*
  enucleated, 112
Androgen, 178
  metabolism, 81
  hormones. *See* Steroid hormones
Androstenedione, 76, 84
Antibody-cellular requirement, 133
  formation, 133
  IgG, 127–133
  IgM, 127–133
Antigenic identity, 283
Antigens, 127
  neutralization, 128
  phage, 127
  retention in macrophage, 134
Antiserum, against
  albumin, 93
  phage, 129
  RNAse, 93
Ascite tumor cells, 164. *See* Cell free system
Ascospores, 260
  germination, 267
Autoradiography, 90
  electron microscopic, 92
  high resolution, 94
  low power, 94
Avian RNA virus, 339
Avidin synthesis, 270
  assay, 273
Axial structure. *See* Secondary axis

Balbiani ring, 289
Balbiani ring RNA, 290–294
  hybridization to DNA, 290, 296
Base complementarity, 208, 222
Beating tissue, 151, 229–246
  tube, 240, 244
Blastocyst implantation, 164
Blastoderms, 137, 138. *See also* Nodal piece
  inoculation, 140
Blastula, 8–12
Bone marrow cells, 335
  leukemic, 335

Bone marrow (BME), 183
effect on PE differentiation, 183
exposure to, 183–198
sensitizing action on PE, 197
threshold, optimum time, 184
Brain nuclear RNA, 238
Brain RNA, 183–185, 229
promoted brain formation, 196
role in ectoderm differentiation, 192
Brain tissue, 192
BSA gradient, 134

Cardiac tissue. *See* Beating tissue
Cell culture, 43
HeLa cells, 21
monolayers, 80
Cell cycle
G_1 phase, 51
G_2 phase, 11, 22
S phase, 11, 22, 23, 24
Cell determination, 22
Cell free system, 4, 80, 115, 130
Krebs ascites system, 5, 7
rabbit reticulocyte lysate, 29, 271
Cellular differentiation, 248. *See* Differentiation
Cellular DNA, 347
Centromeric DNA, 308
Cervical matrix, 252
Cesium chloride gradients. *See* Equilibrium density centrifugation
Chick embryo, 138
Chironomids, 289, 290
Chordal tissue, 233
Chordamesoderm, 153, 184, 225, 339–348
Chloroquine, 335
Chromatin, 21, 22, 170, 305, 324
acceptor sites, 281
complex, 333
liver–chromatin, 324
transcript, 227
transcript competition hybridization, 227
tumor–chromatin, 329
Chromosomal organization, 259
puffs, 305
Chromosomal proteins
nonhistone, 324
regulatory function, 324
regulatory molecules, 328
Chromosomal RNA. *See also* Nuclear RNA
hybridization to DNA, 312–316
hybrid spacing, 314
rate of hybridization, 312
repetitive sequences, 320
Chromosomes, 111, 203
embryonic, 9
Cilia, 98, 279
formation, 270–283. *See* Progesterone
Ciliogenesis, 274
Collagenase, 255, 256
Collagen synthesis, 13, 116, 117. *See also* (m)RNA
Colonial morphology, 259
Conalbumin, 27
Corticosterone, 73, 170
Cortisol (effect) 73, 83, 168, 170, 332
Cot. curve, 308
Culture media, 47
Cyclic AMP
dibutyryl, 167, 177
Cycloheximide, 80
Cytodifferentiation, 27, 28, 243
Cytoplasmic RNA 230, 290. *See also* RNA classes
stability, 294
developmental difference with nRNA, 229

Dactinomycin, 252
Dehydroepiandrosterone, 76–81
N-Demethylrifampicin, 214
Dentine matrix, 252
Depletion experiments, 17–18
DNA reiteration effect, 18
Depressor RNA, 335
hybridization, 333
Development, embryonic, 9
D-RNA content during, 11
Differential gene activity, 110, 111
Differentiation, 9, 110, 302, 339
spontaneous, 282
Dihydrotestosterone, 170, 178
Dimethylsulfate, labeling of RNA, 157, 255
RNA methylating, 161
Diphenylamine reaction, 204, 260
Diploid cells, 330, 331
DNA-effect of hydroxyurea, 5
degree of divergency, 18

mutagenic effect, 267
open loop, closed helix, 335
redundancy degree, 11–21
replication, 5, 110
transcription, 110, 324. *See* Transcription
DNA–DNA hybridization
ovalbumin DNA, 36–40. *See* Reverse transcriptase
DNA-like DNA, 11–25
relations to cell differentiation, 11–25
accumulation, 12
conservation, 13
hybridization to DNA, 12–13
DNA polymerase. *See* Reverse transcriptase
activity during development, 348
antibody neutralization, 346
characterization, 341
DNA dependent, 213, 221
isolation, 341
normal cells, 341
viral cores, 341
RNA dependent, 35, 135, 248, 346–349
origin and function of activity, 346
template specificity, 214
DNA–RNA hybridization. *See* RNA species
additional hybrid, 326
DNAse, 172, 226, 232. *See also* RNAse treatment
Double helix, 168
Double stranded RNA, 67
IAA effect, 69–71
Drosophila genome, 308

Ecdysone, 168
Ectodermal inductor, 186
preparation, 186
Electron microscopy, 94–109, 232, 273, 316–323
Electrophoresis
Laemmli-method, 7
PAA gel, 7, 67
starch gel, 62
Embryo-development, 114
ectoderm, 12, 21
endoderm, 12, 14
mesoderm cells, 12
Embryogenesis, 112, 330
Embryonic cells, 115
Enamel matrix, 252
Endoplasmic reticulum, 297
Endometrial cells, 171
Endometrium, 94
Enzyme, tissue specific, 43
Epidermal structure, 153
Epigenetic factor, 248
Epinephrine, 168
Episomal RNA, 205, 206, 221
Epithelial-layers, 94
mesenchymal interactions, 247
Epithelium, 247–258, 279
Equilibrium density centrifugation, 216
Erythrocytes, 150
Estradiol, 28, 155, 168, 271
active dose, 161
contamination, 155–165, 174, 271
effect on base ratio, 177
hypotropic action, 156
Estrogen, 26, 73, 90, 170, 282, 332, 335
Estrone, 84, 168
Euchromatin, 333
Exogenous RNA, 73–85, 90–109
genotropic nature, 108
Extracellular matrix, 247–258
RNA, 252
Extracellular organelles, 253
Extraction procedure, 62, 74, 139, 173, 260, 272

Ferritin, 314
RNA attachment, 314
Fertilization
protein synthesis, 4
Fibroblasts, 116, 340
Ficoll gradients, 134
Filtrate RNA, 239–244
Folic acid synthetase, 225

Gastrula, 12–21
Gene
acquisition, 20
amplification, 21, 29, 348
control, 306, 330
number, diploid, 113
recognition, 306
regulation, 152, 167. *See* Regulation
variable expression, 330
Gene activity-expression, 304. *See* Regulation
modulation by RNA, 256
Gene derepression, 324, 330–328

control, 336
during oncogenesis, 337
fetal genes, 337
molecular mechanism, 333
Gene repressors, 333
Genetic analysis, 265
changes, mediated by DNA, RNA, 259
diversity, 330
inheritance, 221
Generation time, 135
Genome
functional organization, 311
loss, 111
picture, 311
Germ cell formation, 151
Globin, 114
Globin mRNA, 114–120
life time, 116
Gonadotropin-induced RNA, 73
effect on endocrine tissues, 73
FSH, 80, 85
LH, 79, 85
role in steroid metabolism, 78–85
Gonads, 73–85
Glucose-6-P-cyclase, 260
Glucose-6-P-dehydrogenase, 108, 167. 179
β-Glucuronidase, 92, 108
Golgi apparatus, 163
Granules, ribosome like, 253. *See* Autoradiography
Growth hormone, 168

Haploid ovum, 331
Heart nuclear RNA
differentiation in PNP, 237
Heart mRNA, 140, 171, 229
blastoderm development, 145
concentration effect, 239
DNAse treatment, 232
from calf, 240
isolation, 231
mediated heart formation, 239
HeLa cells, 292. *See* Cell culture
Hemagglutination, 128
Hepatoma cells, 44–51
growth phase, 47–50
replication, 49
Heterocaryotic, 266
Heterochromatin, 111, 333
Heterogeneous RNA, 291
half-life, 294
late, early, 292–297
Heterozygotes, 112
Hexokinase activity, 179
Histone, 168, 335
displacements, 333
Histone, mRNA, 5–10
activity, 6–8
cell free synthesis, 5
delayed, 6
identification, 5
immediate, 6
natural, 5–6
Hn RNA. *See* Nuclear RNA
Hormonal
heterospecific stimulation, 281
regulation, 26–41
target tissues, 41, 86
Hormone
genome interaction hypothesis, 168
histone interaction, 168
peptide hormone, 168
receptor, 168
Hormone stimulated RNA, 271
Hybridization. *See* DNA, RNA
Hydroxyapatite, 18, 313
3β-Hydroxysteroid dehydrogenase, 76
Hydroxyproline, 117
Hypertrophy. *See* Estradiol

Immunocompetent cells, 131
Immunodeficiency diseases, 135
Immunogenicity, 129
Immunogenic RNA. *See* Informational RNA
Immunoglobulins. *See* Antibody
Immunoprecipitation, 29, 128
Immune response, 127, 130
Indoleacetic acid, 61–72
effect on peroxidase development, 64
effect on RNA stability, 65
Induced reversion, 265
Informational RNA, 131–136
cellular transfer, 135
in vitro test, 133
messenger nature, 132
RNA–antigen complex, 131
Initation of DNA replication, 222
Injected in RNA, 119
translation, 119

Injection in oocytes, 110–120
    technique, 118, 273
Inositol, 259
Insulin, 168
Integrator gene, 305
Intercellular communication, 247
Iodine metabolism, 86–89
    incorporation, 87
Iso-DNA, 260
    effect on genetic reversion, 260
Iso-RNA, 260

Kidney RNA, 183–185
    effect on BME, KE treated PE, 188, 193
    effect on steroid synthesis, 76
    mesenchyme formation, 196
    role in ectoderm differentiation, 191
Kleinschmidt technique, 309

Lac repressor, 335
Lactic dehydrogenase, 243
Lampbrush stage, 331
Larvae, 13–25
Latent virus, 347
Leydig cells, 78–81
    immature, 79
    mature, 79
Ligases, 221
Liver-RNA, 91, 171–173
    distribution, 94–98
Lymphocytes, 127, 335, 337
    activation, 330, 333
Lymphoid cells, 134
Lysozyme, 27

Macromolecules. *See* DNA, RNA
    entry into plant cells, 63
Macrophages, 127, 133–136
    adherent, non-adherent cells, 133
Maturation
    4 S RNA, 294
Mature sperm, 331
Maternal RNA hypothesis, 5
    distribution during cleavage, 9
    distribution in cytoplasm, 9
Matrix vesicles, 247–253
    isolation, 253–256
Mesenchyme, 192, 229, 233, 247–258
Mesoderm, 151
    structure, 153
Mesothelium, 192

Methylated albumin, (MAK) 129–131
Methylcholanthrene, 335
Microtubules, 4
mRNA–export, 302
    injection. *See* Injection
    stability. *See* Oocytes, ovalbumin mRNA
    species. *See* Adrenal RNA, Albumin RNA, Brain RNA, DNA like RNA, Globin RNA, Heart RNA, Histone RNA, Kidney RNA, Liver RNA, Ovalbumin RNA, Oviduck RNA, Prostatic RNA, Salivary gland RNA, Seminal vesicle RNA, Spermatid RNA, Testicular RNA, Uterine RNA
Muscle layer, 96
Mutant crosses, 165
Mutation mechanism, 222
Myoblast cells, 43
Myofibrils, 243
Myotome, 192

Neophasm, 331, 332
Neoplastic transformation, 339, 348
Nephric tubules, 150, 151
Neural epithelium, 150, 192
Neural tissue, 229, 233
    structure, 153
Neuroid, 229, 238, 240
Neurospora mutant, 259–267
    RNA, DNA isolation, 260
Neurulae, 13–25
Newt species, 183–198
Nodal piece, 137–154
    post-nodal piece, 137–154, 232
    differentiation analysis, 152
    induced development, 146. *See* Testis RNA, Bone marrow isolation
    mRNA action upon. *See* Brain, Heart, Testis RNA
Nonhistone proteins. *See* Chromosomal proteins
Notochord, 151, 192, 229, 238–240
Nuclear organizer, 112
Nuclear origin, 295
Nuclear RNA, 14, 229–230, 305, 316
    artifactual degradation, 320
    competition hybridization with cRNA, 318–320
    hybridization to DNA, 15, 317
    labeling kinetics, 310

polysomal mRNA precursor, 316
post-transcriptional modification, 320
precursor relationship to cRNA, 319
repetitive sequences, 320
Nucleolus, 112
Nucleoprotein, 153. *See* Ribonucleoprotein

Onconavirus, 221
Oncogenic viruses, 336
Ondontoblasts, 250
Oocytes
genome, 331
mRNA stability in, 119
translational capacity, 119
translational efficiency, 119
Oogenesis, 112
Operator locus, 336
Operator RNA, 336
Orcinol reaction, 260
Organ culture
adrenals, 74
bone marrow, PE, 185
thyroid glands, 86
Organ regeneration, 330
Ovalbumin gene number, 39
Ovalbumin mRNA, 29–35
content, 29
stability, 40
Ovalbumin polysomes, 26
anti-ovalbumin binding, 29–33
isolation, 31
Ovalbumin–synthesis, 26, 31
transcription, 29–35
Ovary removal, 91
Oviduct, 26
chick, 270–283
Xenopus, 270
Oviduct morphology, 274
oviduct RNA. *See* Oviduct RNA
progesterone effect, 274
Oviduct nuclear RNA, 271
Oviduct RNA
effect on avidin synthesis, 270–283
effect on oviduct morphology, 274
heterologous, homologous, 280
injection in oviduct, 273
isolation, 272

Para nitrobenzoic acid, 226
Parenchyma cells, 128
Peptide fingerprints, 8
Peritonial exudate, 128–136
Peroxidase, 61
isoenzyme, 61
repression. *See* Repressor factor
Phagocytosis, 135
Phosphoglucomutase, 260
Phospholipids, 163
Pigment cells, 43
Pinocytosis, 98
Post-mitochondrial supernatant, 7
Post-nodal piece. *See* Nodal piece
Poly A attached mRNA, 100, 139, 229, 231, 264, 271
adsorption to millipore, 34, 91, 139, 229, 231, 271, 295
binding to poly dT cellulose, 34
Polyamines, 174
Polyanions, nuclear, 333
Polycistronic, 306
product, 311
Poly-cytidylic acid, 176
adenylic acid, 176
guanylic acid, 176
uridylic acid, 176
Polynucleotide phosphorylase, 203–205, 221–225
activity, 205
bacterial species, 206
rifampicin, 207
XDP incorporation, 208
Polyoma DNA, 335
Polyornithine, 174
Polyphenol oxidase activity, 64
Polyribonucleoproteins, 128. *See* Ribonucleoproteins
Polysomes, 13, 301
Polythene chromosomes, 308
banding pattern, 289
Pregnenolone, 74–83
Presumptive ectoderm, 183
BME, kidney extract induction, 187–188
differentiation, 187. *See* Brain, Kidney RNA
exposure time to BME, 196
time dependent induction, 187
Processing
post transcriptional, 306

Producer genes, 305, 309
Progesterone, 26, 81, 164, 270, 332
effect on oviduct, 271, 281
receptor complex, 281
Prolactin, 171
Proline incorporation, 117
Pronase, treatment of
adrenal RNA, 84
informational RNA, 131
repressor factor, 61–68
RNA antigen-complex, 128
seminal vesicle RNA, 172
thyroid glands, 87. *See also* Uterine RNA
Pronephros, 192
Pronucleus, 4
Prostatic RNA, 178
DNA hybridization, 178
Protamine, 335
Protein synthesizing system. *See* Cell free system
Protocollagen hydroxylase, 117
Provirus (Temin's hypothesis), 203, 347
protovirus, 203, 336
Pulsating tissue. *See* Beating tissue
Purines, 205, 222
Pyridoxine, 262
mutants, 259
Pyrimidines, 205, 222

Reassociation
mixed suspensions, 252
Receptor gene, 305
Recipient cells
segregation, 227
Recipient strains, 261
Regulation model
Britten, Davidson, 304–329
Jacob Manod, 305
Regeneration. *See* Organ regeneration
Reiteration frequencies, 17
Reovirus, 221
Repetitive-DNA, 16–25, 308
RNA, 305
sequences, 304–311
size, 309
Reticulocyte, 114. *See* Cell free system
Repressor-activity, 67, 68
characterization, 65, 71
effect on peroxidase activity, 61–72
factor, 61–72
fractionation, 67
heating effect, 67
IAA effect, 67–68
rate of decay, 68
specificity, 63
Reverse transcriptase, 135, 203, 309, 225, 267
from Rous sarcoma virus, 35
isolation, 210–213
potassium ion effect, 211
rifampicin, actinomycin D effect, 214
Reversion frequency, 259, 264
Revertant-mutations, 265
stability, 265
Ribonucleoprotein, 13, 183–184
Ribosomal proteins, 204
Ribosomal RNA, 204, 244, 291, 305
controlling sequences in gene system, 113
correlation with cell growth, 113
diffusion, 301
distribution in cytoplasmic zones, 300
DNA hybridization, 112
migration rate, 302
mutations, 112, 113
Rifampicin, 206–221
RNA
appearance lag period, 292
base ratio, 177, 204
classes, 289, 290
cytoplasmic appearance, 290
duplexes, 336 *See also* Extraction, Injection
migration in cytoplasm, 294
nature, 294
precursors, 256
replication, 221
role in transformation, 264
transport, 165. *See also* Activator RNA, Allo-RNA, chromosomal RNA, Episomal RNA, Informational RNA, Iso-RNA, Maternal RNA, Messenger RNA, Nuclear RNA, Operator RNA, Ribosomal RNA, Soma RNA, Transfer RNA, Transforming RNA, Viral RNA
RNA-antigen complex, 127–136
preservation, 134
RNA–DNA complex, 343–347
DNA origin, 346

RNA polymerase, 24, 41, 70, 305, 335
DNA dependent, 205, 208
steroid effect, 170
RNAse activity, 93
RNAse, degradation of
adrenal RNA, 84
extracellular RNA, 252
heart mRNA, 240
oviduct RNA, 279
repressor factor, 61–68
RNA–DNA complex, 342
testis RNA, 140–145
thyroid RNA, 87
transforming RNA–DNA hybrid, 226
uterine RNA, 157

Salivary gland RNA, 289–295
heterogeneous, 289
Sandwich technique, 184, 185
Satellite DNA, 21, 22
Sea urchin, 112
fertilized eggs, unfertilized eggs, 4
Secretory-apparatus, 47
activity, 274
granules, 179
Secondary axis, 144–151
embryo, 153
embryonic induction, 248
Seminal vesicle, 170–175
Seminal vesicle RNA, 172
fractionation, 172
testosterone like activity, 172
Sensor gene, 305
Sequence organization, 331
in *Drosophila* genome, 331
in rat genome, 331
Showdomycin, 203–205, 225
Soma RNA's, 152
Somatic cell, 336
genetic analyses, 336
Somites, 151, 229, 238–240
Spermatid-DNA, 152
RNA, 152
Spinal cord, 192
Stationary growth phase, 204
Steroid hormone-action, mode of, 270
binding protein, 41
contamination, 181. *See* Estradiol
receptor protein, 332
synthesis pattern, 74
Stilbestrol, 272–279
Sucrose gradient, 7, 92, 99, 172, 209, 343
Superinduction, 29, 75. *See also* Sulfonamide resistance

Tadpoles, 113
Tailbud, 13
Testicular
explants, 74
microsomes, 80
Testicular RNA, 137
biological potentiality, 137
blastoderm development, 144
concentration, effect, 141–144, 151
difference from soma RNA's, 137
DNA hybridization, 152
effect on steroid synthesis, 76
isolation, 139
NP (PNP) development, 146–154
Testosterone, 73, 76, 170, 171, 332–335
Theophyllin, 167, 177
Thyroid cells, 43
glands, 86
Thyroid RNA, 86–89
thyrotropin like activity, 86–89
Thyrotropin, 86
Tooth-epithelia, 247
extracellular matrices, 250
mesenchyme, 247–248
morphogenesis, 248
Transcription, 110
control, 304
restriction, 13, 23–24.
Transfer RNA, 305
Transformants, 204
clones, 227
clone characterization, 227
Transformation, 203, 259
Transformed DNA
homology to genome DNA, 219
integration into genome DNA, 220, 266
molecular weight, 219
Transformed information
incorporation in host genome, 227
propagation, 227
Transforming-activity, 225
viruses, 339
Transforming RNA, 203–222
annealing to DNA, 226
DNA contamination, 209

DNA hybrid, 216, 222
effect on base ratio, 208
isolation, 204
replication, 204, 223
synthesis of complementary DNA, 216, 222, 223
template, 205
Translation, 110
control in development, 114
Triiodothyronine, 86
Trophic hormone, tissue specific, 85
Trypsin, 186
Tryptic peptides, 7, 29
Tubule
cells, 229
tissue, 238–240
Tumor production, 340
Tumor RNA virus, 339, 340
envelope, core, 340
weakly, strong, non-transforming, 340

Unbalanced growth, 11–21
Uterine epithelium, 158
modification by estradiol, 158
Uterine growth
stimulation by various RNA's, 175
Uterine horn, 155–166, 173–176
Uterine ribosome activity, 156
Uterine RNA, 90–109, 173–176, 155–166
action mechanism, 90
chromatographic analysis, 158
degradation, 99
effect on protein synthesis, 174
estradiol contamination. *See* Estradiol
estradiol like effects, 173
fate of injected label, 161
injection in uterine horn, 155, 157, 163
localization, 90
preparation, 157, 179
Uterus, protein synthesizing activity, 101
artificial RNA effect, 176
RNA effect, 101

Vesicles, 247–256
Vesicular tissue, 229
Viral
injection, 336
oncogenes, 332
Viral DNA, 347
Viral RNA
DNA hybridization, 346
Virus, induced transformation, 248, 340
cell properties, 340
Virus progeny, 341
Virus replication, 340–341
Vitelline membrane, 230

Walker tumor, 324

X-chromosome, 336
Xenopus eggs, 110, 118
Xenopus oocytes, 13–23, 114–120
translational control, 114

Zygote, 111
genes, 111